Classical
and
Dissipative
Quantum
Systems

Classical and
Dissipative
Quantum
Systems

MOHSEN RAZAVY
University of Alberta, Canada

ICP   Imperial College Press

Published by

Imperial College Press
57 Shelton Street
Covent Garden
London WC2H 9HE

Distributed by

World Scientific Publishing Co. Pte. Ltd.
5 Toh Tuck Link, Singapore 596224
USA office: 27 Warren Street, Suite 401-402, Hackensack, NJ 07601
UK office: 57 Shelton Street, Covent Garden, London WC2H 9HE

British Library Cataloguing-in-Publication Data
A catalogue record for this book is available from the British Library.

CLASSICAL AND QUANTUM DISSIPATIVE SYSTEMS

ISBN-13 978-1-86094-525-0
ISBN-10 1-86094-525-2
ISBN-13 978-1-86094-530-4 (pbk)
ISBN-10 1-86094-530-9 (pbk)

Printed in Singapore

To the memory of my father

M.T. Modarres Razavy

Preface

The aim of this book is to elucidate the origin and nature of dissipative forces and to present a detailed account of various attempts to study the phenomena of dissipation in classical mechanics and in quantum theory. From the early days of the old quantum theory it was realized that the quantization of the velocity-dependent dissipative forces are problematic, partly because of the inconsistencies in the mathematical formulation and partly due to the problem of interpretation. Recent works on the quantum theory of measurement, the theory of collision of heavy ions and the macroscopic quantum tunneling has necessitated a more critical examination of the nature of the frictional forces and their derivations from conservative many-body systems. In this book we discuss the basic concepts of these forces without discussing any particular application. There are monographs available dealing with these applications. For instance *Quantum Theory of Dissipative Systems* (Second Edition), by U. Weiss (World Scientific, 1999) gives an excellent account of the role of dissipative forces in condensed matter physics.

The scope of this book is also limited to the discussion of regular motions since the very important subject of chaotic dissipative motion deserves a completely different approach. Furthermore the emphasis of the present work is on the solvable models where the ideas of symmetries and the conservation laws, the way that the time asymmetry arises in the equations of motion, and the classical-quantum correspondence can be discussed without the ambiguity which is often associated with approximate solutions.

This text is divided into three parts. In the first part we present a detailed coverage of the classical dissipative systems using the canonical formalism. This includes a description of the inverse problem of analytical dynamics, i.e. determination of Lagrangians and Hamiltonians from the equations of motion specifically for the cases involving dissipative forces. Important theoretical concepts such as the Noether theorem and the minimal coupling rule in the presence of frictional forces are presented and applied to simple examples. In addition to the phenomenological frictional forces, a large part is devoted to the problem of derivation of frictional forces from solvable many-body problems. Chapters 2 through 11 cover the classical description of the subject.

In the second part of the book, Chapters 12-17 lays the groundwork for the problem of quantization of classical systems with phenomenological velocity-dependent forces. Various attempts to find a consistent theory satisfying the basic postulates of quantum theory, and their successes and failures are examined. This is followed by a more basic approach in which we try to derive a wave equation for the motion of the damped system, again from a closed many-body system.

Finally in the third part of the book, Chapters 17-18, we discuss a number of problems where the starting point is quantum mechanical, but only in some cases we can associate a classically damped system to these quantal problems.

In selecting references I have tried to cite a number of interesting albeit forgotten works. Unfortunately the task of achieving a complete list of important contributions is an impossible one. I apologize to the authors of the papers that I have overlooked or have failed to give the proper credit to. In writing this monograph I have benefited greatly from the help of my colleagues and my former students. I should thank my wife for sympathetic understanding of the ups and downs of writing a book.

Edmonton, Alberta, Canada, September 2004.

Contents

ix

Chapter 1

Introduction

In classical dynamics the second law of motion is used both as a definition of the force and also as the equation for predicting the position and the momentum of the particle as a function of time. This dual role of the equation of motion has been criticized by a number of eminent scientists [1]. In order to investigate the nature of the force law in many cases we are helped by the so called "theories of forces" [2], where theories like gravitational and electrodynamics specify the force function. Based on these and other well-defined theories we can derive the interaction between complicated system of particles and fields from the inter-particle forces. For instance we can determine the dipole-dipole interaction from the Coulomb force, or the radiation reaction force from the coupling of an electron to the electromagnetic field.

The idea that the force function is dependent on the position of the particle and not on time nor on its velocity has been suggested by a group of philosophers of science [3]. Let us quote Nigel's observation in this regard [4]:

"In point of fact, the force-function employed in many of the familiar applications of the equations of motion is specified in a manner analogous to the Newtonian hypothesis, in so far as it does not contain the time-variable explicitly. Indeed, though there are numerous cases for which the time-variable enters explicitly into the force function (as in the case of damped vibrations), it is commonly assumed that the explicit presence of the time-variable can in principle be eliminated if the initial system of interacting bodies is suitably enlarged by including other bodies into it. For reasons that will be presently apparent, what is called the "principle of causality" (as distinguished from special causal laws) in fact usually construed in classical physics as the maxim that should be force-function for a given physical system contain the time-variable explicitly, the system is to be enlarged in such a manner as to allow a specification of the force-function in which the time-variable does not appear. And it is a matter of historical fact that in the main the search for such enlarged systems that do not coincide with the entire cosmos has been successful."

Thus certain forces such as the damping force exerted on a particle while it is moving in a viscous medium can be postulated as a velocity-dependent force proportional to a given power of the velocity of the particle. The dependence of the force law on velocity in this case is determined from the observed state of the motion. On the other hand one may consider the damping to be due to the collision of the particle in question with the smaller particles forming the environment, e.g. gas molecules. In the latter case the force of friction can be derived and its dependence on the momentum of the particle can be determined.

This idea of dividing a large conservative system into two parts, one part being the heat bath and the other forming the dissipative system is a useful one, particularly in quantum theory. In this way we can avoid certain difficulties in the formulation of the problem, but in turn, nearly in all cases we can only solve the problem approximately.

Usually we assume that the interaction between the two parts causes the transfer of energy from the dissipative system to the heat bath. However we can generalize the models of dissipation to include those cases where the coupling between the two parts requires the transfer of mass or the number of particles from one part to the other.

In quantum theory we borrow the idea of force (or preferably the potential) from the classical dynamics. When the force is conservative and derivable from a potential function the formulation of the quantum analogue of the classical motion is straightforward. But if we try to quantize a non-conservative classical system we find inconsistencies between the equations of motion and the canonical commutation relations. In addition the result depends on our choice of the Hamiltonian or Lagrangian, since for non-conservative systems the Hamiltonian, in general, does not represent the energy of the system.

We define dissipative forces in classical dynamics as any and all types of interaction where the energy is lost when the motion takes place (usually in the form of heat to a heat bath) [5]. Frequently the magnitude of the force, f, on a particle or a body may be closely represented, over a limited range of velocity by a power law, $f_d = av^\nu$, where v is the velocity of the particle or body and a and ν are constants.

Depending on the value of ν we have the following types of dissipative forces [5]:

(1) Frictional force. This work is usually required to slide one surface over another, and once the motion is started the magnitude of the force is independent of the speed. Thus in this case $\nu = 0$.

(2) Viscous force. When the force is proportional to the speed of the particle, i.e. $\nu = 1$ we call the force to be viscous force (see also § 2.1).

(3) Newtonian dissipative force. For high speed motion of an object in the air, the force is proportional to the square of the velocity, i.e. $\nu = 2$.

In this book we will follow the present usage of these terms and occasionally use frictional force and viscous force to indicate dissipative forces of the general type av^{ν} with ν, a positive but not necessarily an integer, taking different values.

Bibliography

[1] B. Russell, *The Principles of Mathematics*, (W.W. Norton, London, 1996) pp. 482-488.

[2] E.J. Konopinski, *Classical Description of Motion*, (W.H. Freeman, San Francisco, 1969) p. 38.

[3] M. Jammer, *Concepts of Force*, (Harvard University Press, Cambridge, Massachusetts, 1957) Chapter 11.

[4] E. Nigel in *The World of Physics*, 1 vol. 3. edited by J.H. Weaver (Simon and Schuster, ew York, 1987) p. 741.

[5] D.A. Wells, *Theory and Problems of Lagrangian Dynamics*, (McGraw-Hill, New York, 1967) Chapter 6.

Chapter 2

Phenomenological Equations of Motion for Dissipative Systems

The simplest way of describing a damped motion in classical dynamics is by adding a resisting force, generally velocity-dependent, to the Newtonian equation of motion. In principle we can derive the damped motion by coupling the system to a heat bath or to the motion of other particles (or to a field) and then eliminate the degrees of freedom of these other particles (or the field) to obtain the equation of the system that we want to study.

We will begin our discussion with an examination of various resistive force laws of common occurrence in physics, some of them are found phenomenologically and others are derived from the fundamental laws governing mechanics, electrodynamics, and hydrodynamics.

2.1 Frictional Forces Linear Velocity

Let us consider a bullet of mass m moving in the air where the coefficient of friction is λ. The equation for the force acting on this particle is given by [1]

$$\mathbf{F} = m\mathbf{g} - m\lambda\mathbf{v}|\mathbf{v}|^{n-1}, \qquad (2.1)$$

where $\mathbf{v}$ denotes the velocity of the particle. For speeds less than $24\ \frac{m}{s}$, $n \approx 1$ (Stoke's law of resistance) and for speeds between $30\ \frac{m}{s}$ and $330\ \frac{m}{s}$, n is approximately equal to 2 (Newton's law of resistance).

5

In the case of the Stoke law of resistance we can use the following simple picture to show that the drag force is linear in velocity:

Consider an object e.g. a plate moving with a velocity V in a direction normal to the surface of the plate in a gas which is at a very low pressure. We assume that the velocity of gas molecules, v, is much larger than V. The rate that the gas molecules hit the plate is proportional to the relative velocity of the incoming molecules and the plate. On the two sides of the plate the relative velocities are $v + V$ and $v - V$ respectively. The pressure is proportional to the product of average momentum transfer per molecule and the rate at which these molecules hit the plate. Since the momentum transfer is also proportional to the relative velocity, therefore the pressures on the two sides of the plates are

$$P_1 \propto (v + V)^2, \quad P_2 \propto (v - V)^2. \tag{2.2}$$

The net pressure P is the difference between P_1 and P_2

$$P = P_1 - P_2 \propto 4vV, \tag{2.3}$$

therefore the force of friction on the plate is proportional to its speed V and opposes the motion.

A more elaborate derivation of this result which takes the quantum nature of the particles into account is the piston model proposed by Gross [2] [3]. This simple model has been used to explain the reason as to how the damping force arises in heavy-ion collisions. Again let us examine the force exerted on a rigid wall when the wall is moving with a velocity V through a gas of fermions. The flux $dj_L(\mathbf{p})$ of the particles with relative momenta $\mathbf{p} = m\mathbf{u}$ hitting the wall from the left is given by

$$dj_L(\mathbf{p}) = \frac{1}{(2\pi\hbar)^3} \frac{p_x}{m} g(\mathbf{p} + m\mathbf{V}) d^3p, \tag{2.4}$$

where $g(\mathbf{p}) = g(-\mathbf{p})$ is the isotropic single particle probability. Since the wall is assumed to be rigid, the reflection coefficient is one and the flux undergoing reflection at the wall is

$$dj'_L(\mathbf{p}) = dj_L(\mathbf{p}) \left[1 - g(\mathbf{p}' + m\mathbf{V}) \right]. \tag{2.5}$$

The term in the square bracket in (2.5) includes the Pauli blocking factor, i.e. the final state for a relative component $\mathbf{p}'$, differing from $\mathbf{p}$ by the reversal of the x component must be unoccupied. The expression for the flux reflected from the rigid wall at right can be found by changing $\mathbf{p}$ to $\mathbf{p}'$. Since the momentum transfer to the wall for a single collision is given by $\Delta p_x = 2p_x$, therefore the force exerted per unit area of the wall is

$$
\begin{aligned}
f &= \int 2p_x \left[dj'_L(\mathbf{p}) - dj_R(\mathbf{p}) \right] \\
&= \frac{2}{m(2\pi\hbar)^3} \int p_x^2 \left[g(\mathbf{p} + m\mathbf{V}) - g(\mathbf{p} - m\mathbf{V}) \right] d^3p. \tag{2.6}
\end{aligned}
$$

By expanding $g\left(\mathbf{p}' \pm m\mathbf{V}\right)$ in powers of V in (2.6) we observe that the leading term in the force f is proportional to V.

The pioneering work of Stokes in hydrodynamics is an important example of the derivation of dissipative force laws. Stokes showed that to the first order in Reynolds number this drag force on a sphere of radius R is

$$F = 6\pi\eta Rv, \tag{2.7}$$

where η is the dynamical viscosity and v is the velocity of the sphere moving in the fluid [4]-[6]. The Stokes law agrees with experiments only for very small Reynolds numbers. Thus if we express the Reynolds number, $\mathcal{R}$, in terms of the diameter of the sphere and velocity then we have

$$\mathcal{R} = \frac{2vR\rho}{\eta}. \tag{2.8}$$

The Reynolds number must be less than one for the validity of the Stokes law [7]. Such a small Reynolds numbers can occur in highly viscous fluids or for very small spheres such as droplets of mist in atmosphere.

We can calculate this drag force to the second order in the Reynolds number (or the velocity of the sphere) and find that [5] [8]

$$F = 6\pi\eta Rv \left(1 + \frac{3vR}{8\nu}\right). \tag{2.9}$$

Here ν is the kinematic viscosity $\nu = \frac{\eta}{\rho}$ and ρ is the density of the fluid. This expression first derived by Oseen [8] extends the range of validity of the drag force law to Reynolds number $\mathcal{R} \leq 5$, a result that has been confirmed by experiment [7].

The drag force can also be calculated for a plane circular disk of radius R moving perpendicular to its plane. This force turns out to be [5]

$$F = 16\eta Rv. \tag{2.10}$$

Another interesting problem is that of the motion of a spherical drop of fluid of viscosity η' moving under gravity in a fluid of viscosity η [5]. In this case the drag force is given by

$$F = 2\pi v\eta R \left(\frac{2\eta + 3\eta'}{\eta + \eta'}\right). \tag{2.11}$$

This expression reduces to (2.7) as $\eta' \to \infty$ (i.e. a solid sphere) and to

$$F = 4\pi\eta Rv, \tag{2.12}$$

when $\eta' \to 0$ (i.e. the case of a gas bubble). For a motion subject to the drag force F, Eq. (2.11), the terminal velocity u_T is given by

$$u_T = \frac{2R^2 g \left(\rho - \rho'\right)\left(\eta + \eta'\right)}{3\eta \left(2\eta + 3\eta'\right)}. \tag{2.13}$$

2.2   Raleigh's Oscillator

Let us consider a tuning fork vibrating in vacuum with a given frequency. We want to know how the vibration of this tuning fork will be changed if we immerse it in a viscous fluid, e.g. in air or in a gas. Raleigh observed that the fork and the air surrounding it constitute a single system whose parts cannot be treated separately [9]. However he noted that we can simplify the problem when the effect of the medium on the fork, during few periods is very small, and only becomes important by accumulation. Thus he was led to consider the effect as a perturbation, where the disturbing force is periodic (roughly with the periodicity of the fork) and may be divided into two parts. One part is proportional to the acceleration $(m' - m)\ddot{x}$, where $(m' - m)$ is a small mass, and the other part is proportional to the velocity $\dot{x}$ and opposes the motion. The first part can be absorbed in $m\ddot{x}$, with the result that the actual mass m is replaced by the effective mass m'. Thus we find the equation of motion

$$m'\ddot{x} + m'\lambda\dot{x} + m'\omega_0^2 x = 0, \tag{2.14}$$

where λ and ω_0^2 are defined in such a way that $m'\lambda\dot{x}$ is the frictional and $m'\omega_0^2 x$ is the harmonic force acting on the fork.

2.3   One-Dimensional Motion and Bopp Transformation

For a one-dimensional motion (say along x-axis) and for $n = 1$ Newton's equation of motion is

$$m\frac{d^2x}{dt^2} + m\lambda\frac{dx}{dt} = f(x). \tag{2.15}$$

In this equation $f(x)$ represents all of the other conservative forces which act on the particle and are derivable from potentials. For instance for a simple pendulum which is oscillating in a liquid Eq. (2.15) takes the following form:

$$m\frac{d^2x}{dt^2} + m\lambda\frac{dx}{dt} = -m\omega_0^2 x. \tag{2.16}$$

Denoting the time derivative of x by $\dot{x}$, the initial conditions of the motion are $x(0) = x_0$ and $\dot{x}(0) = \dot{x}_0$. With these initial conditions Eq. (2.16) has a simple solution, and if λ is sufficiently small while the particle is oscillating, the amplitude of x decreases exponentially and asymptotically we have $\dot{x}(\infty) = x(\infty) = 0$, i.e. after a very long time we know both the position and the momentum of the particle precisely.

The classical equation of motion for an harmonically bound electron coupled to the electromagnetic field (Abraham-Lorentz equation [10]-[15]) is an equation similar to (2.15) and (2.16):

$$m\frac{d^2 X}{dt^2} = -kX + \left(\frac{2e^2}{3c^3}\right)\left(\frac{d^3 X}{dt^3}\right), \tag{2.17}$$

where the coefficient of $\left(d^3 X/dt^3\right)$ can be expressed in terms of a very small constant having the dimension of time:

$$\tau = \left(\frac{2e^2}{3mc^3}\right) = 6.26 \times 10^{-24} s. \tag{2.18}$$

According to classical electrodynamics, an electron moving in an external electromagnetic field experiences the Lorentz force [16]

$$m\frac{d\mathbf{v}}{dt} = e\left(\mathbf{E} + \frac{1}{c}\mathbf{v} \wedge \mathbf{B}\right). \tag{2.19}$$

If we assume that the electron is uniformly charged sphere of radius r_0, each part of this charged sphere repels every other part with a Coulomb force, and this repulsion is responsible for the self force $\mathbf{F}_S$

$$\mathbf{F}_S = \int \left(\mathbf{E} + \frac{1}{c}\mathbf{v} \wedge \mathbf{B}\right)\rho dV. \tag{2.20}$$

In this expression $\mathbf{E}$ and $\mathbf{B}$ are the fields produced by the electron itself and ρ is the charge density. A lengthy calculation yields the following result, first derived by Lorentz:

$$
\begin{aligned}
\mathbf{F}_S &= -\frac{2}{3c^2}\sum_{n=0}^{\infty}\frac{(-1)^n}{n!c^n}\frac{d^n \mathbf{a}}{dt^n}\int\frac{\rho(\mathbf{r})\rho(\mathbf{r}')}{|\mathbf{r}-\mathbf{r}'|}|\mathbf{r}-\mathbf{r}'|^n d^3 r d^3 r' \\
&= -\frac{4}{3c^2}\left(\frac{1}{2}\int\frac{\rho(\mathbf{r})\rho(\mathbf{r}')}{|\mathbf{r}-\mathbf{r}'|}d^3 r d^3 r'\right)\times\frac{d\mathbf{v}}{dt} + \frac{2e^2}{3c^3}\frac{d\mathbf{a}}{dt} \\
&\quad -\frac{2e^2}{3c^2}\sum_{n=2}^{\infty}\frac{(-1)^n}{n!c^n}\frac{d^n \mathbf{a}}{dt^n}\mathcal{O}\left(r_0^{n-1}\right),
\end{aligned}
\tag{2.21}
$$

where $\mathbf{a}$ is the acceleration of the electron. The first term on the right side which is proportional to acceleration represents the electromagnetic mass of the electron which we denote by m'. Thus if we ignore all of the terms proportional to r_0 and its higher powers, then we have the equation of motion

$$m\frac{d\mathbf{v}}{dt} = -m'\frac{d\mathbf{v}}{dt} + \frac{2e^2}{3c^3}\frac{d\mathbf{a}}{dt} + \mathbf{F}_{ext}, \tag{2.22}$$

where $\mathbf{F}_{ext}$ is the external force [17].

The relativistic generalization of this equation first obtained by von Laue in 1909, in covariant notation is [18]

$$(m + m')\frac{dv^{\mu}}{ds} = \frac{2e^2}{3c^3}\left[\frac{da^{\mu}}{ds} - \frac{1}{c^2}\sum_{\nu}a_{\nu}a^{\nu}v^{\mu}\right] + F^{\mu}, \tag{2.23}$$

where again in (2.23) higher order derivatives of v^μ have been ignored. In this relation s denotes the proper time, and $a^\nu = dv^\nu/ds$.

Finally Dirac in 1932 found the exact relativistic form of this equation without ignoring higher order terms [19]. This classical Lorentz-Dirac equation in the presence of the external electromagnetic field is given by

$$ma^\mu = \frac{e}{c}F^{\mu\nu}v_\nu + \left(\frac{2e^2}{3c^3}\right)\left[\frac{da^\mu}{ds} - \frac{1}{c^2}\sum_\nu a^\nu a_\nu v^\mu\right],\tag{2.24}$$

where $F^{\mu\nu}$ is related to the 4-potential by

$$F^{\mu\nu} = \frac{\partial A^\nu}{\partial x_\mu} - \frac{\partial A^\mu}{\partial x_\nu}.\tag{2.25}$$

and a^μ and v^μ are the μ component of the acceleration and velocity respectively and s is the proper time.

When the radiating electron is harmonically bound, we write Eq. (2.17) as

$$\left(\frac{d^2X}{dt^2}\right) - \tau\left(\frac{d^3X}{dt^3}\right) + \nu^2 X = 0,\tag{2.26}$$

where $k = m\nu^2$. This equation can be transformed to a second order equation for the damped harmonic oscillator by the following transformations [20]

$$x = \left[\frac{dX}{dt} - \left(\frac{1}{a}\right)\left(\frac{d^2X}{dt^2}\right)\right],\tag{2.27}$$

and

$$y = \left[X - \left(\frac{1}{2a}\right)\frac{dX}{dt} - \left(\frac{1}{2a^2}\right)\left(\frac{d^2X}{dt^2}\right)\right],\tag{2.28}$$

where

$$a = \frac{\nu}{\sqrt{(\lambda\tau)}} = \frac{1}{\tau} + \lambda,\tag{2.29}$$

and λ is the real positive root of the equation

$$\lambda(1 + \lambda\tau)^2 - \nu^2\tau = 0.\tag{2.30}$$

By differentiating (2.28) we find the coupled set of equations

$$\dot{y} + \frac{1}{2}\lambda y - \frac{(1 + \frac{3}{4}\lambda\tau)}{1 + \lambda\tau}x = 0,\tag{2.31}$$

and

$$\dot{x} + \frac{1}{2}\lambda x + \frac{\lambda}{\tau}(1 + \lambda\tau)y = 0.\tag{2.32}$$

If we eliminate x or y between these equations we obtain the equation of motion for x which is the same as Eq. (2.16) with $\omega_0^2 = \frac{\nu^2}{a\tau}$.

2.4   The Classical Theory of Line Width

The spectral lines observed in a spectrograph have an observable width and this is related to the width of the entrance slit. By reducing the width of the slit further we reach a limiting value which is determined by diffraction. The line width can be reduced further by increasing the resolving power of the instrument. Under the best set of conditions we find that each line consists of a distribution of the frequencies and is not monochromatic, and this is because of the finite lifetime of the excited state [21].

Let us study the classical theory of the line width emitted by an isolated stationary atom [12]. We start with Eq. (2.17), i.e. we assume that the radiating electron is harmonically bound $\omega_0^2 = k/m$ [22]. We write (2.17) as

$$\frac{d^2X}{dt^2} - \tau \frac{d^3X}{dt^3} + \omega_0^2 X = 0. \tag{2.33}$$

This equation can be solved by taking X to be of the form $\exp(-\alpha t)$ and substituting this in Eq. (2.33) to find a cubic equation for α;

$$\tau\alpha^3 + \alpha^2 + \omega_0^2 = 0. \tag{2.34}$$

The cubic equation (2.34) has a real negative root, (the runaway solution [12]) and two other complex roots, one being the complex conjugate of the other. Since τ is very small, an accurate solution of (2.34) can be found by assuming that $\tau\omega_0 << 1$, and then we have

$$\alpha = \frac{\Gamma}{2} \pm i\left(\omega_0 + \Delta\omega\right), \tag{2.35}$$

where

$$\Gamma = \omega_0^2 \tau, \tag{2.36}$$

is the decay constant and

$$\Delta\omega = -\frac{5}{8}\omega_0^3 \tau^2, \tag{2.37}$$

is the level shift. The energy of the oscillator which is proportional to X^2 decays as $e^{-\Gamma t}$ because of the radiation damping. Thus the emitted radiation is like a wave train of an effective length c/Γ. But the wave train of a finite length cannot be monochromatic, and we can find the shape of the frequency spectrum through the Fourier transform of $\ddot{X}$ which is proportional to the electric field $E(\omega)$;

$$E(\omega) \propto \int_0^\infty e^{-\alpha t} e^{i\omega t} dt = \frac{1}{\alpha - i\omega}. \tag{2.38}$$

The distribution of the intensity, $I(\omega)$, that would be obtained from a spectrograph of infinite resolving power is proportional to $|E(\omega)|^2$ and is given by the Lorentzian distribution [23]

$$I(\omega) = I_0 \left(\frac{\Gamma}{2\pi}\right) \frac{1}{\left(\omega - \omega_0 - \Delta\omega\right)^2 + \frac{1}{4}\Gamma^2}, \tag{2.39}$$

where I_0 is the total energy radiated. If we express the classical line width $\Gamma/2$ for an oscillator in terms of the wavelength we get a universal constant

$$\Delta\lambda = \frac{2\pi c}{\omega_0^2}\Gamma = 2\pi c\tau = 1.2 \times 10^{-4} \quad \text{Angstrom.} \tag{2.40}$$

In Chapter 18 we will study the quantum theory of the natural line width.

2.5   Frictional Forces Quadratic in Velocity

The drag force on an object which moves with high speed through a gas (e.g. the motion of a meteoroid through atmosphere [24]) can be calculated in a simple way if we assume that the mean free path of the molecule is large compared to the linear dimensions of the object. Let us denote the mass of the object by M, its effective cross-sectional area by S and the density of the gas by ρ. For simplicity we assume that the collisions are inelastic and the molecules stick to the object. Then the law of conservation of momentum implies that

$$Mv = (M + \rho Svdt)(v + dv), \tag{2.41}$$

from which it follows that the force of friction is proportional to the square of velocity

$$\frac{dv}{dt} = -\frac{\rho S}{M}v^2. \tag{2.42}$$

For the general case the last equation can be written as

$$\frac{dv}{dt} = -\Gamma\frac{\rho S}{M}v^2, \tag{2.43}$$

where Γ is a dimensionless constant.

The atmospheric drag force on an artificial satellite which enters the Earth atmosphere also depends quadratically on the velocity of the satellite, i.e.

$$F = -\frac{1}{2}C\rho Sv^2, \tag{2.44}$$

where ρ is the air density which varies exponentially with height above Earth's surface, S is the effective cross section of the satellite and C is the drag coefficient based on the shape of satellite. The energy loss due to the dissipative force is reflected into the orbital motion as a minor contraction of the orbit.

Air drag tends to make the elliptic orbit closer to the circular orbit by constantly reducing the apogee distance while having a minor effect on the perigee distance [25]. The motion of a projectile in the atmosphere with a quadratic drag force is an interesting example of the modification of the orbit caused by the dissipative forces [26].

2.6 Non-Newtonian and Nonlocal Dissipative Forces

Dissipative forces which explicitly depend on acceleration, i.e. $\mathbf{F} = \mathbf{F}(\mathbf{r}, \dot{\mathbf{r}}, \ddot{\mathbf{r}}, t)$ are called non-Newtonian [27]. These forces violate some of the basic principles of the Newtonian dynamics. For instance, in the presence of these forces, the total acceleration is not given by the vector sum of accelerations produced by each individual force. Equation (2.24) describing the relativistic motion of an electron is an example of a non-Newtonian force law.

When a particle interacts with a system of particles, or when an extended object moves in a resistive medium, then the effective force on the particle or on the extended object is nonlocal. This nonlocality can be spatial or temporal. Examples of the latter type of nonlocality are given in Chapter 8, where the effective force felt by the particle at the time t is given by

$$F(x, t) = \int_0^t K(t - t') x(t') dt'. \qquad (2.45)$$

Here the motion is assumed to be one-dimensional.

Nonlocal forces which occur in the description of the motion of an extended object depend on the shape and structure of the object and have the general form

$$\mathbf{F}(\mathbf{r}, \dot{\mathbf{r}}, t) = \int \mathbf{Q}(\mathbf{r}, \dot{\mathbf{r}}; \mathbf{r}', \dot{\mathbf{r}}', t) d^3 r', \qquad (2.46)$$

where the integral is taken over the volume of the object. Under certain conditions we can expand the integrand in (2.46) in powers of velocity with the result that

$$\mathbf{F}(\mathbf{r}, \dot{\mathbf{r}}, t) \approx \mathbf{f}(\mathbf{r}) - \lambda_1(\mathbf{r})\dot{\mathbf{r}} - \lambda_2(\mathbf{r})|\dot{\mathbf{r}}|\dot{\mathbf{r}} - \cdots \qquad (2.47)$$

The analogue of this type of nonlocal force (or potential) is important in the quantum theory of nuclear structure where it is obtained from the coupling of the object to a large number of different channels (see Chapters 18 and 19).

Bibliography

[1] See for instance J.B. Marion, *Classical Dynamics of Particles and Systems*, (Academic Press, New York, 1965) p. 65.

[2] D.H.E. Gross, Nucl. Phys. A240, 472 (1975).

[3] W.E. Schröder and J.R. Huizenga in *Treatise on Heavy-Ion Scattering, vol. 2, Fusion and Qusi-Fusion Phenomena*, Edited by D.A. Bromley, (Plenum Press, New York, 1984) p. 192.

[4] G. Stokes, Tran. Cambridge Phil. Soc. Soc. vol. 9 (1850).

[5] L.D. Landau and E.M. Lifshitz, *Fluid Mechanics*, (Pergamon Press, London, 1959) pp. 67-68.

[6] H. Lamb, *Hydrodynamics*, Sixth Edition (Cambridge University Press, 1932) §342-343.

[7] W. Kaufmann, *Fluid Mechanics*, (McGraw-Hill, New York, N.Y. 1963) p. 236.

[8] C.W. Oseen, Arkiv Mat. Astron. Fys. vol. 6. no. 29 (1910).

[9] J.W.S. Rayleigh, *The Theory of Sound*, vol I (Dover Publications, New York, N.Y. 1945) p. 45.

[10] M. Abraham, Ann. Physik, 10, 105 (1903).

[11] H.A. Lorentz, *The Theory of Electrons and Its Applications to the Phenomena of Light and Radiant Heat*, Second Edition (Dover Publications, New York, 1952).

[12] J.D. Jackson, *Classical Electrodynamics*, Third Edition (John Wiley & Sons, New York, 1999) Chap. 16.

[13] F. Rohrlich, *Classical Charged Particles*, (Addison-Wesley, Reading, 1965) p. 214.

[14] B. Leaf, Phys. Rev. 127, 1369 (1962).

[15] B. Leaf, Phys. Rev. 132, 1321 (1963).

[16] See for instance A.O. Barut, *Electrodynamics and Classical Theory of Fields and Particles*, (The McMillan Company, New York, 1964).

[17] For an interesting argument due to Kirchhoff regarding equations of motion with higher derivatives see *Concepts of Force*, (Harvard University Press, Cambridge, Massachusetts, 1957) p. 223.

[18] M. von Laue, Annalen der Physik, 28, 436 1909).

[19] P.A.M. Dirac, Proc. Roy. Soc. A167, 148 (1932).

[20] F. Bopp, Zeit. Angw. Phys. 14, 699 (1962).

[21] W. Heitler *The Quantum Theory of Radiation*, Third Edition (Oxford University Press, London, 1954) p. 33.

[22] R.H. Good Jr. and T.J. Nelson, *Classical Theory of Electric and Magnetic Fields*, (Academic Press, New York, 1971) p. 509.

[23] A. Corney, *Atomic and Laser Spectroscopy*, (Oxford University Press, London, 1977) Chapter 8.

[24] C. Kittel, W.D. Knight and M.A. Rutherman, *Berkeley Physics Course* volume 1 (McGraw-Hill, New York, N.Y. 1962) p. 188.

[25] See for instance R. Deutsch, *Orbital Dynamics of Space Vehicles*, (Prentice-Hall, Englewood Cliffs, N.J. 1963) p. 207.

[26] D. Hesteness, *New Foundations for Classical Mechanics*, (D. Reidel Publishing Company , 1986, Dordrecht) p. 140.

[27] R.M. Santilli, *Foundations of Theoretical Mechanics II*, (Springer-Verlag, New York, N.Y. 1983) p. 2.

Chapter 3

Lagrangian Formulations

In this chapter we want to review the canonical formulation of the simplest forms of dissipative systems. The methods of construction of the Lagrangian and the Hamiltonian for the dissipative systems have been discussed in a number of papers and books [1]-[8]. Once a Lagrangian or a Hamiltonian for the system is found we can obtain conserved quantities associated with the motion as well as the solution, for the latter either by using the conventional technique of integrating the equations of motion or by applying methods based on the calculus of variation. But the main purpose of the Hamiltonian (or the Lagrangian) formulation is to find the quantum analogue of the dissipative motion.

3.1   Rayleigh and Lur'e Dissipative Functions

The simplest way of incorporating dissipative forces in the Lagrangian formalism is by introducing Raleigh's dissipation function $\mathcal{F}$, where this $\mathcal{F}$ is defined in such a way that $-\frac{\partial \mathcal{F}}{\partial \dot{x}_j}$ is the j-th component of the dissipative force [9] [10]. The equations of motion in this case are given by

$$\frac{d}{dt}\left(\frac{\partial L}{\partial \dot{x}_j}\right) - \frac{\partial L}{\partial x_j} + \frac{\partial \mathcal{F}}{\partial \dot{x}_j} = 0, \quad j = 1, 2, \cdots \tag{3.1}$$

For instance if the force of friction is linear in velocity

$$\mathbf{f} = -m\lambda\dot{\mathbf{r}}, \tag{3.2}$$

then $\mathcal{F}$ is given by

$$\mathcal{F} = \frac{m}{2}\lambda\dot{\mathbf{r}}^2. \tag{3.3}$$

In this approach $2\mathcal{F}$ represents the rate of energy dissipation due to friction since

$$dW_f = -\mathbf{f} \cdot d\mathbf{r} = -\mathbf{f} \cdot \mathbf{v}dt = m\lambda\dot{\mathbf{r}}^2. \tag{3.4}$$

One can generalize Rayleigh's dissipation function to include nonlinear damping forces [10]. Lur'e has formulated this problem by considering the damping force with the components

$$f_j = -k_j(x_1, x_2, \cdots x_N)g_j(\dot{x}_i), \tag{3.5}$$

where k_j s are positive functions of their arguments and

$$\dot{x}_j g_j(\dot{x}_i) \geq 0. \tag{3.6}$$

Here x_j s are the Cartesian components of the N-dimensional configuration space. We can transform the coordinates $x_j(t)$ by the transformation

$$x_j = x_j(q_1, q_2 \cdots q_N), \tag{3.7}$$

to the generalized coordinates $q_j(t)$. Then the generalized force becomes

$$Q_s^L = -\sum_{j=1}^{N} k_j(x_1, x_2, \cdots x_N)g_j(\dot{x}_i)\frac{\partial x_j}{\partial q_s}. \tag{3.8}$$

From Eq. (3.7) it follows that

$$\dot{x}_j = \sum_{s=1}^{N} \frac{\partial x_j}{\partial q_s}\dot{q}_s, \tag{3.9}$$

and therefore

$$\frac{\partial \dot{x}_j}{\partial \dot{q}_s} = \frac{\partial x_j}{\partial q_s}. \tag{3.10}$$

Substituting (3.10) in (3.8) we find Q_s^L to be

$$Q_s^L = -\sum_{j=1}^{N} k_j(x_1, x_2, \cdots x_N)g_j(\dot{x}_i)\frac{\partial \dot{x}_j}{\partial \dot{q}_s}. \tag{3.11}$$

Now Lur'e defines the dissipation function $\mathcal{F}^L$ by

$$\mathcal{F}^L = \sum_{j=1}^{N} k_j(x_1, x_2, \cdots x_N)\int_0^{\dot{x}_j} g_j(y)dy. \tag{3.12}$$

From this definition and Eq. (3.11) we have

$$\frac{\partial \mathcal{F}^L}{\partial \dot{q}_s} = -Q_s^L, \quad s = 1, 2, \cdots N. \tag{3.13}$$

This result should be compared with Eq. (3.3) for the Rayleigh function. The latter case is a special case of (3.12) if we choose

$$g_j(\dot{x}_j) = \dot{x}_j, \quad k_j = \text{constant.} \tag{3.14}$$

Let us consider the special case of a one-dimensional motion where the force of friction is of the form

$$f = -k|\dot{x}|^n. \tag{3.15}$$

In this relation n is an even integer and k is a positive constant. Then from Eqs (3.11) and (3.12) we find

$$\mathcal{F}^L = \frac{k}{n+1}|\dot{x}|^{n+1}, \tag{3.16}$$

and

$$Q^L = -k\dot{x}^n \text{sgn}(\dot{x}), \tag{3.17}$$

where the function $\text{sgn}(\dot{x})$ is defined by

$$\text{sgn}(\dot{x}) = \begin{cases} 1 & \text{for } \dot{x} > 0 \\ -1 & \text{for } \dot{x} < 0 \end{cases}. \tag{3.18}$$

The Rayleigh and the Lur'e dissipative functions will account for the damping forces proportional to a given power of velocity, but not to the radiation reaction force which is proportional to the time derivative of the acceleration.

There are physical systems such as Drude model for the motion of a valence electron where both of these forces are present [11]. For these systems we replace $\mathcal{F}$ by $\mathcal{L} = \mathcal{F} + \mathcal{G}$ where

$$\mathcal{F} = \frac{1}{2}\sum_{i,j=1}^{3} \dot{x}_i R_{ij} \dot{x}_j, \tag{3.19}$$

and

$$\mathcal{G} = \frac{1}{2}\sum_{i,j=1}^{3} \ddot{x}_i P_{ij} \ddot{x}_j. \tag{3.20}$$

The equations of motion in this case are given by the generalized Euler-Lagrange equations:

$$\left\{\frac{\partial \mathcal{L}}{\partial \dot{x}_i} - \frac{d}{dt}\left(\frac{\partial \mathcal{L}}{\partial \ddot{x}_i}\right)\right\} - \left\{\frac{\partial L}{\partial x_i} - \frac{d}{dt}\left(\frac{\partial L}{\partial \dot{x}_i}\right)\right\} = f_i, \tag{3.21}$$

where

$$L = T - V(x_i), \tag{3.22}$$

and f_i is the i-th component of the external time-dependent force. Note that L has the dimension of energy, but $\mathcal{L}$ has the dimension of power. For small displacements we can take V to be a quadratic function of the coordinates,

$$\frac{1}{2}\sum_{i,j=1}^{3} x_i V_{ij} x_j, \tag{3.23}$$

T as the sum of the quadratic terms in velocities,

$$\frac{1}{2} \sum_{i,j=1}^{3} \dot{x}_i M_{ij} \dot{x}_j, \tag{3.24}$$

and in addition we can have a "gyroscopic" term,

$$\frac{1}{2} \sum_{i,j=1}^{3} \dot{x}_i G_{ij} x_j. \tag{3.25}$$

Here the matrix elements M_{ij} and V_{ij} are real and symmetrical, while G_{ij} s are asymmetrical, but all are constants. Using these quadratic forms we find the linear equations of motion for x_i s

$$\sum_{j=1}^{3} \left\{ -P_{ij}\frac{d^3 x_j}{dt^3} + M_{ij}\frac{d^2 x_j}{dt^2} + B_{ij}\frac{dx_j}{dt} + V_{ij}x_j \right\} = f_i, \quad i = 1, 2, 3, \tag{3.26}$$

where

$$B_{ij} = R_{ij} + G_{ij}. \tag{3.27}$$

The theory of damped heavy ion scattering offers an interesting example of the application of the above-mentioned Lagrangian formulation. While we expect that in the domain of nuclear scattering the problem has to be treated quantum mechanically, but under the conditions where semi-classical approximations such as WKB are valid, we can justify a classical description for the motion of the nuclei.

As a specific case let us consider the classical collision of two rigid spherical nuclei of radii R_P and R_T, where P and T refer to the projectile and target nuclei respectively. We formulate this scattering using the generalized coordinates q_i s instead of x_i in Eq. (3.1), and write the Raleigh dissipation function as [12]

$$\mathcal{F} = \frac{1}{2} \sum_{i,j} k_{ij} \dot{q}_i \dot{q}_j, \tag{3.28}$$

where k_{ij} is the friction tensor.

We note that there are four degrees of freedom for the collision. These are: the distance r between the centers of the projectile and the target, their relative orientation angles θ_P and θ_T, and the angle of orientation of the total system θ. We assume that the interaction potentials which consists of the nuclear and Coulomb are only functions of r. Thus the Lagrangian depends on r, $\dot{r}$ and $\dot{\theta}$,

$$L = \frac{\mu}{2}\left(\dot{r}^2 + r^2\dot{\theta}^2\right) + \frac{1}{2}I_P\dot{\theta}_P^2 + \frac{1}{2}I_T\dot{\theta}_T^2 - V_{Coul}(r) - V_N(r). \tag{3.29}$$

Here I_P and I_T denote the moments of inertia of the projectile and the target about their centers, i.e.

$$I_P = \frac{2}{5}M_P R_P^2, \quad I_T = \frac{2}{5}M_T R_T^2, \tag{3.30}$$

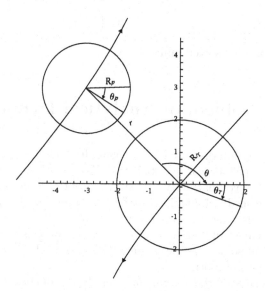

Figure 3.1: Scattering of the projectile P from the target T showing the degrees of freedom of the motion. Both P and T are assumed to be rigid spheres.

μ is the reduced mass

$$\mu = \frac{M_P M_T}{(M_P + M_T)},$$

(3.31)

and V_{Coul} and V_N are Coulomb and nuclear potentials respectively.

Associated with these four degrees of freedom we can introduce a symmetric 4×4 matrix for the friction tensor k_{ij}. However to fit the observed data it seems that we need a coefficient of friction k_r for the radial motion, and a coefficient k_t for the relative sliding motion of the two nuclear surfaces [12] [13]. Thus $\mathcal{F}$ can be written as

$$\mathcal{F} = \frac{1}{2} \left\{ k_r \dot{r}^2 + k_t \left(\frac{r}{R_P + R_T} \right)^2 \left[R_P \left(\dot{\theta}_P - \dot{\theta} \right) + R_T \left(\dot{\theta}_T - \dot{\theta} \right) \right]^2 \right\}.$$

(3.32)

The equations of motion derived from (3.29) and (3.32) are:

$$\mu \ddot{r} - \mu r \dot{\theta}^2 + k_r \dot{r} + \frac{\partial V_{Coul}}{\partial r} + \frac{\partial V_N}{\partial r} = 0,$$

(3.33)

$$\mu r^2 \ddot{\theta} + I_P \ddot{\theta}_P + I_T \ddot{\theta}_T = 0,$$

(3.34)

$$I_P \ddot{\theta}_P = -k_t \left(\frac{r}{R_P + R_T} \right)^2 R_P \left(\dot{\theta}_P - \dot{\theta} \right),$$

(3.35)

and

$$I_T \ddot{\theta}_T = -k_t \left(\frac{r}{R_P + R_T} \right)^2 R_T \left(\dot{\theta}_T - \dot{\theta} \right). \tag{3.36}$$

These equations can only be solved numerically.

3.2   Inverse Problem of Analytical Dynamics

While the simple formulation discussed in §3.1 is useful for solving certain classical problems, for the purpose of quantizing dissipative systems and for the questions relating to the compatibility of the commutation relations with the equations of motion we need to find a Lagrangian describing the damped motion and at the same time defining the canonical coordinates. In what follows we will see that for a given set of equations of motion we can construct infinitely many Lagrangians or in some cases we cannot find a single Lagrangian which will give us the equations of motion.

The non-uniqueness of the Lagrangian and Hamiltonian formalism for conservative as well as dissipative systems have been discussed extensively. Excellent surveys of the essential works in construction of q-equivalent Lagrangians, i.e. Lagrangians generating the same equation of motion in coordinate space are given by Dodonov *et al.* [14] and by Santilli [5]. Here we will briefly mention conditions under which Lagrangians for a given set of equations of motion do exist and a method for the construction of these Lagrangians.

In classical dynamics for a conservative system the simple expression of

$$L \left(\dot{x}_i, t \right) = T \left(x_i, \dot{x}_i \right) - V(x_i, t), \tag{3.37}$$

or in terms of generalized coordinates $q_j(t)$ s and generalized velocities $\dot{q}_j(t)$ s [4]

$$L \left(q_i, \dot{q}_i, t \right) = T \left(q_i, \dot{q}_i \right) - V(q_i, t), \tag{3.38}$$

is generally regarded as the Lagrangian of the system [15]. Here $T \left(x_i, \dot{x}_i \right)$ is the kinetic and $V(x_i, t)$ is the potential energy of the system and x_i s and $\dot{x}_i$ s are the $2N$ coordinates and velocities of the particles in the system and if we define $L \left(x, \dot{x}, t \right)$ by (3.37), then it is unique. However we may define $L \left(x, \dot{x}, t \right)$ as the generator of the equation of motion, i.e. a twice differentiable function of the coordinates x_i, and of velocities $\dot{x}_i$ in such a way that when this L is substituted in the Euler-Lagrange equation

$$\frac{\delta L}{\delta x_i} \equiv \frac{\partial L}{\partial x_i} - \frac{d}{dt} \left(\frac{\partial L}{\partial \dot{x}_i} \right) = 0, \quad i = 1, 2, \cdots N, \tag{3.39}$$

the equations of motion are obtained. That the first definition (3.37) is a special case of L satisfying Eq. (3.39) for conservative systems is well-known [15]. When the system is not conservative then we may or may not have Lagrangians and when we have, then the Lagrangian is not unique.

For the equations of motion of second order the question of existence of the Lagrangian and its dependence on the number of degrees of freedom has been studied by a number of authors [3] [16]-[18]. If the motion is one-dimensional, we can always find a Lagrangian whose variational derivative yields the equation of motion. When the number of degrees of freedom is equal or greater than two, then we may or may not be able to have a Lagrangian formulation [17].

As an example of the systems with no Lagrangian let us consider the coupled system with the equations of motion

$$\begin{cases} m_1\ddot{x}_1 = A\left(x_1 - x_2\right)^\alpha \\ m_2\ddot{x}_2 = B\left(x_2 - x_1\right)^\beta \end{cases}. \tag{3.40}$$

This system, for arbitrary values of A, B, α and β has no Lagrangian. As a second example of a motion which cannot be derived from a Lagrangian we can cite the oscillations of two coupled oscillators with different frequencies and different damping constants:

$$\begin{cases} \ddot{x}_1 + \lambda_1\dot{x}_1 + \omega_1^2 x_1 - c_2 x_2 = 0 \\ \ddot{x}_2 + \lambda_2\dot{x}_2 + \omega_2^2 x_2 - c_1 x_1 = 0 \end{cases}, \tag{3.41}$$

where c_1 and c_2 are constants.

The non-uniqueness of the Lagrangians function for dynamical systems will be studied in detail later. But first let us examine the conditions for the existence of a Lagrangian function and how it can be constructed [3] [16] [19].

The laws governing the motion of a dynamical system can be formulated according to Hamilton's principle [20] and this principle states that there is a function (or functions) $L\left(x_1, \cdots x_N; \dot{x}_1 \cdots \dot{x}_N, t\right)$ and this L satisfies the following condition:

If the system at the instant t_1 has the coordinates $x_i^{(1)}$ and at t_2 has the coordinates $x_i^{(2)}$ then this condition implies that the functional differential

$$\begin{aligned} \delta S &= \delta \int_{t_1}^{t_2} L\left(x_i, \dot{x}_i, t\right) dt = \int_{t_1}^{t_2} \sum_{j=1}^N \frac{\delta L\left(x_i, \dot{x}_i, t\right)}{\delta x_j} \delta x_j(t) dt \\ &= \int_{t_1}^{t_2} \sum_{j,k} \mu_{jk}\left(x_i, \dot{x}_i, t\right) \left[\ddot{x}_k - \frac{1}{m} f_k\left(x_i, \dot{x}_i, t\right)\right] \delta x_j(t) dt, \\ & \qquad j = 1, 2, \cdots N, \end{aligned} \tag{3.42}$$

must vanish identically for any continuous variation of the path consistent with the requirement

$$\delta x_j\left(t_1\right) = \delta x_j\left(t_2\right) = 0. \tag{3.43}$$

In Eq. (3.42) μ_{jk} is a positive definite matrix which may be viewed as an integrating factor [3] [19] [21]. Thus the requirement of the extremum of action S implies the vanishing of the integrand between any two arbitrary instances t_1 and t_2. Since we have imposed the condition of positive definiteness on the integrating factor μ_{jk} and since the variation $\delta x_j(t)$ is arbitrary (apart from the continuity condition and the variational conditions of (3.43)), therefore we conclude that

$$\ddot{x}_k(t) = \frac{1}{m} f_k\left(x_i, \dot{x}_i, t\right),\qquad (3.44)$$

which is Newton's second law of motion.

The Lagrangian function determined from the requirement of $\delta S \equiv 0$ is not unique. We can add the total time derivative of any function $\Phi\left(x, \dot{x}, t\right)$ to the Lagrangian without affecting the resulting equations of motion (3.44).

For the special case of $\mu_{jk} = \delta_{jk}$ we find the Lagrangian to be the same as that given by (3.37).

The second variation of S with respect to $\delta x_j(t)$ and $\delta x_k(t')$ must satisfy the identity

$$\frac{\delta^2 S}{\delta x_j\left(t\right)\delta x_k\left(t'\right)} = \frac{\delta^2 S}{\delta x_k\left(t'\right)\delta x_j\left(t\right)}.\qquad (3.45)$$

By imposing this condition on (3.42) we find the following equations which μ_{jk} has to satisfy:

$$\mu_{jk} = \mu_{kj},\qquad (3.46)$$

$$\frac{\partial \mu_{kj}}{\partial \dot{x}_l} = \frac{\partial \mu_{jl}}{\partial \dot{x}_k} = \frac{\partial \mu_{lk}}{\partial \dot{x}_j},\qquad (3.47)$$

$$\hat{\mathcal{D}}\mu_{jk} + \frac{1}{2m}\sum_l\left(\mu_{jl}\frac{\partial f_l}{\partial \dot{x}_k} + \mu_{kl}\frac{\partial f_l}{\partial \dot{x}_j}\right) = 0,\qquad (3.48)$$

and

$$\hat{\mathcal{D}}\sum_k\left(\mu_{ik}\frac{\partial f_k}{\partial \dot{x}_j} - \mu_{jk}\frac{\partial f_k}{\partial \dot{x}_i}\right) - 2\sum_k\left(\mu_{ik}\frac{\partial f_k}{\partial \dot{x}_j} - \mu_{jk}\frac{\partial f_k}{\partial \dot{x}_i}\right) = 0,\qquad (3.49)$$

where in Eqs. (3.48) and (3.49) $\hat{\mathcal{D}}$ denotes the following operator

$$\hat{\mathcal{D}} = \frac{\partial}{\partial t} + \sum_k\left[\dot{x}_k\frac{\partial}{\partial x_k} + \frac{1}{m}f_k\left(x_i, \dot{x}_i, t\right)\frac{\partial}{\partial \dot{x}_k}\right].\qquad (3.50)$$

These are the Helmholtz conditions [3] [16].

We can construct the Lagrangian from the equations of motion $m\ddot{x}_k = f_k\left(x_i, \dot{x}_i, t\right)$ by solving a linear partial differential equation. To this end we first write the following identity,

$$\begin{aligned}
\frac{\delta L}{\delta x_k} &= \frac{\partial L}{\partial x_k} - \frac{d}{dt}\left(\frac{\partial L}{\partial \dot{x}_k}\right) \equiv \frac{\partial L}{\partial x_k} - \frac{\partial^2 L}{\partial t\partial \dot{x}_k} \\
&\quad - \sum_j \dot{x}_j\frac{\partial^2 L}{\partial x_j\partial \dot{x}_k} - \sum_j \ddot{x}_j\frac{\partial^2 L}{\partial \dot{x}_j\partial \dot{x}_k} = 0,
\end{aligned}$$

$$(3.51)$$

and then we substitute for $\ddot{x}$ from the equation of motion to obtain

$$\frac{\partial L}{\partial x_k} - \frac{\partial^2 L}{\partial t \partial \dot{x}_k} - \sum_j \dot{x}_j \frac{\partial^2 L}{\partial x_j \partial \dot{x}_k} - \frac{1}{m} \sum_j \frac{\partial^2 L}{\partial \dot{x}_j \partial \dot{x}_k} f_j(x_i, \dot{x}_i, t) = 0. \qquad (3.52)$$

This equation together with the set of equations (3.46)-(3.49) determines both $L(x, \dot{x}, t)$ and $\mu(x, \dot{x}, t)$.

For a one-dimensional motion Eq. (3.52) simplifies and later we will use it to find the Lagrangian for simple damping systems. But there are other ways of constructing the Lagrangian for motions confined to one dimension [7] [22]. Let us start with the equation of motion

$$m\ddot{x} = f(x, \dot{x}), \qquad (3.53)$$

and write the Lagrangian for this motion as

$$L(x, \dot{x}) = \dot{x} \int^{\dot{x}} \frac{1}{v^2} G(v, x) dv, \qquad (3.54)$$

where $G(v, x)$ is a function to be determined later. By substituting $L(x, \dot{x})$ in the Euler-Lagrange equation we find

$$\frac{d}{dt}\left(\frac{\partial L}{\partial \dot{x}}\right) - \frac{\partial L}{\partial x} = \frac{\ddot{x}}{\dot{x}}\left(\frac{\partial G(\dot{x}, x)}{\partial \dot{x}}\right) + \frac{\partial G(\dot{x}, x)}{\partial x} \equiv m\ddot{x} - f(\dot{x}, x). \qquad (3.55)$$

From Eq. (3.55) it follows that $G(\dot{x}, x)$ is the solution of the following differential equation

$$\frac{1}{m\dot{x}} f(x, \dot{x}) \frac{\partial}{\partial \dot{x}} G(\dot{x}, x) = -\frac{\partial}{\partial x} G(\dot{x}, x), \qquad (3.56)$$

provided that

$$\frac{1}{\dot{x}}\frac{\partial}{\partial \dot{x}} G(\dot{x}, x) \neq 0. \qquad (3.57)$$

Now let us consider two examples using this method of construction of the Lagrangian. When the equation of motion for a damped system is of the form

$$\ddot{x} + \frac{\dot{x}}{\left(\frac{dg(\dot{x})}{d\dot{x}}\right)} \frac{dV(x)}{dx} = 0, \qquad (3.58)$$

then the Lagrangian which is found by solving Eq. (3.56) is;

$$L(x, \dot{x}) = \dot{x} \int^{\dot{x}} \frac{1}{v^2} F[z(x, v)] dv, \qquad (3.59)$$

where F is an arbitrary function of its argument

$$z(v, x) = g(v) + V(x). \qquad (3.60)$$

However the condition

$$\frac{1}{v} \frac{dg(v)}{dv} \frac{dF(z)}{dz} \neq 0, \tag{3.61}$$

must be satisfied. Thus for the very simple case of $\ddot{x} + \lambda \dot{x} = 0$ and for $F(z) = z$ we find

$$L(x, \dot{x}) = \dot{x} \ln \dot{x} - \lambda x. \tag{3.62}$$

As a second example let us consider the damped harmonic oscillator

$$\ddot{x} + \lambda \dot{x} + \omega_0^2 x = 0. \tag{3.63}$$

By calculating $G(\dot{x}, x)$ we find that $G(\dot{x}, x) = F(z)$, where now

$$z = \frac{1}{2} \frac{\left[x + \left(\frac{\lambda}{2} - i\omega\right) \dot{x}\right]^{\nu^*}}{\left[x + \left(\frac{\lambda}{2} + i\omega\right) \dot{x}\right]^{\nu}} = z^*, \tag{3.64}$$

and

$$\nu^2 = \frac{\frac{\lambda}{2} + i\omega}{\frac{\lambda}{2} - i\omega}, \tag{3.65}$$

with $\omega^2 = \omega_0^2 - \frac{\lambda^2}{4}$. In the following section we study some special cases where L can be obtained analytically.

3.3   Some Examples of the Lagrangians for Dissipative Systems

(1) The simplest choice of μ_{jk} is $\mu_{jk} = \delta_{jk}$. Then Eqs. (3.48) and (3.49) imply that $\frac{\partial f_k}{\partial \dot{x}_j} = 0$ for all k s and j s. Hence f_k can be a function of the coordinate and time but not of velocity.

(2) For the one-dimensional motion Eqs. (3.49) and (3.51) reduce to

$$\frac{\partial \mu}{\partial t} + \dot{x} \frac{\partial \mu}{\partial x} + \frac{1}{m} \left[\frac{\partial \mu}{\partial \dot{x}} f(x, \dot{x}, t) + \mu \frac{\partial f(x, \dot{x}, t)}{\partial \dot{x}} \right] = 0, \tag{3.66}$$

and

$$\frac{\partial L}{\partial x} - \frac{\partial^2 L}{\partial t \partial \dot{x}} - \dot{x} \frac{\partial^2 L}{\partial x \partial \dot{x}} - \frac{1}{m} \frac{\partial^2 L}{\partial \dot{x}^2} f(x, \dot{x}, t) = 0. \tag{3.67}$$

We note that μ is a dimensionless function and L has the dimension of energy. If L in Eq. (3.67) is known, then μ can be obtained from L. Thus if we substitute for μ from

$$\mu = \frac{1}{m} \left(\frac{\partial^2 L}{\partial \dot{x}^2} \right), \tag{3.68}$$

in Eq. (3.66) and use Eq. (3.67) we find that (3.68) is a solution of (3.66).

For analytically solvable cases of Eq. (3.67) when the motion is dissipative

we have the following cases:
(a) If μ is only a function of time then from (3.66) we have

$$\frac{d\mu(t)}{dt} + \frac{\mu(t)}{m}\left(\frac{\partial f}{\partial \dot{x}}\right) = 0, \tag{3.69}$$

or f has to have the general form of

$$f = -m\lambda g(t)\dot{x} - \frac{\partial V(x)}{\partial x}, \tag{3.70}$$

where

$$\mu(t) = m\lambda \int^t g(t')\,dt'. \tag{3.71}$$

In Eq. (3.71) λ is a constant and $g(t)$ is an arbitrary function of time.

A Lagrangian which will reduce to $L = T - V$ in the limit of $\lambda \to 0$ is of the form

$$L = \left[\frac{1}{2}m\dot{x}^2 - V(x,t)\right]\exp\left(\lambda \int^t g(t')\,dt'\right). \tag{3.72}$$

The particular case of $g(t) = 1$ corresponds to the damping force linear in velocity where the force is defined by (2.15).
(b) When μ depends on x only, i.e. $\mu = \mu(x)$, then Eq. (3.66) reduces to

$$\dot{x}\frac{d\mu(x)}{dx} + \left(\frac{\mu(x)}{m}\right)\frac{\partial f}{\partial \dot{x}} = 0. \tag{3.73}$$

Thus the force $f(x, \dot{x})$ is of the form

$$f(x, \dot{x}) = -m\gamma\dot{x}^2 - \frac{\partial V(x)}{\partial x}, \tag{3.74}$$

and from Eq. (3.73) we find $\mu(x)$ to be

$$\mu(x) = \exp(2\gamma x). \tag{3.75}$$

In Eq. (3.74) we have written the x-dependent part of the force f in terms of the potential $V(x)$. From Eq. (3.68) and (3.75) we obtain the following expression for the Lagrangian

$$L = \frac{1}{2}m\dot{x}^2 \exp(2\gamma x) - W(x). \tag{3.76}$$

Now if we substitute (3.74) and (3.76) in Eq. (3.67) for L, we find $W(x)$ to be

$$W(x) = \int^x e^{-2\gamma y}\frac{\partial V(y)}{\partial y}\,dy, \tag{3.77}$$

which agrees with $L = T - V$ in the limit of $\gamma \to 0$.

As another example let us obtain the Lagrangian for the motion of a particle subject to the resistive force $m\beta\dot{x}^\nu$ where ν is a positive number $1 <$

$\nu < 2$. We can formulate this problem as that of the motion of a particle with a velocity-dependent mass moving in a constant potential. Let us write the equation of motion as

$$m\mu''(\dot{x}) \frac{d\dot{x}}{dt} + m\beta = 0, \tag{3.78}$$

where primes denote derivatives with respect to $\dot{x}$, and

$$\mu''(\dot{x}) = \dot{x}^{(-\nu)}. \tag{3.79}$$

The Lagrangian for this motion is given by [23]

$$L(\dot{x}, x) = m\mu(\dot{x}) - m\beta x. \tag{3.80}$$

The Euler-Lagrange equation for $L(\dot{x}, x)$, Eq. (3.80) yields the equation of motion (3.78).

The motion of a rocket is another example of the motion of a system with variable mass moving in a viscous medium. Consider a rocket with the initial mass $m_v + m_f$, where m_v is the mass of the rocket and m_f is the mass of fuel. Assuming a constant exhaust rate and denoting the exhaust velocity by u, we have the equation of motion [24];

$$\frac{dv}{dt} + \lambda v = -u \frac{d}{dt} \ln[m(t)], \tag{3.81}$$

where

$$m(t) = \begin{cases} m_v + m_f \left(1 - \frac{t}{T}\right) & \text{for } 0 \leq t \leq T \\ m_v & \text{for } t > T \end{cases}. \tag{3.82}$$

Using (3.82) we can write the right hand side of (3.81) as

$$-u \frac{d}{dt} \ln[m(t)] = \frac{\alpha u}{1 - \alpha t} = F(t), \tag{3.83}$$

where

$$\alpha = \frac{1}{T} \left(\frac{m_f}{m_v + m_f} \right). \tag{3.84}$$

Equation (3.81) can be integrated to yield $v(t)$ [24]

$$v(t) = u \exp\left[\frac{\lambda}{\alpha}(1 - \alpha t)\right] \left[Ei\left(-\frac{\lambda}{\alpha}\right) - Ei\left(-\frac{\lambda}{\alpha}(1 - \alpha t)\right) \right], \tag{3.85}$$

where

$$Ei(y) = -\int_{-y}^{\infty} \frac{e^{-z}}{z} dz, \tag{3.86}$$

is the exponential integral function [25]. The position of the rocket as a function of time can be found by integrating $v(t)$ with respect to t.

We can construct the Lagrangian for this system using equation (3.68) with $\mu(t) = e^{\lambda t}$. Thus we obtain the following expression for the Lagrangian

$$L = \left[\frac{1}{2}\dot{x}^2 + xF(t) \right] e^{\lambda t}. \tag{3.87}$$

3.4 Non-Uniqueness of the Lagrangian

If $L(x_i, \dot{x}_i, t)$ is the Lagrangian describing a dissipative system then the functional derivative of L, i.e. $\frac{\delta L}{\delta x_i}$, Eq. (3.51), must vanish. Now if we multiply L by a constant number, α, and add the total time derivative of a function of x_i and t to it we find a new Lagrangian $\bar{L}$,

$$\bar{L} = \alpha L + \frac{dg(x_i, t)}{dt}. \qquad (3.88)$$

This Lagrangian is equivalent to L, i.e. $\frac{\delta \bar{L}}{\delta x_i} = 0$ gives us the same equation of motion as Eq. (3.51). However the Lagrangian $\bar{L}$ is not the most general Lagrangian for the motion. We can show this in a simple way for one-dimensional motion of a particle. But this method, in general, cannot be extended to multi-dimensional motions. We start with Eq (3.51) for a single generalized coordinate x, and solve it for $\ddot{x}$, assuming that the coefficient of $\ddot{x}$ is not zero,

$$\ddot{x} = \left(\frac{\partial^2 L}{\partial \dot{x}^2}\right)^{-1} \left[\frac{\partial L}{\partial x} - \frac{\partial^2 L}{\partial \dot{x} \partial t} - \frac{\partial^2 L}{\partial \dot{x} \partial x} \dot{x}\right]. \qquad (3.89)$$

Now we assume that $\bar{L}$ is equivalent to L, i.e. $\frac{\delta \bar{L}}{\delta x} = 0$ yields the same equation of motion as $\frac{\delta L}{\delta x} = 0$

$$\ddot{x} = \left(\frac{\partial^2 \bar{L}}{\partial \dot{x}^2}\right)^{-1} \left[\frac{\partial \bar{L}}{\partial x} - \frac{\partial^2 \bar{L}}{\partial \dot{x} \partial t} - \frac{\partial^2 \bar{L}}{\partial \dot{x} \partial x} \dot{x}\right]. \qquad (3.90)$$

Next we define Λ called "fouling function" by [26]

$$\Lambda = \left(\frac{\partial^2 \bar{L}}{\partial \dot{x}^2}\right) \left(\frac{\partial^2 L}{\partial \dot{x}^2}\right)^{-1}, \qquad (3.91)$$

so that

$$\frac{\partial \bar{L}}{\partial x} - \frac{\partial^2 \bar{L}}{\partial \dot{x} \partial t} - \frac{\partial^2 \bar{L}}{\partial \dot{x} \partial x} \dot{x} = \Lambda \left(\frac{\partial L}{\partial x} - \frac{\partial^2 L}{\partial \dot{x} \partial t} - \frac{\partial^2 L}{\partial \dot{x} \partial x} \dot{x}\right). \qquad (3.92)$$

For any $x(t)$ which is the solution of the equation of motion (3.92) is satisfied. Now we want to show that Λ is a constant of motion. First we note that

$$\frac{d\Lambda}{dt} = \frac{\partial \Lambda}{\partial \dot{x}} \ddot{x} + \frac{\partial \Lambda}{\partial x} \dot{x} + \frac{\partial \Lambda}{\partial t}, \qquad (3.93)$$

and for the moment we do not assume that $x(t)$ is a solution of $\frac{\delta L}{\delta x} = 0$. By taking the partial derivatives of Λ, Eq. (3.91), with respect to x and t, and substituting the results in (3.93) and also eliminating $\ddot{x}$ between (3.93) and (3.89) we find

$$\frac{d\Lambda}{dt} = \frac{\partial \Lambda}{\partial \dot{x}} \left(\frac{\partial^2 L}{\partial \dot{x}^2}\right)^{-1} \left[\frac{\delta L}{\delta x} + \frac{\partial L}{\partial x} - \frac{\partial^2 L}{\partial \dot{x} \partial t} - \frac{\partial^2 L}{\partial \dot{x} \partial x} \dot{x}\right]$$

$$- \left(\frac{\partial^2 L}{\partial \dot{x}^2} \right)^{-2} \left(\frac{\partial^2 \bar{L}}{\partial \dot{x}^2} \right) \left(\frac{\partial^3 L}{\partial \dot{x}^2 \partial x} \dot{x} + \frac{\partial^3 L}{\partial \dot{x}^2 \partial t} \right)$$

$$+ \left(\frac{\partial^2 L}{\partial \dot{x}^2} \right)^{-1} \left(\frac{\partial^3 \bar{L}}{\partial \dot{x}^2 \partial x} \dot{x} + \frac{\partial^3 \bar{L}}{\partial \dot{x}^2 \partial t} \right).$$

$$(3.94)$$

We also take the partial derivative of Eq. (3.92) with respect to $\dot{x}$

$$\left(\frac{\partial^3 \bar{L}}{\partial \dot{x}^2 \partial x} \dot{x} + \frac{\partial^3 \bar{L}}{\partial \dot{x}^2 \partial t} \right) = \frac{\partial \Lambda}{\partial \dot{x}} \left[\frac{\partial^2 L}{\partial \dot{x} \partial t} + \frac{\partial^2 L}{\partial \dot{x} \partial x} \dot{x} - \frac{\partial L}{\partial x} \right]$$

$$+ \Lambda \left(\frac{\partial^3 L}{\partial \dot{x}^2 \partial x} \dot{x} + \frac{\partial^3 L}{\partial \dot{x}^2 \partial t} \right). \qquad (3.95)$$

From Eqs. (3.94) and (3.95) it follows that

$$\frac{d\Lambda}{dt} = \frac{\partial \Lambda}{\partial \dot{x}} \left(\frac{\delta L}{\delta x} \right), \qquad (3.96)$$

and this is true whether $\delta L / \delta x$ is zero or not.

When $x(t)$ is a solution of Euler-Lagrange equation, $\frac{\delta L}{\delta x} = 0$, then $\frac{d\Lambda}{dt} = 0$ and Λ is a constant of motion. Conversely for any constant of motion Λ, there does exist a Lagrangian $\bar{L}$ such that [26]

$$\frac{\delta \bar{L}}{\delta x} = \Lambda \frac{\delta L}{\delta x}. \qquad (3.97)$$

As a specific example let us consider the simple case of a particle moving in a viscous medium with linear damping, where the Lagrangian is

$$L = \frac{1}{2} m \dot{x}^2 e^{\lambda t}. \qquad (3.98)$$

The generalized Lagrangian $\bar{L}$ can be found from Eq. (3.98),

$$\frac{\partial^2 \bar{L}}{\partial \dot{x}^2} = \Lambda \frac{\partial^2 L}{\partial \dot{x}^2} = m \Lambda e^{\lambda t}. \qquad (3.99)$$

Since Λ is a constant of motion, we can find it from L. By noting that

$$\frac{d}{dt} \left(\frac{\partial L}{\partial \dot{x}} \right) = 0, \quad \text{therefore} \quad m \dot{x} e^{\lambda t} = \text{constant}, \qquad (3.100)$$

we have a possible choice of Λ

$$\Lambda = \frac{m \dot{x} e^{\lambda t}}{p_0}, \qquad (3.101)$$

with p_0 being a constant ($p_0 \neq 0$). From Eqs. (3.99) and (3.101) we find

$$\bar{L} = \frac{1}{6 p_0} m^2 \dot{x}^3 e^{2\lambda t} + g_1(x, t) \dot{x} + g_2(x, t), \qquad (3.102)$$

where $g_1(x,t)$ and $g_2(x,t)$ are functions of their arguments. By substituting $\bar{L}$ in the Euler-Lagrange equation we get the result

$$\frac{\partial g_1(x,t)}{\partial t} = \frac{\partial g_2(x,t)}{\partial x}, \tag{3.103}$$

that is $g_1(x,t)\dot{x} + g_2(x,t)$ is the total time derivative of a function $g(x,t)$.

In a similar way we find the Lagrangian

$$\bar{L} = \frac{1}{24}m^2\dot{x}^4 e^{4\gamma x} + \frac{dg(x,t)}{dt}, \tag{3.104}$$

for the damping force $\left(-\gamma\dot{x}^2\right)$ acting on the particle. Here we have used $\Lambda = \frac{1}{2}m\dot{x}^2 e^{2\gamma x}$ as the constant of motion.

The fouling method can be used to construct the Lagrangian for a given frictional force law provided that a first integral of motion is known [7]. Let $\Lambda(x,\dot{x})$ be a first integral for a one-dimensional motion in a viscous medium and let $F(\Lambda)$ be an arbitrary but differentiable function of its argument, $\Lambda(x,\dot{x})$, then the Lagrangian for this motion is expressible as (see also Eq. (3.54))

$$L = \dot{x}\int^{\dot{x}} \frac{G\left(\Lambda(x,v)\right)}{v^2}dv. \tag{3.105}$$

To show that L is a solution of the Euler-Lagrange equation we calculate $\frac{\partial L}{\partial x}$ and $\frac{d}{dt}\left(\frac{\partial L}{\partial \dot{x}}\right)$, and we simplify the results using the fact that $\frac{dF(\Lambda)}{dt} = 0$. By substituting these in (3.39) we observe that L is indeed a Lagrangian.

If we apply this method to the problem of the damped harmonic oscillator, $\ddot{x} + \lambda\dot{x} + \omega_0^2 x = 0$, we find that $\Lambda(x,\dot{x})$ is a solution of the partial differential equation

$$\frac{d\Lambda}{dt} = \dot{x}\frac{\partial\Lambda}{\partial x} - \left(\lambda\dot{x} + \omega_0^2 x\right)\frac{\partial\Lambda}{\partial \dot{x}} = 0. \tag{3.106}$$

This equation can be solved by the method of characteristics, i.e. by solving the set of ordinary differential equations [27]

$$\frac{dx}{\dot{x}} = -\frac{d\dot{x}}{\lambda\dot{x} + \omega_0^2 x} = d\Lambda, \tag{3.107}$$

and these yield the result

$$\Lambda(x,\dot{x}) = \exp\left\{\frac{1}{2}\ln\left(\dot{x}^2 + \lambda x\dot{x} + \omega_0^2 x\right) - \frac{\lambda}{2\omega}\tan^{-1}\left[\frac{1}{2\omega}\left(\frac{2\dot{x}}{x} + \lambda\right)\right]\right\}. \tag{3.108}$$

Thus the general form of the time-independent Lagrangian for the damped oscillator is expressed by the integral Eq. (3.105) where $\Lambda(x,\dot{x})$ is given by (3.108).

3.5 Acceptable Lagrangians for Dissipative Systems

By solving the set of differential equations (3.52) we find a finite or infinite set of Lagrangians L_i. But not all of these L_i's are acceptable, some because of their singular or defective nature, and others because they violate other physical requirements. We have already obtained an explicitly time-dependent Lagrangian for the damped harmonic oscillator,

$$\ddot{x} + \lambda \dot{x} + \omega_o^2 x = 0, \tag{3.109}$$

for which from Eqs. (3.70) and (3.72) we have

$$L = \frac{m}{2} \left(\dot{x}^2 - \omega_0^2 x^2 \right) e^{\lambda t}. \tag{3.110}$$

Among the explicitly time-independent Lagrangians we can obtain the following Lagrangian if we choose $F(\Lambda) = -\Lambda$ in (3.105) (see Chapter 7 Eq. (7.56))

$$
\begin{aligned}
L_2(x, \dot{x}) =\ & -\frac{1}{2} \ln \left[\dot{x}^2 + \lambda x \dot{x} + \left(\omega^2 + \frac{\lambda^2}{4} \right) x^2 \right] \\
& + \left(\frac{1}{2\omega} \right) \left(\frac{2\dot{x}}{x} + \lambda \right) \arctan \left[\left(\frac{1}{2\omega} \right) \left(\frac{2\dot{x}}{x} + \lambda \right) \right].
\end{aligned} \tag{3.111}
$$

This Lagrangian is defective in the sense that it does not reduce to the undamped Lagrangian when $\lambda \to 0$. A time-independent Lagrangian which has the correct form as $\lambda \to 0$ is obtained if we choose $F(\Lambda) = \frac{1}{2} \exp(2\Lambda)$ in (3.105) [7]

$$
\begin{aligned}
L(x, \dot{x}) =\ & \dot{x} \int^{\dot{x}} \left(\frac{1}{2} \right) \left[\left(1 + \frac{\lambda x}{y} + \frac{\omega^2 x^2}{y^2} \right) \right. \\
& \left. \times \ \exp \left[- \left(\frac{\lambda}{\omega} \right) \arctan \left\{ \left(\frac{1}{2\omega} \right) \left(\frac{2y}{x} + \lambda \right) \right\} \right] \right] dy. \tag{3.112}
\end{aligned}
$$

Bibliography

[1] P. Cardirola, Nuovo Cimento 18, 393 (1941).

[2] E. Kanai, Prog. Theor. Phys. 3, 440 (1948).

[3] P. Havas, Nuovo Cimento Supp. 5, 363 (1957).

[4] H. Goldstein, *Classical Mechanics*, Second Edition (Addison-Wesley Reading, 1980) Chapter 8.

[5] R.M. Santilli, *Foundations of Theoretical Mechanics*, vol. 1 (Springer-Verlag, New York, 1978).

[6] M. Razavy, Z. Phys. B26, 201 (1977).

[7] J.A. Kobussen, Acta Phys. Austriaca, 59, 293 (1979).

[8] P. Caldirola, Rend. Ist. Lomb. Sc., A 93, 439 (1959).

[9] L. Meirovitch, *Methods of Analytical Dynamics*, (McGraw-Hill, New York, 1970) p. 88.

[10] R.M. Rosenberg, *Analytical Dynamics of Discrete Systems*, (Plenum Press, 1977) p. 229.

[11] B.R. Gossick, *Hamilton's Principle and Physical Systems*, (Academic Press, New York, 1967) p. 102.

[12] W.E. Schröder and J.R. Huizenga in *Treatise on Heavy-Ion Scattering, vol. 2, Fusion and Qusi-Fusion Phenomena*, Edited by D.A. Bromley, (Plenum Press, New York, 1984) p. 138.

[13] J.R. Birkelund, J.R. Huizenga, J.N. De and D. Sperber, Phys. Rev. Lett. 40, 1123 (1978).

[14] V.V. Dodonov, V.I. Man'ko and V.D. Skarzhinskiy in *Quantization, Gravitation and Group Methods in Physics*, Edited by A.A. Komar (Nova Science, Commack, 1988) p. 57.

[15] L.A. Pars, *A Treatise on Analytical Dynamics*, (John Wiley & Sons, New York, 1965).

[16] H. Helmholtz, J. reine angew. Math. 100, 137 (1887).

[17] J. Douglas, Trans. Am. Math. Soc. 50, 71 (1940).

[18] M. Henneaux, Ann. Phys. (NY) 140, 45 (1982).

[19] V. Dodonov, V.I. Man'ko and V.D. Skarzhinsky, Lebedev Physical Institute Preprint No (216), Moscow (1978).

[20] L.D. Landau and E.M. Lifshitz, *Mechanics*, (Pergamon Press, London, 1960) p. 2.

[21] V. Dodonov, V.I. Man'ko and V.D. Skarzhinsky, Hadronic J. 4, 1734 (1981).

[22] S. Okubo Phys. Rev. A 23, 2776 (1981).

[23] J. Geicke, unpublished (2000).

[24] I. Campos, J.L. Jimenez and G. del Valle, Eur. J. Phys. 24, 469 (2003).

[25] I.S. Gradshetyn and I.M. Ryzhik, *Tables of Integrals, Series and Products*, (Academic Press, New York, N.Y. 1965) p. 925.

[26] R.A. Matzner and L.C. Shepley, *Classical Mechanics*, (Prentice Hall, Englewood Cliffs, New Jersey, 1991) Chapter 5.

[27] C.R. Chester, *Techniques in Partial Differential Equations*, (McGraw-Hill, New York, 1971) Chapter 8.

Chapter 4

Hamiltonian Formulation

For conservative as well as certain dissipative motions the Lagrange method provides an elegant and concise formulation of dynamics. However as we have seen in the last chapter, not all forms of dissipative systems can be described by means of a Lagrangian. Whereas in the Lagrangian description, for a system of N degrees of freedom, there are N differential equations of second order, in the Hamiltonian formulation of the same system we have $2N$ first order differential equations.

As in the case of conservative systems, for the dissipative systems the Hamiltonian formulation has the following advantages:
(1) The equal footing of the canonical variables provides for a larger choice of transformations.
(2) The Hamiltonian is a function of the variables alone and not of their derivatives.
(3) For quantizing the system the Hamiltonian form is more convenient.
As we will see in the following sections, we can introduce the Hamiltonian in a number of ways, and for a dissipative motion, just as in the Lagrangian formulation, there will be no unique Hamiltonian form.

4.1 Inverse Problem for the Hamiltonian

In the last chapter we observed that we can solve the inverse problem of classical dynamics by finding the solution of a linear partial differential equation for the Lagrangian L if the forces acting on the system are known. We turn now to an investigation of the Hamiltonian formulation, where for a system of N degrees of freedom, we find $2N$ first order differential equations. But unlike the previous inverse problem, i.e. determination of the Lagrangian, the partial differential equation for obtaining the Hamiltonian is nonlinear and is of second order. As in the case of Lagrangian, the Hamiltonian for a given force law is not unique. However for conservative systems we can choose the Hamiltonian to be the generator of motion as well as a specific constant of motion, viz, the energy of the system. But when dissipative forces are present, the second condition cannot be met. Having already defined the canonical momentum by

$$p_j = \frac{\partial L}{\partial x_j}, \quad j = 1, 2, \cdots N, \tag{4.1}$$

we define the Hamiltonian $H(x_j, p_j, t)$ by the relation

$$H(x_j, p_j, t) = \left(\sum_k \dot{x}_k p_k - L \right)_{\dot{x}_k = \dot{x}_k(p_j)}, \tag{4.2}$$

where the subscript $\dot{x}_k = \dot{x}_k(p_j)$ means that we replace all $\dot{x}_j$ s on the right hand side of (4.2) by p_k s using Eq. (4.1).

We can also write the action S in terms of the Hamiltonian

$$S(x_j, p_j) = \int_{t_1}^{t_2} \left[\sum_j \{ \dot{x}_j(p_k) p_j \} - H(x_j, p_j, t) \right] dt. \tag{4.3}$$

Here x_j s and p_j s are assumed to be independent variables [1] [2]. By setting the functional derivatives of $S(x_j, p_j)$ with respect to x_j and p_j equal to zero we find the Hamilton canonical equations. For a system of particles interacting via the potential $V(x_1 \cdots x_N, t)$ with the Lagrangian (3.37), the Hamiltonian (4.2) reduces to

$$H = \sum_{j=1}^N \frac{p_j^2}{2m_j} + V(x_1, x_2 \cdots x_N, t) = \sum_{j=1}^N \frac{p_j^2}{2m_j} + V(x_1, x_2 \cdots x_N, t). \tag{4.4}$$

But we can also define H as a function of p_j s, x_j s and t in such a way that by eliminating p_j s between Hamilton's canonical equations

$$\dot{p}_j(t) = -\frac{\partial H(x_k, p_k, t)}{\partial x_j}, \tag{4.5}$$

$$\dot{x}_j(t) = \frac{\partial H(x_k, p_k, t)}{\partial p_j}, \tag{4.6}$$

we get the equations of motion

$$m\ddot{x}_j(t) = f_j(x_1, x_2 \cdots x_N; \dot{x}_1, \dot{x}_2 \cdots \dot{x}_N, t), \quad j = 1, 2 \cdots N. \tag{4.7}$$

This definition of H contains (4.4) as a special case. Using the latter definition we can extend the Hamiltonian formulation to dissipative systems, where unlike Eq. (4.4), H does not represent the energy of the system, and p_j, in general is not the same as the mechanical momentum $m_j \dot{x}_j$. Here as in the case of Lagrangian formulation there are infinitely many Hamiltonians generating the same equation of motion in coordinate space. We call these Hamiltonians q-equivalent, implying that when H is substituted in the canonical equations (4.5) and (4.6) and p_j s are eliminated we obtain Eq. (4.7) [3]-[5]. We note that in this formulation p_j s play the role of dummy variables. In classical dynamics these q-equivalent Hamiltonians are all acceptable for the variational formulation and the solution of the equations of motion. But we can impose other conditions on the Hamiltonian function. For instance we may require that $H(x_j, p_j, t)$ should generate the equations of motion in phase space in which case only a subset of q-equivalent Hamiltonians will be acceptable.

Examples of q-equivalent Hamiltonian for an under-damped harmonic oscillator are given later in this chapter, but first let us consider the partial differential equation for the most general Hamiltonian when the force $f(x, \dot{x})$ does not depend explicitly on time. We note from (4.5) and (4.6) that for one-dimensional motion

$$\ddot{x}(t) = \left[\frac{\partial^2 H}{\partial t \partial p} + \left(\frac{\partial^2 H}{\partial x \partial p} \right) \left(\frac{\partial H}{\partial p} \right) - \left(\frac{\partial^2 H}{\partial p^2} \right) \left(\frac{\partial H}{\partial x} \right) \right]. \tag{4.8}$$

Substituting this expression and (4.6) in the equation of motion $m\ddot{x} = f(x, \dot{x})$ we find

$$m \left[\frac{\partial^2 H}{\partial t \partial p} + \left(\frac{\partial^2 H}{\partial x \partial p} \right) \left(\frac{\partial H}{\partial p} \right) - \left(\frac{\partial^2 H}{\partial p^2} \right) \left(\frac{\partial H}{\partial x} \right) \right] = f \left(x, \frac{\partial H}{\partial p} \right). \tag{4.9}$$

For instance when $f(x, \dot{x})$ is given by (3.74) we obtain the equation for H to be

$$m \left[\left(\frac{\partial^2 H}{\partial x \partial p} \right) \left(\frac{\partial H}{\partial p} \right) - \left(\frac{\partial^2 H}{\partial p^2} \right) \left(\frac{\partial H}{\partial x} \right) \right] + \frac{\partial V(x)}{\partial x} + m\gamma \left(\frac{\partial H}{\partial p} \right)^2 = 0, \tag{4.10}$$

and in this case one of the solutions of (4.10) is given by

$$H = \frac{1}{2m} p^2 e^{-2\gamma x} + \int^x e^{2\gamma y} \frac{\partial V(y)}{\partial y} dy. \tag{4.11}$$

We can generalize (4.9) when there are a number of degrees of freedom. In this general case (4.9) is replaced by

$$\frac{\partial^2 H}{\partial t \partial p_k} + \sum_j \left[\left(\frac{\partial^2 H}{\partial x_j \partial p_k} \right) \left(\frac{\partial H}{\partial p_j} \right) - \left(\frac{\partial^2 H}{\partial p_j \partial p_k} \right) \left(\frac{\partial H}{\partial x_j} \right) \right] = \frac{1}{m} f_k \left(x_i, \frac{\partial H}{\partial p_i} \right). \tag{4.12}$$

Since Eqs. (4.9) and (4.12) are nonlinear partial differential equations their most general solutions are not known. However for very special cases we can find solutions for (4.9).

4.2   Hamiltonians for Simple Dissipative Systems

Let us start with the simplest case when the frictional force is linear in velocity and there are no conservative forces, i.e. $f = -m\lambda\dot{x}$. In this case we can find a number of q-equivalent Hamiltonians.

(1) If we choose H to be independent of x, then for a one-dimensional motion (4.9) reduces to [6] [7]

$$\frac{\partial}{\partial p}\left(\frac{\partial H}{\partial t} + \lambda H\right) = 0,$$
(4.13)

and the most general solution of this equation is

$$H(p,t) = h(p)\exp(-\lambda t),$$
(4.14)

where $h(p)$ is a differentiable but otherwise an arbitrary function of p.

(2) If we assume that H can be written as the sum of two terms

$$H(p,x) = h_1(p) + h_2(x),$$
(4.15)

then the solution of (4.9) is

$$H(p,x) = \mathcal{E}\exp\left(\frac{p}{p_0}\right) + \lambda p_0 x,$$
(4.16)

where $\mathcal{E}$ and p_0 are constants.

(3) Finally Dodonov and collaborators have considered a Hamiltonian which is of the form $H(t, xp)$. For this Hamiltonian Eq. (4.9) reduces to

$$\frac{\partial^2 H}{\partial t\partial\xi} + \left(\frac{\partial H}{\partial\xi}\right)^2 + \lambda\left(\frac{\partial H}{\partial\xi}\right) = 0,$$
(4.17)

where $\xi = xp$. By changing $\frac{\partial H}{\partial\xi}$ to u we find the differential equation satisfied by u

$$\frac{du}{dt} + \lambda u + u^2 = 0.$$
(4.18)

This equation can be integrated and the result is

$$u = \frac{\partial H}{\partial\xi} = \frac{\lambda}{\exp[\lambda(t - \alpha)] - 1},$$
(4.19)

or

$$H = \lambda\int^{xp}\frac{\lambda d\xi}{\exp[\lambda(t - \alpha(\xi))] - 1},$$
(4.20)

where $\alpha(\xi)$ is an arbitrary function of its argument.

Since the phase space formulation plays an essential role in quantum-classical correspondence, we need to study the non-uniqueness of the Hamiltonian formalism for the damped motion in phase space. For this group of problems it is more difficult to construct qp-equivalent Hamiltonians (see below).

For the simple systems with Stoke's type resistive force, i.e. a force proportional to $\dot{x}(t)$, Eq. (2.15), we can find explicitly time-dependent Lagrangian and Hamiltonian or time-independent Hamiltonian. Let us consider the Lagrangian function given by Cardirola [8] and by Havas [9] (see Eq. (3.72))

$$L = \frac{m}{2}\left[\dot{x}^2 + \int^x f(x')\,dx'\right]\exp(\lambda t), \tag{4.21}$$

where $\dot{x}(t)$ is the velocity of the particle and $f(x)$ is the conservative force acting on it. Substituting this Lagrangian in Euler-Lagrange differential equation, i.e. [1]

$$\frac{\partial L}{\partial x} - \frac{d}{dt}\left(\frac{\partial L}{\partial \dot{x}}\right) = 0, \tag{4.22}$$

we find the equation of motion (2.15). We note that in this formulation the canonical momentum p is explicitly time-dependent and is given by

$$p = \left(\frac{\partial L}{\partial \dot{x}}\right) = m\dot{x}e^{\lambda t}. \tag{4.23}$$

The Hamiltonian is found from (4.21) and (4.23);

$$H(x,p,t) = (p\dot{x} - L)_{\dot{x}=\dot{x}(p,t)} = \frac{p^2}{2m}e^{-\lambda t} - e^{\lambda t}\int^x f(x')\,dx'. \tag{4.24}$$

In particular, for a harmonically bound particle $V(x) = \frac{1}{2}m\omega_0^2 x^2$ we have

$$H_1(x,p,t) = \frac{p^2}{2m}e^{-\lambda t} + \left(\frac{m}{2}\right)e^{\lambda t}\omega_0^2 x^2. \tag{4.25}$$

For these systems the Hamiltonian function is not unique. For instance another Hamiltonian which does not explicitly depend on time and generates the same motion in coordinate space as (4.25) is:

$$H_2(x,p) = -\left(\frac{\lambda}{2}\right)xp - \ln\cos(\omega px) + \ln x, \tag{4.26}$$

where $\omega^2 = \omega_0^2 - \frac{\lambda^2}{4}$. This Hamiltonian corresponds to the Lagrangian given by Eq. (3.111).

Let us discuss some of the similarities and differences of the Hamiltonians $H_1(x,p,t)$ and $H_2(x,p)$. Both of these Hamiltonians generate the same equations of motion in coordinate space, viz,

$$m\ddot{x} + m\lambda\dot{x} + m\left(\omega^2 + \frac{\lambda^2}{4}\right)x = 0, \tag{4.27}$$

but if we eliminate x and $\dot{x}$ between the Hamilton canonical equations

$$\dot{p} = -\frac{\partial H}{\partial x}, \quad \text{and} \quad \dot{x} = \frac{\partial H}{\partial p}, \tag{4.28}$$

then we find different equations for p.

From the Hamiltonian $H_1(x, p, t)$ and the Hamilton canonical equation we can find the equation of motion for the canonical momentum;

$$m\ddot{p} - m\lambda\dot{p} + m\left(\omega^2 + \frac{\lambda^2}{4}\right)p = 0. \tag{4.29}$$

It is important to note that the equations for x and p are not the same and therefore p, the canonical momentum, is not the same as the mechanical momentum $m\dot{x}$, nor it satisfies the equation for p derived from $H_2(x, p)$.

Thus $H_1(x, p, t)$ and $H_2(x, p)$ are q-equivalent Hamiltonians, but they give rise to different motions in phase space. Both $H_1(x, p, t)$ and $H_2(x, p)$ are not invariant under time-reversal transformation $t \to -t$, and this is expected since the equation of motion (4.27) is not invariant under this transformation. However $H_2(x, p)$ is invariant under time translation transformation, i.e. $t \to t + t_0$ whereas $H_1(x, p, t)$ is not. In fact for $H_2(x, p)$ we have

$$\frac{dH_2}{dt} = \frac{\partial H_2}{\partial t} + \{H_2, H_2\} = 0. \tag{4.30}$$

This result implies that H_2 is a constant of motion and therefore cannot be the energy of the system.

We can also regard $H_1(x, p, t)$ as the Hamiltonian describing an oscillating system whose mass is increasing with time

$$m(t) = m\exp(\lambda t). \tag{4.31}$$

For example consider an empty bucket attached with a rope to a fixed point and is oscillating while it is collecting raindrops. If the mass of this system increases exponentially like (4.31) then its motion can be described by $H_1(x, p, t)$.

Different Hamiltonians that generate the same equation of motion in co-ordinate space, i.e. q-equivalent Hamiltonians are related by canonical transformations. For instance consider the two Hamiltonians given by (Eqs. (4.16) and (4.24))

$$H_1 = \frac{p^2}{2m}e^{-\lambda t}, \quad \text{and} \quad H_2 = \mathcal{E}_2\exp\left(\frac{P}{p_0}\right) + \lambda p_0 X. \tag{4.32}$$

Both of these give us the equation of motion $m\ddot{x} + m\lambda\dot{x} = 0$, and therefore are q-equivalent. The generator of the canonical transformation relating these two is given by [1] [6]

$$F_3(p, X, t) = X(\lambda p_0 t - p) + \frac{1}{\lambda}\left(1 - e^{-\lambda t}\right)\left[\mathcal{E}\exp\left(\frac{p}{p_0}\right) - \frac{p^2}{2m}\right]. \tag{4.33}$$

Using this function $F_3(p, X, t)$ we can connect the old coordinate and the old momentum, x and p to the new ones X and P;

$$P = -\frac{\partial F_3}{\partial X} = p - \lambda p_0 t, \tag{4.34}$$

$$x = -\frac{\partial F_3}{\partial p} = X - \frac{1}{\lambda}\left(1 - e^{-\lambda t}\right)\left[\frac{\mathcal{E}}{p_0}\exp\left(\frac{p}{p_0}\right) - \frac{p}{m}\right], \tag{4.35}$$

and

$$H_2 = H_1 + \frac{\partial F_3}{\partial t}. \tag{4.36}$$

Since

$$\frac{\partial F_3}{\partial t} = \lambda X p_0 + e^{-\lambda t}\left[\mathcal{E}\exp\left(\frac{p}{p_0}\right) - \frac{p^2}{2m}\right], \tag{4.37}$$

we have

$$H_1(x, p, t) + \frac{\partial F_3}{\partial t} = \lambda p_0 X + \mathcal{E}e^{-\lambda t}\exp\left(\frac{P + \lambda p_0 t}{p_0}\right) = H_2(X, P). \tag{4.38}$$

We note that the new coordinate X is a combination of the old coordinate x and the old momentum p, as well as time.

Other q-equivalent Hamiltonians can be found from the equivalent Lagrangians, for example from (3.98) and (3.104) we obtain;

$$H = \frac{1}{3mp_0}(2p_0 p)^{\frac{3}{2}} e^{-\lambda t}, \tag{4.39}$$

and

$$H = \frac{m^2}{8}\left(\frac{6p}{m^2}\right)^{\frac{4}{3}} e^{\frac{-4\gamma x}{3}}. \tag{4.40}$$

These Hamiltonians are not just q-equivalents, but they are as well qp-equivalents of the Hamiltonians $H_1(p, t)$ for linear damping and of

$$H = \frac{1}{2m}p^2 e^{-2\gamma x}, \tag{4.41}$$

for quadratic damping.

4.3 Ostrogradsky's Method

We have already seen that the equations of motion for a radiating electron has a damping term proportional to $\left(\frac{d^3 x}{dt^3}\right)$ and possibly higher derivatives. For dissipative dynamical systems where forces do depend on higher derivatives of acceleration, we can construct the Hamiltonian function using a method due to

Ostrogradsky [10]-[12].

The starting point in this construction is to obtain the extremum of the action integral

$$S = \int L\left(x, x^{(1)}, x^{(2)}, \cdots x^{(j)}, t\right) dt. \tag{4.42}$$

where $x^{(j)}$ denotes the j-th time derivative of x. Here for the sake of simplicity we consider a one-dimensional motion, but the method can be generalized to more dimensions in a straightforward way. By requiring that the functional S in (4.42) be stationary, i.e. $\delta S \equiv 0$, we find that the integrand L must satisfy the differential equation

$$\frac{\partial L}{\partial x} - \frac{d}{dt}\left(\frac{\partial L}{\partial x^{(1)}}\right) + \cdots + (-1)^j \frac{d^j}{dt^j}\left(\frac{\partial L}{\partial x^{(j)}}\right) = 0. \tag{4.43}$$

To construct the Hamiltonian we first introduce the canonical momenta:

$$p_1 = \frac{\partial L}{\partial x^{(1)}} - \frac{d}{dt}\left(\frac{\partial L}{\partial x^{(2)}}\right) + \cdots + (-1)^{j-1}\frac{d^{j-1}}{dt^{j-1}}\left(\frac{\partial L}{\partial x^{(j)}}\right), \tag{4.44}$$

$$p_2 = \frac{\partial L}{\partial x^{(2)}} - \frac{d}{dt}\left(\frac{\partial L}{\partial x^{(3)}}\right) + \cdots + (-1)^{j-2}\frac{d^{j-2}}{dt^{j-2}}\left(\frac{\partial L}{\partial x^{(j)}}\right), \tag{4.45}$$

$$\cdots\cdots\cdots\cdots\cdots\cdots\cdots\cdots$$

$$p_j = \frac{\partial L}{\partial x^{(j)}}. \tag{4.46}$$

We also define the canonical coordinates q_j s by

$$q_1 = x, \quad q_2 = x^{(1)}, \quad \cdots \quad q_j = x^{(j-1)}. \tag{4.47}$$

In terms of these p s and q s we can write the Hamiltonian function $H(q_1, \cdots q_j; p_1 \cdots p_j, t)$ as

$$H = -L + p_1 q_2 + p_2 q_3 + \cdots + p_{j-1} q_j + p_j x^{(j)}, \tag{4.48}$$

where in the last term of (4.48), $x^{(j)}$ should be replaced in terms of other p s and q s by solving $p_j = \frac{\partial L}{\partial x^{(j)}}$ for $x^{(j)}$.

Let us now consider the small variations $\delta q_1, \cdots \delta q_j, \delta p_1 \cdots \delta p_j$ of q s and p s and the corresponding variation in H,

$$\delta H = -\sum_{i=0}^{j-1} \frac{\partial L}{\partial x^{(j)}} \delta q_{j+1} - \frac{\partial L}{\partial x^{(j)}} \delta x^{(j)} + \sum_{i=1}^{j-1} p_j \delta q_{i+1} + p_j \delta x^{(j)}$$

$$+ \sum_{i=1}^{j-1} q_{i+1} \delta p_i + x^{(j)} \delta p_j. \tag{4.49}$$

From Eqs. (4.43)-(4.46) we have

$$\frac{\partial L}{\partial x} = \dot{p}_1, \quad \frac{\partial L}{\partial x^{(1)}} = \dot{p}_2 + p_1, \quad \frac{\partial L}{\partial x^{(2)}} = \dot{p}_3 + p_2, \quad \frac{\partial L}{\partial x^{(j)}} = p_j. \tag{4.50}$$

Substituting these in (4.49) we find

$$\delta H = -\sum_{i=1}^{j} \dot{p}_i \delta q_i + \sum_{i=1}^{j} \dot{q}_i \delta p_i. \tag{4.51}$$

Now we compare (4.51) with the variation of H, i.e.

$$\delta H\left(q_1 \cdots q_j; p_1, \cdots p_j, t\right) = \sum_{i=1}^{j} \left(\frac{\partial H}{\partial q_i} \delta q_i + \frac{\partial H}{\partial p_i} \delta p_i\right), \tag{4.52}$$

and we obtain the Hamilton canonical equations

$$\dot{q}_i = \frac{\partial H}{\partial p_i}, \quad \dot{p}_i = -\frac{\partial H}{\partial q_i}. \tag{4.53}$$

As an example of the application of this method let us consider a slightly different model for radiating electron than the one given by Eq. (2.33). Here we want to find the Hamiltonian for the three-dimensional motion of an electron subject to the potential $V(q, t)$, where the equations of motion are given by [13] [14]

$$m\left[\left(\frac{d^2 x_i}{dt^2}\right) - \tau\left(\frac{d^3 x_i}{dt^3}\right) + \left(\frac{\tau}{2}\right)^2 \left(\frac{d^4 x_i}{dt^4}\right)\right] = -\frac{\partial V}{\partial x_i}, \quad i = 1, 2, 3. \tag{4.54}$$

These equations can be derived from the Lagrangian

$$L = -\left[\frac{\tau^2}{8}\sum_i m\left(\frac{d^2 x_i}{dt^2}\right)^2 + V\right]\exp\left[-\frac{2t}{\tau}\right]. \tag{4.55}$$

Since the equation of motion is of the fourth order we can construct the generalized canonical variables by Ostrogradsky method [10] [11]

$$q_i = x_i, \quad p_i = \frac{\partial L}{\partial \dot{q}_i} - \frac{d}{dt}\left(\frac{\partial L}{\partial \ddot{q}}\right),$$

$$Q_i = \dot{x}_i, \quad P_i = \frac{\partial L}{\partial \ddot{q}_i} - \frac{d}{dt}\left(\frac{\partial L}{\partial \frac{d^3 q}{dt^3}}\right). \tag{4.56}$$

With the help of these canonical variables we can write the Hamiltonian as

$$H = \sum_i p_i Q_i - \frac{2}{m\tau^2}\exp\left(\frac{2t}{\tau}\right)\sum_i P_i^2 + \exp\left(-\frac{2t}{\tau}\right)V(q, t). \tag{4.57}$$

For other attempts to formulate the Lagrangian and the Hamiltonian for a radiative electron see Infeld [15] and also Englert [16]. In the latter work, using Ostrogradsky's method the Hamiltonian for the third order equation of motion is constructed and then quantized.

4.4   Complex or Leaky Spring Constant

The complex classical Lagrangian that we want to consider is modeled after the complex optical potentials used in nuclear physics [17]. Let us study a two-dimensional harmonic oscillator which is subject to a damping force linear in velocity, i.e.

$$\frac{d^2\mathbf{r}}{dt^2} + 2\alpha\omega\frac{d\mathbf{r}}{dt} + \omega^2\left(1+\alpha^2\right)\mathbf{r} = 0. \tag{4.58}$$

The x and y components of this motion are

$$x(t) = r_0 \cos(\omega t)e^{-\alpha\omega t}, \quad y(t) = r_0 \sin(\omega t)e^{-\alpha\omega t}, \tag{4.59}$$

where we have used the initial conditions $x(0) = r_0$ and $y(0) = 0$. Now let us consider a complex time variable τ defined by the relation

$$\tau = (1 - i\alpha)t, \tag{4.60}$$

and then write the equation of motion as

$$\frac{d^2 z}{d\tau^2} = -\omega^2 z, \tag{4.61}$$

or

$$\frac{d^2 z}{dt^2} = -\Omega^2 z = -\omega^2(1-i\alpha)^2 z, \quad \text{Im } \Omega^2 = -2\alpha\omega^2 < 0. \tag{4.62}$$

Of the two solutions of (4.61) the one which is of the form $e^{-i\omega\tau}$ is acceptable, i.e. z remains finite as $t \to \infty$. If we write

$$z(t) = x(t) + iy(t) = r_0 \exp(-i\omega t - \alpha t), \tag{4.63}$$

we observe that $x(t)$ and $y(t)$ are the same as those given by (4.59).
    Equation (4.61) has a simple Lagrangian

$$L = \left(\frac{m}{2}\right)\left[\left(\frac{dz}{d\tau}\right)^2 - \omega^2 z^2\right], \tag{4.64}$$

from which we can determine the Hamiltonian and also the classical action.

4.5   Dekker's Complex Coordinate Formulation

Another method using complex coordinates was suggested by Dekker [18] [19], which unlike the previous model has no direct connection to the optical potential model.

Consider the under-damped harmonic oscillator with the equation of motion

$$\ddot{x} + \lambda\dot{x} + \left(\omega^2 + \frac{\lambda^2}{4}\right)x = 0, \tag{4.65}$$

and let us introduce the complex coordinate $q(t)$ by

$$q(t) = \frac{1}{\sqrt{\omega}}\left[p(t) + \left(\frac{\lambda}{2} - i\omega\right)x(t)\right], \tag{4.66}$$

where

$$p(t) = \dot{x}. \tag{4.67}$$

By differentiating $q(t)$ and using Eq. (4.67) we note that $q(t)$ is the solution of the first order differential equation

$$\dot{q}(t) + i\omega q(t) + \frac{\lambda}{2}q(t) = 0. \tag{4.68}$$

Assuming that (4.68) rather than (4.65) is the equation of motion, then the equation for $q(t)$ can be derived from the Lagrangian

$$\mathsf{L} = \frac{i}{2}[q^*(t)\dot{q}(t) - q(t)\dot{q}^*(t)] - \left(\omega - i\frac{\lambda}{2}\right)q^*(t)q(t). \tag{4.69}$$

Next we find the conjugate to the complex coordinate $q(t)$ which we denote by $\pi(t)$

$$\pi(t) = \frac{\partial\mathsf{L}}{\partial\dot{q}} = \frac{i}{2}q^*(t). \tag{4.70}$$

From the expression for the Lagrangian L and the canonical momentum $\pi(t)$ we find the Hamiltonian H in terms of $\pi(t)$ and $q(t)$;

$$\mathsf{H} = H_1 + i\Gamma = -\left(i\omega + \frac{\lambda}{2}\right)\pi(t)q(t). \tag{4.71}$$

We can express this Hamiltonian in terms of the original variables $x(t)$ and $p(t)$. For this we note that $q(t)$ is given by Eq. (4.66) and $\pi(t)$ is

$$\pi(t) = \frac{i}{2\sqrt{\omega}}\left[p(t) + \left(i\omega + \frac{\lambda}{2}\right)x(t)\right]. \tag{4.72}$$

Now by substituting for $\pi(t)$ and $q(t)$ in (4.71) we find that

$$\mathsf{H} = \frac{1}{2}\left\{p^2 + \left(\omega^2 + \frac{\lambda^2}{4}\right)x^2 + \lambda(xp + px)\right\} - \frac{i}{2}\lambda, \tag{4.73}$$

is the complex Hamiltonian of the system. From this Hamiltonian we obtain the equation of motion (4.65) for $x(t)$ and a similar equation for $p(t)$;

$$\ddot{p} + \lambda\dot{p} + \left(\omega_0^2 + \frac{\lambda^2}{4}\right)p = 0. \tag{4.74}$$

4.6 Hamiltonian Formulation of the Motion of a Particle with Variable Mass

We have seen that the Kanai-Caldirola Hamiltonian generates the classical equation of motion for the coordinate of the particle when the motion takes place in a viscous medium with a drag force proportional to the velocity of the particle. However a different interpretation can be given to this Hamiltonian, viz, that it generates the equation for motion of a particle with a variable (time-dependent) mass [20]-[22].

We write the Hamiltonian as

$$H = H(x, p; \theta, m; t), \tag{4.75}$$

where in this generalized form we consider the mass m (measured in units of energy, i.e. mc^2, with $c = 1$), and the "proper time" θ as conjugate variables. Thus p and m should be treated as momenta and x and θ as their conjugates. Thus we have the following canonical equations for such a Hamiltonian:

$$\dot{x} = \frac{\partial H}{\partial p}, \quad \dot{\theta} = \frac{\partial H}{\partial m}, \tag{4.76}$$

and

$$\dot{p} = -\frac{\partial H}{\partial x}, \quad \dot{m} = -\frac{\partial H}{\partial \theta}. \tag{4.77}$$

According to this formulation a potential which depends on x provides a force, thus causing a change in momentum, and similarly a potential which depends on θ causes a change in the mass of the particle.

Let us consider the specific example of a harmonically bound particle with variable mass

$$m = m_0 e^{\lambda t}. \tag{4.78}$$

The Hamiltonian in this case is

$$H = \frac{p^2}{2m} + \frac{1}{2}m\omega_0^2 x^2 - \lambda m\theta. \tag{4.79}$$

From this Hamiltonian we find the canonical equations of motion

$$\dot{x} = \frac{\partial H}{\partial p} = \frac{p}{m}, \tag{4.80}$$

$$\dot{\theta} = \frac{\partial H}{\partial m} = -\frac{p^2}{2m^2} + \frac{1}{2}\omega_0^2 x^2 - \lambda\theta, \tag{4.81}$$

$$\dot{m} = -\frac{\partial H}{\partial \theta} = \lambda m, \tag{4.82}$$

and

$$\dot{p} = -\frac{\partial H}{\partial x} = -m\omega_0^2 x. \tag{4.83}$$

Equation (4.82) can be integrated and the result is given by (4.78). By differentiating (4.80) with respect to t and substituting for $\dot{p}$ and $\dot{m}$ from (4.82) and (4.83) we find the equation of motion for x;

$$\ddot{x} + \lambda \dot{x} + \omega_0^2 x = 0. \tag{4.84}$$

Similarly we obtain the equation of motion for p

$$\ddot{p} + \lambda \dot{p} + \omega_0^2 p = 0. \tag{4.85}$$

Thus H given by (4.75) is qp-equivalent to Dekker's complex Hamiltonian and gives us the correct equations for the motion in phase space.

4.7 Variable Mass Oscillator

A system with variable mass for which the equations of motion is solvable and can be tested in laboratory is that of an oscillator where the mass is decreasing uniformly and is subject to a frictional force proportional to velocity.

Let us consider a container filled with sand and attached to a fixed point by means of a spring with a spring constant K. The equation of motion of this system is [23]

$$m(t)\frac{d^2 x}{dt^2} + b\frac{dx}{dt} + Kx = 0, \tag{4.86}$$

where

$$m(t) = m_0 - \beta t, \tag{4.87}$$

and b is the coefficient of friction. Thus $\beta = -dm/dt$ is the rate with which the system is losing its mass. For the equation of motion (4.86) with $m(t)$ defined by (4.87) we find a Hamiltonian similar to H defined by (4.79), i.e.

$$H = \frac{1}{2m(t)}p^2 + \frac{1}{2}Kx^2 - b\theta. \tag{4.88}$$

Using the canonical equations (4.80)-(4.83) and eliminating p and $\dot{p}$ between the resulting equations we obtain (4.86). Also by multiplying Eq. (4.86) by dx/dt we can find the rate of change of mechanical energy of the system;

$$\frac{d}{dt}(E_k + E_p) = \left(\frac{1}{2}\frac{dm(t)}{dt} - b\right)\left(\frac{dx}{dt}\right)^2 = 0, \tag{4.89}$$

where $E_k = (m(t)/2)\dot{x}^2$ and $E_p = (K/2)x^2$ are the kinetic and potential energies respectively. Equations (4.86)-(4.87) can be solved analytically. We write the solution as the product of an amplitude and a time-dependent phase [23]

$$x(t) = A_0 f(t) \sin[h(t) + \phi], \tag{4.90}$$

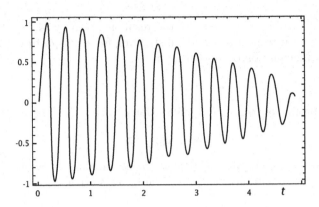

Figure 4.1: Damped oscillations of a system with variable mass described by Eqs. (4.86) and (4.87).

where A_0 and ϕ are constants, and we substitute $x(t)$, Eq. (4.90) in (4.86) and equate the coefficients of $\sin[h(t)]$ and $\cos[h(t)]$ separately equal to zero to find

$$m(t)\left[\ddot{f} - \left(\dot{h}\right)^2 f\right] + b\dot{f} + Kf = 0, \qquad (4.91)$$

and

$$m(t)\left[2\dot{f}\dot{h} + f\ddot{h}\right] + bf\dot{h} = 0. \qquad (4.92)$$

The solutions of these differential equations are given by:

$$f(t) = \left(1 - \frac{\beta t}{m_0}\right)^\alpha, \qquad (4.93)$$

and

$$h(t) = \frac{2\mu}{\beta}\left[\tan^{-1}\sqrt{\frac{1 - a - \frac{\beta t}{m_0}}{a}} - \sqrt{\frac{1 - a - \frac{\beta t}{m_0}}{a}}\right], \qquad (4.94)$$

where

$$\alpha = \frac{b}{2\beta} + \frac{1}{4}, \qquad (4.95)$$

$$\mu = \frac{1}{2}\sqrt{\left(b + \frac{\beta}{2}\right)\left(b + \frac{3}{2}\beta\right)}, \qquad (4.96)$$

and

$$a = \frac{\mu^2}{m_0 K}. \qquad (4.97)$$

In Fig. 4.1 the damped oscillations of a container with variable mass connected to a spring is shown. For this calculation the constants, $b = 0.1$, $\beta = 0.2$, $K = 2.5$ and $m_0 = 1$ have been used.

4.8 Bateman's Damped-Amplified Harmonic Oscillators

A different two-dimensional model which as a whole is conservative is the Bateman's system which is described by the Lagrangian [24]-[27]

$$L = m\dot{x}\dot{y} + \frac{m}{2}\lambda(x\dot{y} - \dot{x}y) - m\left(\omega^2 + \frac{\lambda^2}{4}\right)xy. \qquad (4.98)$$

From this Lagrangian we find the classical equations of motion to be

$$m\ddot{x} + m\lambda\dot{x} + m\left(\omega^2 + \frac{\lambda^2}{4}\right)x = 0, \qquad (4.99)$$

and

$$m\ddot{y} - m\lambda\dot{y} + m\left(\omega^2 + \frac{\lambda^2}{4}\right)y = 0. \qquad (4.100)$$

We note that the equation for y is the time-reversed of the x motion. Thus while the x oscillator is losing energy, the y oscillator is gaining, so that the total energy of the system is conserved. This also follows from the fact that

$$H = (p_x\dot{x} + p_y\dot{y} - L)_{\dot{x}(p_x,p_y),\dot{y}(p_x,p_y)}, \qquad (4.101)$$

is a constant of motion. We can write H in terms of the canonical momenta p_x and p_y as

$$H = \frac{p_x p_y}{2m} - \frac{\lambda}{2}(x p_x - y p_y) + m\omega^2 xy, \qquad (4.102)$$

and since H does not depend on time explicitly, therefore $dH/dt = 0$.

The Bateman approach for constructing the Lagrangian for a dissipative system has been extended to include the classical motion of a radiating electron by Englert [16]. Again let us consider the motion of a harmonically bound radiating electron Eq. (2.17) which we write as

$$m\frac{d^2x}{dt^2} = -m\omega_0^2 x + m\tau\left(\frac{d^3x}{dt^3}\right). \qquad (4.103)$$

We can write the Lagrangian either as

$$L = -\frac{1}{2}m\tau(\ddot{x}\dot{y} - \dot{x}\ddot{y}) + m\dot{x}\dot{y} - m\omega_0^2 xy, \qquad (4.104)$$

or alternatively the equivalent Lagrangian which is asymmetric in x and y

$$L_1 = -m\tau\dot{x}\ddot{y} + m\dot{x}\dot{y} - m\omega_0^2 xy. \qquad (4.105)$$

Both of these Lagrangians generate the equation of motion (4.103).

4.9   Dissipative Forces Quadratic in Velocity

In the last section we studied the one-dimensional motion of a particle when it is affected by a force linear in velocity. Now let us consider the motion of a particle of mass m in a medium where in addition to the conservative force a Newtonian resistive force $m\gamma\dot{x}^2$ is acting on it. Here the equation of motion is given by [28]

$$m\left(\frac{d^2x}{dt^2}\right) + m\gamma\left(\frac{dx}{dt}\right)^2 = f(x). \qquad (4.106)$$

Again the Hamiltonian for this system is not unique, but a simple time-independent Hamiltonian for this motion is given by

$$H(x,p) = \left(\frac{p^2}{2m}\right)e^{-2\gamma x} - \int^x f(y)e^{2\gamma y}dy. \qquad (4.107)$$

Note that (4.106) remains invariant under time-reversal transformation. If the motion is such that $\dot{x}$ changes its sign, then for a dissipative motion we must replace $(dx/dt)^2$ by $\left(\frac{dx}{dt}\right)\left|\frac{dx}{dt}\right|$ (see §12.6).

4.10   Resistive Forces Proportional to Arbitrary Powers of Velocity

As we have already mentioned in Section (2.1) in the case of a bullet in the air the resistance can be proportional to $\dot{x}^\nu$, ($\nu \approx 2$). In the following discussion we assume ν to be a number greater than one, but not equal to two
The equation of motion of the particle in the absence of a conservative force is

$$m\left(\frac{d^2x}{dt^2}\right) + m\beta\left(\frac{dx}{dt}\right)^\nu = 0, \qquad (4.108)$$

or if x changes sign we have

$$m\left(\frac{d^2x}{dt^2}\right) + m\beta\left(\frac{dx}{dt}\right)\left|\frac{dx}{dt}\right|^{\nu-1} = 0. \qquad (4.109)$$

For Eq. (4.108) the Hamiltonian is given by the implicit relation

$$p(H,x) = \int^H [C(H) - \beta(2-\nu)x]^{\frac{1}{(\nu-2)}}\, dH, \quad \nu \neq 2, \qquad (4.110)$$

where C is an arbitrary function of H. The simplest choice is that of C being a constant independent of H. Then from (4.110) we find

$$H = p\left[C - \beta(2-\nu)x\right]^{\frac{1}{(2-\nu)}} = pf(x). \qquad (4.111)$$

Noting that since H does not explicitly depend on time and hence is a constant of motion, the Hamilton canonical equations yield

$$\dot{x} = \frac{H}{p} \quad \text{and} \quad \dot{p} = \beta p \dot{x}^{(\nu-1)}. \tag{4.112}$$

By eliminating p between these two relations we obtain Eq. (4.108).

While Eq. (4.110) gives us the most general although an implicit Hamiltonian, we can use the Lagrangian formulation Eq. (3.80) to construct the Hamiltonian. The canonical momentum p found from the Lagrangian is [29]

$$p = \frac{\partial L}{\partial \dot{x}} = m\mu'(\dot{x}) = (1-\nu)\dot{x}^{1-\nu}, \quad 1 < \nu < 2, \tag{4.113}$$

where prime denotes derivative with respect to the argument and $\mu''(\dot{x})$ is defined by Eq. (3.79). The Hamiltonian of the system is found from the definition of H;

$$H = p\dot{x} - L = m[\dot{x}\mu'(\dot{x}) - \mu(\dot{x})] + m\beta x. \tag{4.114}$$

By eliminating $\dot{x}$ between (4.113) and (4.114) we have

$$H = \left(\frac{m}{2-\nu}\right)\left(\frac{1-\nu}{m}p\right)^{\frac{\nu-2}{\nu-1}} + m\beta x, \quad 1 < \nu < 2. \tag{4.115}$$

The above Hamiltonian can also be obtained from (4.110) if we choose $C(H)$ to be

$$C(H) = \frac{2-\nu}{m}H, \tag{4.116}$$

and after carrying out the integration over H, solve for H in terms of p.

4.11 Universal Lagrangian and Hamiltonian

In the previous chapter we studied the Helmholtz condition for the existence of a Lagrangian for the set of equations of motion

$$m\ddot{x}_k - f_k(x_i, \dot{x}_i, t) \equiv G(x_i, \dot{x}_i, \ddot{x}_i, t) = 0, \quad k = 1, 2 \cdots N, \tag{4.117}$$

and we noticed that in general, for dissipative systems we have to introduce an integrating factor $\mu(x_i, \dot{x}_i, t)$ in order to satisfy the Helmholtz conditions. We can inquire about the possibility of constructing Lagrangian (or Hamiltonian) functions for open systems where, like simple conservative systems, there are no integrating factors. This possibility has been studied by Cawley [30] who has called such a Lagrangian a "universal Lagrangian". We introduce the Lagrangian $L_u(x_i, z_i, \dot{x}_i, \dot{z}_i, y_i, t)$ with $2N$ additional coordinates z_i and y_i and additional velocities $\dot{z}_i$ by

$$L_u(x_i, z_i, \dot{x}_i, \dot{z}_i, y_i, t) = \sum_{k=1}^{N}\left[f_k(x_i, \dot{x}_i, t)z_k + m\dot{x}_k\dot{z}_k + \frac{1}{2}y_k z_k^2\right]. \tag{4.118}$$

The Euler-Lagrange equation for L yields the N equations of motion (4.117) together with a set of N subsidiary conditions

$$z_k = 0, \quad k = 1, \cdots N. \tag{4.119}$$

If we try to construct a Hamiltonian for (4.117) we obtain N first class constraints:

$$\frac{\partial L_u}{\partial \dot{y}_k} = p_{y_k} \approx 0, \quad k = 1, \cdots N, \tag{4.120}$$

where the wavy equality sign "$\approx$" denotes the weak equality. By weak equality we mean that these constraints should not be used before working out the Poisson brackets [31]. For the momenta p_{z_k} from (4.117) we find

$$p_{z_k} = m\dot{x}_k, \quad k = 1, \cdots N. \tag{4.121}$$

Using these momenta, we obtain the Hamiltonian

$$H_u = \sum_{k=1}^{N} \left[\frac{p_{z_k} p_{x_k}}{m} - f_k \left(x_i, \frac{1}{m} p_{z_i}, t \right) z_k - \frac{1}{2} y_k z_k^2 + \phi_k(t) p_{y_k} \right], \tag{4.122}$$

where ϕ_k s are arbitrary functions of time. Expressing the time derivative in terms of the Poisson bracket

$$\dot{g}_k = \{g_k, H_u\}, \tag{4.123}$$

we get the following relations:

$$\dot{p}_{y_k} \approx \{p_{y_k}, H_u\} \approx \frac{1}{2} z_k^2 \approx 0, \quad k = 1, \cdots N, \tag{4.124}$$

$$z_k \approx 0, \quad k = 1, \cdots N, \tag{4.125}$$

and

$$z_k^2 \approx 0, \quad k = 1, \cdots N. \tag{4.126}$$

We observe that Eqs. (4.125) are secondary constraints and the consistency conditions are [31] [32]:

$$\dot{z}_k \approx \{z_k, H_u\} = \frac{1}{m} p_{x_k} \approx 0, \quad k = 1, \cdots N. \tag{4.127}$$

Let us apply this method to the problem of motion of a particle subject to an applied force $F(x, t)$ in addition to the damping force $-m\lambda \dot{x}$. According to the present formulation the Lagrangian of the system is

$$L_u(x, \dot{x}, z, \dot{z}, y, t) = \left[m\dot{x}\dot{z} + (F(x, t) - m\lambda \dot{x}) z + \frac{1}{2} y z^2 \right]. \tag{4.128}$$

From this Lagrangian we find

$$p_z = m\dot{x}, \tag{4.129}$$

and

$$p_x = m\dot{z} - m\lambda z. \tag{4.130}$$

Knowing the momenta p_x and p_z and L we obtain the Hamiltonian H_u

$$H_u = \frac{1}{m}p_z p_x + (\lambda p_z - F(x,t))\, z - \frac{1}{2}yz^2. \tag{4.131}$$

This Hamiltonian can be related to the Birkhoff's Hamiltonian [33] defined in the following way: Let us write the classical equations of motion as first order coupled equations

$$\frac{dq_k}{dt} = X_k\,(q_i, t)\,, \quad k = 1, \cdots 2N. \tag{4.132}$$

The Birkhoff Hamiltonian has the form

$$H_B = \sum_{k=1}^{2N} X_k\,(q_i, t)\, p_k, \tag{4.133}$$

where p_k s are the $2N$ momenta conjugate to the coordinates q_k. Equations (4.132) are one-half of the Hamilton canonical equations. The rest are given by

$$\frac{dp_k}{dt} == -\frac{\partial H_B}{\partial q_k} = -\sum_{j=1}^{2N} \frac{\partial X_j\,(q_i, t)}{\partial q_k} p_j, \quad k = 1, \cdots 2N. \tag{4.134}$$

From the Hamiltonian H_B we can find the Lagrangian L_B to be

$$L_B = \sum_{k=1}^{2N} \frac{dq_k}{dt} p_k - H_B \equiv 0. \tag{4.135}$$

Next we identify the canonical variables x_j, z_j, p_{x_j} and p_{x_j} by the following relations:

$$q_1, \cdots q_N \rightarrow x_1, \cdots x_N, \tag{4.136}$$

$$q_{N+1}, \cdots q_{2N} \rightarrow p_{z_1}, \cdots p_{z_N}, \tag{4.137}$$

$$p_1, \cdots p_N \rightarrow p_{x_1}, \cdots p_{x_N}, \tag{4.138}$$

$$p_{N+1}, \cdots p_{2N} \rightarrow -z_1, \cdots -z_N, \tag{4.139}$$

$$X_1, \cdots X_N \rightarrow \dot{x}_1\,(p_{z_i})\,, \cdots \dot{x}_N\,(p_{z_i})\,, \tag{4.140}$$

and

$$X_{N+1}, \cdots X_{2N} \rightarrow f_1\left(x_i, \frac{1}{m}p_{z_i}, t\right) \cdots f_N\left(x_i, \frac{1}{m}p_{z_i}, t\right). \tag{4.141}$$

Substituting for q_k s and p_k s in H_B, Eq. (4.133) we find

$$H_B = \sum_{j=1}^{N} \dot{x}_j\,(p_{z_i})\, p_{x_j} - \sum_{j=1}^{N} f_j\left(x_i, \frac{1}{m}p_{z_i}, t\right) z_j. \tag{4.142}$$

This H_B is the same as H_u, except for the two terms,

$$\left(-\frac{1}{2}y_k z_k^2 + \phi(t)p_{y_k}\right),\tag{4.143}$$

which do not affect the equations of motion. For the damped harmonic oscillator we have $N = 1$ and

$$X_1 = \dot{q}_1 = \frac{p_z}{m} = \dot{x}, \quad p_1 = p_x,\tag{4.144}$$

and

$$X_2 = -\lambda q_2 + F\left(q_1, t\right), \quad p_2 = -z.\tag{4.145}$$

By substituting these in H_B, we find the same expression as H_u, Eq. (4.131) without the last term.

4.12 Hamiltonian Formulation in Phase Space of N-Dimensions

The Abraham-Lorentz equation for the non-relativistic motion of a radiating electron is of third order, Eq. (2.17). The Hamiltonian formulation of equations of motion containing higher derivatives of the coordinate(s) of a particle has been studied in connection with nonlocal field theories [11]. The time-reversal invariance imposed on these fields allows only even derivatives of the coordinate(s) to appear in the equation of motion. However there are other generalizations of the Hamiltonian dynamics where odd derivatives can occur in the equations of motion and that the phase space can have odd dimensions [4] [12] [34].

Let us study the phase space of N dimensions where N can be odd or even. The equations of motion for a system of particles can be written as a set of first order differential equations

$$\dot{\eta}_i = X_i\left(\eta_1, \eta_2...\eta_N\right), \quad i = 1, 2, \cdots N,\tag{4.146}$$

where $\dot{\eta}_i = \frac{d\eta_i}{dt}$ and η_i s are the dynamical variables of the system. Next let us suppose that the system is integrable [2] and that $H_1, H_2 \cdots H_{N-1}$ are the $N-1$ constants of motion of this system, then

$$\frac{dH_i}{dt} = \sum_{k=1}^{N}\left(\frac{\partial H_i}{\partial \eta_k}\right)\dot{\eta}_k = 0, \quad i = 1, 2, \cdots N - 1.\tag{4.147}$$

This set of $N - 1$ equations can be solved for the variables $\dot{\eta}_1, \dot{\eta}_2 \cdots \dot{\eta}_{N-1}$ in terms of $\dot{\eta}_N$;

$$\dot{\eta}_j = (-1)^N \frac{d\eta_N}{dt}\left(\frac{\Delta_j}{\Delta_N}\right),\tag{4.148}$$

where

$$\Delta_j = \sum_{[j]} \varepsilon_{i_1, i_2, \cdots i_{N-1}} \frac{\partial H_1}{\partial \eta_{i_1}} \cdots \frac{\partial H_{N-1}}{\partial \eta_{i_{N-1}}}. \tag{4.149}$$

In Eq. (4.149) $\varepsilon_{i_1, i_2, \cdots i_{N-1}}$ is the Levi-Civita tensor, and the summation is over all integers $1, 2 \cdots N$ except j. Thus Δ_j is the determinant obtained from N by $N-1$ matrix elements $\left(\frac{\partial H_j}{\partial \eta_k} \right)$ by deleting the j-th column. Now if we compare (4.146) and (4.148) we find

$$\frac{X_1}{\Delta_1} = \frac{X_2}{-\Delta_2} = \cdots = \frac{X_N}{(-1)^{N-1}\Delta_N} = \frac{1}{M(\eta_1 \cdots \eta_N)}, \tag{4.150}$$

where M is, in general, a function of η_i and is called the multiplier of the system [12]. From Eq. (4.149) it follows that

$$\frac{\partial \Delta_1}{\partial \eta_1} - \frac{\partial \Delta_2}{\partial \eta_2} + \cdots + (-1)^{N-1}\frac{\partial \Delta_N}{\partial \eta_N} = 0, \tag{4.151}$$

or in terms of X_i s we have

$$\sum_{i=1}^{N} \frac{\partial}{\partial \eta_i}(MX_i) = 0. \tag{4.152}$$

This relation can be used to determine $M(\eta_1 \cdots \eta_N)$.

Let us consider the phase space of three dimensions when the forces acting on the system satisfy the relation

$$\sum_{i=1}^{3} \frac{\partial X_i}{\partial \eta_i} = \nabla \cdot \mathbf{X} = 0, \tag{4.153}$$

then the present formulation reduces to Nambu's dynamics [4] [34] . Thus from (4.148) and (4.149) we find $\dot{\eta}_i$ to be

$$\dot{\eta}_i = \sum_{j,k} \varepsilon_{ijk} \frac{\partial H_1}{\partial \eta_j} \frac{\partial H_2}{\partial \eta_k}, \tag{4.154}$$

and (4.151) simplifies to

$$\nabla \cdot (\nabla H_1 \wedge \nabla H_2) = 0, \tag{4.155}$$

where ∇ acts on the coordinates η_1, η_2 and η_3 of the phase space.

In an N-dimensional phase space let us take H_{N-1} to be one of the generators of motion for a set of dynamical variables $\eta_1 \cdots \eta_N$. Then we define the Poisson bracket for any two functions of the dynamical variables ϕ and ψ by the relation

$$\{\phi, \psi\} = \frac{1}{M} \sum_{j,k} \mathcal{J}_{jk} \frac{\partial \phi}{\partial \eta_j} \frac{\partial \psi}{\partial \eta_k}, \tag{4.156}$$

where M is defined by (4.150) and (4.152) and $\mathcal{J}_{jk}$ is the antisymmetric matrix

$$\mathcal{J}_{jk} = \sum_{i_3 \dots i_N} (-1)^N \varepsilon_{jk \, i_3 \dots i_N} \frac{\partial H_1}{\partial \eta_{i_3}} \cdots \frac{\partial H_{N-2}}{\partial \eta_{i_N}}. \tag{4.157}$$

From the definition of $\mathcal{J}_{jk}$ it follows that

$$\mathcal{J}_{jk} + \mathcal{J}_{kj} = 0. \tag{4.158}$$

However the Jacobi identity [1]

$$\sum_s^N \left[\mathcal{J}_{ks} \frac{\partial}{\partial \eta_s} \left(\frac{\mathcal{J}_{lm}}{M} \right) + \mathcal{J}_{ls} \frac{\partial}{\partial \eta_s} \left(\frac{\mathcal{J}_{mk}}{M} \right) + \mathcal{J}_{ms} \frac{\partial}{\partial \eta_s} \left(\frac{\mathcal{J}_{kl}}{M} \right) \right], \tag{4.159}$$

must be verified directly.

By replacing ψ in (4.156) by H_{N-1} we find

$$\{\phi, H_{N-1}\} = \frac{1}{M} \left[\frac{\partial \phi}{\partial \eta_1} \Delta_1 - \frac{\partial \phi}{\partial \eta_2} \Delta_2 + \cdots + (-1)^{N-1} \frac{\partial \phi}{\partial \eta_N} \Delta_N \right] = 0. \tag{4.160}$$

In particular we have

$$\{\eta_i, H_{N-1}\} = (-1)^{i-1} \left(\frac{\Delta_i}{M} \right) = X_i = \dot{\eta}_i, \quad i = 1, \cdots N. \tag{4.161}$$

The first integrals of motion, $H_1, H_2 \cdots H_{N-1}$ are found by writing the equations of motion (4.146) as

$$\frac{d\eta_1}{X_1} = \frac{d\eta_2}{X_2} = \cdots = \frac{d\eta_N}{X_N}, \tag{4.162}$$

and integrating them. Here we assume that X_i s do not depend explicitly on time. This set of equations are the Lagrange's subsidiary equations for the partial differential equations

$$\sum_{i=1}^N X_i \frac{\partial H_k}{\partial \eta_i} = 0, \quad k = 1, 2 \cdots N - 1, \tag{4.163}$$

and H_ks are the first integrals of motion. We also observe that the Poisson brackets of the canonical variables $\eta_1 \cdots \eta_N$ are given by

$$\{\eta_j, \eta_k\} = \frac{1}{M} \mathcal{J}_{jk}, \quad j, k = 1, 2 \cdots N. \tag{4.164}$$

There are others ways of formulating the Hamiltonian of a dynamical system with the generalized canonical equations of motion similar to what we have discussed here. Such formulations can also be used in the case of dissipative classical motions (for instance see reference [35]).

Returning to the problem of a radiating electron of mass m moving in a constant external field, $(-ma)$, we first write the equation of motion as

$$m\tau \frac{d^3\eta_1}{dt^3} - m\frac{d^2\eta_1}{dt^2} = ma, \tag{4.165}$$

where the constant external field is written as ma. We can also write (4.165) as three coupled first order equations:

$$\frac{d\eta_1}{dt} = \eta_2, \quad \frac{d\eta_2}{dt} = \eta_3, \quad \text{and} \quad \frac{d\eta_3}{dt} = \frac{1}{\tau}(\eta_3 + a). \tag{4.166}$$

By integrating these equations, we can construct two constants of motion:

$$H_1 = \eta_2 - \tau\eta_3 + a\tau \ln(\eta_3 + a), \tag{4.167}$$

and

$$H_2 = \frac{1}{\tau}\eta_1 - \frac{a\tau}{2}\left[\ln(\eta_3 + a)\right]^2 - \tau\eta_3 + \left[\tau(\eta_3 + a) - \eta_2\right]\ln(\eta_3 + a). \tag{4.168}$$

These two constants H_1 and H_2 are the generators of the equation of motion (4.166) with the multiplier M given by

$$M = -\frac{1}{(\eta_3 + a)}. \tag{4.169}$$

From Eq. (4.164) we calculate the Poisson bracket for the dynamical variables η_1, η_2 and η_3:

$$\{\eta_1, \eta_2\} = -\tau\eta_3, \quad \{\eta_2, \eta_3\} = 0, \quad \text{and} \quad \{\eta_3, \eta_1\} = \eta_3 + a. \tag{4.170}$$

These relations together with the Hamilton's equations

$$\dot{\eta}_i = \{\eta_i, H_2\}, \tag{4.171}$$

determine the equations of motion (4.166).

4.13 Symmetric Phase Space Formulation of the Damped Harmonic Oscillator

In the previous section we studied the Hamiltonian formulation in an extended phase space. A different but symmetric extension of the phase space, again suitable for studying dissipative motions, can be found by treating the phase space coordinates on the same footing and by introducing canonical momenta conjugate to these coordinates in the Lagrangian formalism [36]. Here we start with the extended Lagrangian $L(q, p, \dot{q}, \dot{p})$ defined by

$$L(q, p, \dot{q}, \dot{p}) = -\dot{q}p - \dot{p}q + L_q(q, \dot{q}) + L_p(p, \dot{p}),\qquad(4.172)$$

where L_q and L_p are the q and p space Lagrangians, i.e. they give correct equations of motion in coordinate and momentum space respectively. From the Lagrangins we find the canonical conjugate variables corresponding to q and p:

$$\pi_q = \frac{\partial L}{\partial \dot{q}} = \frac{\partial L_q}{\partial \dot{q}} - p,\qquad(4.173)$$

and

$$\pi_p = \frac{\partial L}{\partial \dot{p}} = \frac{\partial L_p}{\partial \dot{p}} - q.\qquad(4.174)$$

We note that π_p has the dimension of the coordinate q whereas π_q has the dimension of the canonical momentum p.

From these canonical momenta we obtain the extended phase space Hamiltonian

$$H(q, p, \pi_q, \pi_p) = \dot{q}\pi_q + \dot{p}\pi_p - L = H(p + \pi_q, q) - H(p, q + \pi_p).\qquad(4.175)$$

This formulation can be applied to describe the motion of a damped harmonic oscillator using successive canonical transformations [37]. To this end we start with the problem of the undamped harmonic oscillator for which the extended Hamiltonian is

$$H = \frac{1}{2m}\pi_q^2 + \frac{1}{m}p\pi_q - \frac{m\omega^2}{2}\pi_p^2 - m\omega^2 q\pi_p.\qquad(4.176)$$

Now we make the canonical transformation

$$q \to q, \quad p \to p, \quad \pi_q \to -\pi_q - p, \quad \pi_p \to -\pi_p - q,\qquad(4.177)$$

with the result that

$$H_1 = \left(\frac{\pi_q^2}{2m} + \frac{1}{2}m\omega^2 q^2\right) - \left(\frac{p^2}{2m} + \frac{1}{2}m\omega^2\pi_p^2\right).\qquad(4.178)$$

A second transformation

$$q \to q, \quad p \to p, \quad \pi_q \to \pi_q + \frac{m\lambda q}{2}, \quad \pi_p \to \pi_p - \frac{\lambda p}{2m\omega^2},\qquad(4.179)$$

changes H_1 to H_2, where

$$H_2 = \left[\frac{\pi_q^2}{2m} + \frac{1}{2}m\left(\omega^2 + \frac{\lambda^2}{4}\right)q^2\right] - \left[\frac{p^2}{2m}\left(1 + \frac{\lambda^2}{4\omega^2}\right) + \frac{1}{2}m\omega^2\pi_p^2\right]$$

$$+ \frac{\lambda}{2}(q\pi_q + p\pi_p).\qquad(4.180)$$

Finally we use a third transformation with the generator

$$F_2\left(q, \mathcal{P}_q, p, \mathcal{P}_p, t\right) = q\mathcal{P}_q e^{\frac{-\lambda t}{2}} + p\mathcal{P}_p e^{\frac{\lambda t}{2}}. \tag{4.181}$$

From this last transformation, (4.181), we obtain the following relations between the old and the new canonical variables:

$$Q = \frac{\partial F_2}{\partial \mathcal{P}_q} = qe^{\frac{-\lambda t}{2}}, \quad P = \frac{\partial F_2}{\partial \mathcal{P}_p} = pe^{\frac{\lambda t}{2}}, \tag{4.182}$$

and

$$\pi_q = \frac{\partial F_2}{\partial q} = \mathcal{P}_q e^{\frac{-\lambda t}{2}}, \quad \pi_p = \frac{\partial F_2}{\partial p} = \mathcal{P}_p e^{\frac{\lambda t}{2}}. \tag{4.183}$$

Substituting these in the new Hamiltonian H_3 we find

$$H_3 = H_2\left(Q, \mathcal{P}_q, P, \mathcal{P}_p, t\right) + \frac{\partial F_2}{\partial t}, \tag{4.184}$$

or

$$\begin{aligned} H_3 &= \left[\frac{\mathcal{P}_q^2}{2m} e^{-\lambda t} + \frac{1}{2}m\left(\omega^2 + \frac{\lambda^2}{4}\right)Q^2 e^{\lambda t}\right] \\ &\quad - \left[\frac{P^2}{2m}\left(1 + \frac{\lambda^2}{4\omega^2}\right)e^{-\lambda t} + \frac{1}{2}m\omega^2\mathcal{P}_p^2 e^{\lambda t}\right]. \end{aligned} \tag{4.185}$$

The expression in the first bracket of (4.185) is the Kanai-Caldirola Hamiltonian and the terms in the second bracket give us the classical evolution for the image oscillator [37]. The equation of motion for the image oscillator found from H_3 is

$$\ddot{P} - \lambda P + \left(\omega^2 + \frac{\lambda^2}{4}\right)P = 0, \tag{4.186}$$

and this is the same as that of the equation for y of the Bateman model we obtained earlier (see §(4.8)).

4.14 Dynamical Systems Expressible as Linear Difference Equations

For certain dynamical systems the momentum of the system (or the particle) can change discontinuously as a function of time. This discontinuous change can be described simply in terms of difference equations. Among the well-known examples of this type of motion one can cite the kicked rotator and the Fermi accelerator [38] [39]. Both of these systems have been studied to determine, among other characteristics, conditions for their regular and chaotic behaviors.

Before we start on the Hamiltonian description of motions expressible in terms of difference equations we must emphasize that in these systems the forces acting on the particle are impulsive forces or forces of constraint. These are idealization of stiff elastic forces and therefore we have to be careful when we take the limit of the stiffness going to infinity. In classical mechanics taking this limit is nontrivial, whereas in quantum theory the limit in general does not exist [40].

Classically there are at least two distinct forms of problems where the momentum changes discontinuously:

(1) Those cases where the discrete times at which the momentum changes are predetermined and are independent of the state of motion, e.g. a rotator which is kicked periodically.

(2) For a number of problems the time interval between the events that causes the change in momentum may depend on the state of the motion of the particle. The bouncing motion of a ball between two rigid plane, when these planes are moving relative to each other is an example of the latter case. Here we want to examine the question of the Hamiltonian description of such a motion and later study the problems that we encounter in quantizing such a motion.

The general form of the equation of motion for the momentum p is given by the linear difference equation [41]

$$p(t_{n+1}) = -\rho p(t_n) + 2mV(t_n). \tag{4.187}$$

In this equation ρ is the coefficient of restitution, m is the mass of the particle and $V(t)$ is a given function of time.

The Fermi-Ulam accelerator [42] [43] can be described in the following way: A rigid wall is fixed at $x = 0$ and a second rigid wall is moving and its position is given by $x = L(t)$. Between these walls the particle moves freely, but at the boundaries we have the conditions

$$p \rightarrow -p \quad \text{at} \quad x = 0, \quad p \rightarrow -p + 2m\dot{L}(t) \quad \text{at} \quad x = L(t). \tag{4.188}$$

The behavior of the system is completely specified by giving the values of p, x and t at the n-th bounce from the moving wall. The time between two successive bounces from the moving wall is

$$t_{n+1} - t_n = \frac{m}{|p_n|} \left[L(t_n) + L(t_{n+1}) \right], \tag{4.189}$$

where $p_n = p(t_n)$ is the momentum of the ball after the n-th bounce from the moving wall and this momentum is always negative;

$$p_n = p_{n-1} + 2m\dot{L}(t_n). \tag{4.190}$$

This equation is a special case of (4.187) for $\rho = 1$ and $V(t_n) = \dot{L}(t_n)$. Equations (4.189) and (4.190) form a set of difference equations. Except for a few special forms of $L(t)$, these equations can only be solved numerically. The most general form of $L(t)$ for which we can solve (4.189) and (4.190) recursively is when

$$L(t) = \left(At^2 + Bt + C\right)^{\frac{1}{2}}, \tag{4.191}$$

where A, B and C are arbitrary constants subject to the condition that $L(t)$ has to be real. For this case we have (see also §18.7)

$$t_{n+1} = \frac{B + \left[A + \left(\frac{p_n}{m}\right)^2\right] t_n + 2 \left(\frac{|p_n|}{m}\right) \left(At_n^2 + Bt_n + C\right)^{\frac{1}{2}}}{\left(\frac{p_n}{m}\right)^2 - A}, \tag{4.192}$$

and

$$p_n = p_{n-1} + 2m \frac{At_n + \frac{1}{2}B}{\left(At_n^2 + Bt_n + C\right)^{\frac{1}{2}}}. \tag{4.193}$$

We can find a Hamiltonian description for the equation of motion (4.187) when t_n s are defined by an expression similar to (4.189) or preferably given as a set of numbers independent of p_n s. To this end we write Eq. (4.187) as a differential equation

$$\frac{dp}{dt} = -f(t)p + g(t), \tag{4.194}$$

where $f(t)$ and $g(t)$ are defined by

$$f(t) = (i\pi - \ln \rho) \sum_n \delta(t - t_n), \tag{4.195}$$

and

$$g(t) = -2m \sum_n V(t)\delta(t - t_n). \tag{4.196}$$

Equation (4.194) shows that the impulsive forces are time- and velocity-dependent, and the system, in general, is not conservative.

From the equation of motion (4.194) which does not explicitly depend on the x-coordinate, we can construct the Hamiltonian. Noting that the mechanical momentum $p = m\frac{dx}{dt}$ is not the same as the canonical momentum, P, the Hamiltonian will be a function of x, P and t;

$$H = \frac{P^2}{2m} \exp\left[-\int_0^t f\left(t'\right) dt'\right] - xg(t) \exp\left[\int_0^t f\left(t'\right) dt'\right]. \tag{4.197}$$

The canonical equations of motion found for H are:

$$p = m\frac{dx}{dt} = m\frac{\partial H}{\partial P} = P \exp\left[-\int_0^t f\left(t'\right) dt'\right], \tag{4.198}$$

and

$$\frac{dP}{dt} = -\frac{\partial H}{\partial x} = g(t) \exp\left[\int_0^t f\left(t'\right) dt'\right]. \tag{4.199}$$

By eliminating P between these two equations we find (4.194). In order to relate the Hamiltonian H to the energy of the particle, $E(t)$, we substitute for P in terms of p in H, Eq. (4.197);

$$H = \left[\frac{p^2}{2m} - xg(t)\right] \exp\left[2\int_0^t f(t')\,dt'\right] = E \exp\left[2\int_0^t f(t')\,dt'\right], \quad (4.200)$$

where E is the sum of kinetic and potential energies of the particle. When the coefficient of restitution ρ is one, then $f(t) = i\pi \sum_n \delta(t - t_n)$, and $H = E$.

Bibliography

[1] H. Goldstein, *Classical Mechanics*, Second Edition (Addison-Wesley Reading, 1980) Chapter 8.

[2] J.L. McCauley, *Classicl Mechanics*, (Cambridge University Press, Cambridge, 1997) p. 75.

[3] F.J. Kennedy and E.H. Kerner, Am. J. Phys. 33, 463 (1965).

[4] M. Razavy and F.J. Kennedy, Can. J. Phys. 52, 1532 (1974).

[5] V.V. Dodonov, V.I. Man'ko and V.D. Skarzhinskiy in *Quantization, Gravitation and Group Methods in Physics* Edited by A.A. Komar (Nova Science, Commack, 1988) p. 73.

[6] V.V. Dodonov, V.I. Man'ko and V.D. Skarzhinsky, Hadronic J. 4, 1734 (1981).

[7] V.V. Dodonov, V.I. Man'ko and V.D. Skarzhinsky, Nuovo Cimento B69, 185 (1982).

[8] P. Cardirola, Nuovo Cimento 18, 393 (1941).

[9] P. Havas, Nuovo Cimento Supp. 5, 363 (1957).

[10] M. Ostrogradsky, Mem. Ac. St. Petersbourg, VI, 385 (1850).

[11] A. Pais and G. Uhlenbeck, Phys. Rev. 79, 145 (1950).

[12] E.T. Whittaker, *A Treatise on the Analytical Dynamics of Particles and Rigid Bodies*, Fourth Edition (Cambridge University Press, London, 1965) Chapter X.

[13] P. Caldirola, Rend. Ist. Lomb. Sc., A 93, 439 (1959).

[14] G. Valentini, Nuovo Cimento, XIX, 1280 (1961).

[15] L. Infeld, Acta Physica Hungaricae XVII, 7 (1964).

[16] B-G. Englert, Ann. Phys. (NY), 129, 1 (1980).

[17] M. Razavy, Hadronic J. 10, 7 (1987).

[18] H. Dekker, Z. Phys. B26, 273 (1977).

[19] H. Dekker, Phys. Rev. A16, 2126 (1977).

[20] D.M. Greenberger J. Math. Phys. 20, 762 (1979).

[21] D.M. Greenberger J. Math. Phys. 20, 771 (1979).

[22] J.R. Ray, Am. J. Phys. 47, 626 (1979).

[23] J. Flores, G. Solovey and S. Gil, Am. J. Phys. 71, 721 (2003).

[24] H. Bateman, Phys. Rev. 38, 815 (1931).

[25] P.M. Morse and H. Feshbach, *Methods of Theoretical Physics*, Part I (McGraw-Hill, New York, 1953) p. 298.

[26] H. Dekker, Phys. Rep. 80, 1 (1981).

[27] H. Feshbach and Y. Tikochinsky, Tran. N.Y. Acad. Sci. Ser. II, 38, 44 (1977).

[28] M. Razavy, Phys. Rev. A 36, 482 (1987).

[29] J. Geicke, unpublished, (2000).

[30] R. Cawley, Phys. Rev. A 20, 2370 (1979).

[31] P.A.M. Dirac, *Lectures on Quantum Mechanics*, (Yeshiva University, NewYork, 1965) Lecture No. 1.

[32] R. Cawley, Phys. Rev. Lett. 42, 413 (1979).

[33] G.D. Birkhoff, *Dynamical Systems*, (American Mathematical Society, New York, N.Y. 1927) p. 57.

[34] Y. Nambu, Phys. Rev. D7, 2405 (1973).

[35] S.Q.H. Nguyen and L.A. Turski, J. Phy. A 34, 8281 (2001).

[36] Y. Sobouti and S. Nasiri, Int. J. Mod. Phys. B7, 3255 (1993).

[37] S. Nasiri and H. Safari, Proc. Inst. Math. NAS, Ukraine, 43, 654 (2002).

[38] G.M. Zaslavsky, *Chaos in Dynamics Systems*, (Harwood Academic, New York, 1985) Chapter 3.

[39] G. Casati, B.L. Chirikov, J. Ford and F. Israilev, in *Stochastic Behavior in Classical and Quantum Hamiltonian Systems*, Edited by G. Casati and J. Ford (Springer-Verlag, New York, 1979) p. 334.

[40] N.G. van Kampen and J.J. Lodder, Am. J. Phys. 52, 419 (1984).

[41] M. Razavy, Hadronic J. 17, 515 (1994).

[42] E. Fermi, Phys. Rev. 75, 1169 (1949).

[43] S.M. Ulam, in *Proceedings of the Fourth Berkely Symposium on Mathematical Statistics and Probability*, vol. III (1961).

Chapter 5

Hamilton-Jacobi Formulation

Our next task is to study the simplest forms of the Hamilton-Jacobi equation for damped systems. The idea here is to construct a generating function $F_2 (q_1 \cdots q_N; P_1 \cdots P_N)$ which transforms an arbitrary set of canonical variables $(q_1 \cdots q_N; P_1 \cdots P_N)$ to a new set of canonical variables $Q_i = \beta_i = $ constant and $P_i = \alpha_i = $ constant [1] [2]. Here we assume that the system is integrable, and that the generating function exists globally [3]. The condition of integrability excludes most of the dissipative systems, yet for few simple but interesting cases that we want to consider the Hamilton-Jacobi (H-J) is separable and thus integrable.

The Hamilton-Jacobi equation which is a nonlinear equation of first order admits a great variety of complete integrals, i.e. solutions expressed in terms of N q_j s and t and N parameters $\alpha_1 \cdots \alpha_N$. Now if $S(q_1...q_N; \alpha_1...\alpha_N)$ is a complete integral of the Hamilton-Jacobi equation then the integrals of the Hamilton's equations of motion are given by [2]

$$\frac{\partial S}{\partial \alpha_j} = -\beta_j, \quad \text{and} \quad \frac{\partial S}{\partial q_j} = p_j, \quad j = 1, 2 \cdots N. \tag{5.1}$$

The close connection between the Schrödinger equation and the Hamilton-Jacobi (H-J) equation of classical dynamics enables us to study the semi-classical solution of the wave equation and also use H-J formulation to quantize the classical dissipative motion [4]-[6].

5.1   The Hamilton-Jacobi Equation for Linear Damping

Again let us consider the Hamiltonian for the linearly damped motion in three dimensions;

$$H\left(\mathbf{p}, \mathbf{r}, t\right) = \left(\frac{\mathbf{p}^2}{2m}\right) \exp(-\lambda t) + V(\mathbf{r}) \exp(\lambda t), \tag{5.2}$$

where $V(\mathbf{r})$ is the potential which acts on the particle.

The standard Hamilton-Jacobi equation for this Hamiltonian is given by [1]

$$\left(\frac{1}{2m}\right) (\nabla \zeta)^2 e^{-\lambda t} + V(\mathbf{r}) e^{\lambda t} + \frac{\partial \zeta}{\partial t} = 0. \tag{5.3}$$

A different form of the H-J equation can be found by utilizing the transformation

$$\zeta(\mathbf{r}, t) = S(\mathbf{r}, t) e^{\lambda t}, \tag{5.4}$$

and substituting for ζ in (5.3) to get

$$\left(\frac{1}{2m}\right) (\nabla S)^2 + V(\mathbf{r}) + \lambda S + \frac{\partial S}{\partial t} = 0. \tag{5.5}$$

Both of these forms can be used to obtain solutions for the motion of a particle in the presence of dissipative forces. For instance let us consider the application of the Hamilton-Jacobi equation to the problem of the damped harmonic oscillator. If we write $V(x) = \frac{1}{2} m \omega_0^2 x^2$, then Eq. (5.3) for one-dimensional motion reduces to

$$\left(\frac{1}{2m}\right) \left(\frac{\partial \zeta}{\partial x}\right)^2 e^{-\lambda t} + \frac{1}{2} m \omega_0^2 x^2 e^{\lambda t} + \frac{\partial \zeta}{\partial t} = 0. \tag{5.6}$$

Changing the variable x to y where

$$y = x \exp\left(\frac{\lambda t}{2}\right), \tag{5.7}$$

we transform (5.6) to an equation with variables y and t

$$\left(\frac{1}{2m}\right) \left(\frac{\partial \zeta}{\partial y}\right)^2 + \frac{1}{2} m \omega_0 y^2 = -\left(\frac{\partial \zeta}{\partial t} + \frac{1}{2} \lambda y \frac{\partial \zeta}{\partial y}\right). \tag{5.8}$$

Now we can expand the variables in the usual way of separation used in the Hamilton-Jacobi equation by assuming that ζ is the sum of two terms [1]

$$\zeta(y, t) = -\alpha t + W(y). \tag{5.9}$$

By substituting (5.9) in (5.8) we obtain the following differential equation for $W(y)$

$$\frac{1}{2m} \left[\left(\frac{dW}{dy}\right) - \frac{1}{2} \lambda m y\right]^2 + \frac{1}{2} m \omega^2 y^2 = \alpha, \tag{5.10}$$

where $\omega^2 = \omega_0^2 - \frac{\lambda^2}{4}$. By integrating (5.10) and substituting for $W(y)$ in (5.9) we find $\zeta(y, t)$

$$\zeta(y, t) = -\alpha t + \frac{1}{4}\lambda m y^2 + \int \left\{ 2m \left[\alpha - \frac{m}{2}\omega^2 y^2 \right] \right\}^{\frac{1}{2}} dy. \qquad (5.11)$$

From this equation we can obtain y as a function of time t by noting that [1]

$$\beta = \frac{\partial \zeta(y, t)}{\partial \alpha} = -t + \sqrt{\frac{m}{2\alpha}} \int \frac{dy}{\sqrt{1 - \frac{m\omega^2 y^2}{2\alpha}}}, \qquad (5.12)$$

where β is a constant. Carrying out the integration in (5.12) and then substituting from (5.7) we finally determine x as a function of t

$$x(t) = \sqrt{\frac{2\alpha}{m\omega^2}} e^{\frac{-\lambda t}{2}} \sin[\omega(t + \beta)]. \qquad (5.13)$$

We can also solve Eq. (5.5) for the damped harmonic oscillator. For the Jacobi function S we can separate the time dependent part by writing

$$S(x, t) = W_1(x) + \frac{q^2}{2m\lambda} \left(e^{-\lambda t} - 1 \right), \qquad (5.14)$$

and substitute this in the equation

$$\left(\frac{1}{2m} \right) \left(\frac{\partial S}{\partial x} \right)^2 + V(x) + \lambda S + \frac{\partial S}{\partial t} = 0, \qquad (5.15)$$

to find

$$\left(\frac{1}{2m} \right) \left(\frac{dW_1}{dx} \right)^2 + V(x) + \lambda W_1 = \frac{q^2}{2m}. \qquad (5.16)$$

In Eqs. (5.14) and (5.16) q^2 is the separation constant.

For the harmonic oscillator $V(x) = \frac{1}{2}m\omega_0^2 x^2$ and by using this method, we can obtain a parametric solution for W_1. Thus

$$W_1(z, q) = \frac{q^2}{2m\lambda} \left\{ 1 - \frac{1 + \omega_0^2 z^2}{1 + \lambda z + \omega_0^2 z^2} \exp\left[-\frac{\lambda}{\omega} \tan^{-1} \left(\frac{\lambda + 2\omega_0^2 z}{\omega} \right) \right] \right\}, \qquad (5.17)$$

where

$$x(z, q) = \frac{qz}{m\sqrt{(1 + \lambda z + \omega_0^2 z^2)}} \exp\left[-\frac{\lambda}{2\omega} \tan^{-1} \left(\frac{\lambda + 2\omega_0^2 z}{\omega} \right) \right]. \qquad (5.18)$$

Equation (5.14) is not the only way that we can write S. Another interesting form of S which is complex is given by [7]

$$S_J(x, t) = \frac{m}{2} \left(-\frac{\lambda}{2} \pm i\omega \right) (x - \xi(t))^2 + mx\dot{\xi}(t) + C(t), \qquad (5.19)$$

where $\xi(t)$ and $C(t)$ are solutions of the differential equations:

$$\ddot{\xi}(t) + \lambda \dot{\xi}(t) + \omega_0^2 \xi(t) = 0, \tag{5.20}$$

and

$$\dot{C}(t) + \lambda C(t) + \frac{1}{2} m \left(\dot{\xi}^2(t) - \omega_0^2 \xi^2(t) \right) = 0. \tag{5.21}$$

5.2   Classical Action for an Oscillator with Leaky Spring Constant

In §4.4 we found the Lagrangian for an oscillator with complex spring constant, Eq. (4.64). The classical action S for this motion can be obtained directly from the definition of S;

$$S = \int_{\tau_1}^{\tau_2} L \left(z, \frac{dz}{d\tau}, \tau \right) d\tau. \tag{5.22}$$

By substituting for L and carrying out the integration we find

$$S = \frac{m\omega}{2 \sin [\omega (\tau_2 - \tau_1)]} \left[(z_2^2 + z_1^2) \cos \omega(\tau_2 - \tau_1) - 2z_1 z_2 \right], \tag{5.23}$$

where z_1 and z_2 refer to the coordinates of the particle at τ_1 and τ_2 respectively.

We can determine the position of the particle directly from $S(\tau,0)$ by noting that

$$p = m \frac{dz}{d\tau} = \frac{\partial S(\tau,0)}{\partial z} = m\omega \left[z \cot(\omega\tau) - \frac{z_0}{\sin(\omega\tau)} \right]. \tag{5.24}$$

There is another way that we can formulate the action for this damped harmonic oscillator. Consider the classical equation of motion

$$m \left(\frac{d^2 x}{dt^2} \right) = m \left[-\omega^2 \left(1 + \alpha^2 \right) x + 2\alpha\omega \left(\frac{d\xi(t)}{dt} \right) \right], \tag{5.25}$$

where $\xi(t)$ is given as the solution of the differential equation

$$\ddot{\xi} + 2\alpha\omega\dot{\xi} + \omega^2 \left(1 + \alpha^2 \right) \xi = 0. \tag{5.26}$$

Thus x describes the motion of a damped harmonic oscillator, $\xi(t)$, superimposed on the background of a harmonic oscillator of frequency $\omega\sqrt{1 + \alpha^2}$. The Hamiltonian and the Hamilton-Jacobi equation for the motion (5.25) are:

$$H = \frac{p^2}{2m} + \frac{1}{2} m\omega^2 \left(1 + \alpha^2 \right) x^2 + 2\alpha\omega m \dot{\xi}(x - \xi), \tag{5.27}$$

and

$$\frac{\partial S}{\partial t} + \frac{1}{2m} \left(\frac{\partial S}{\partial x}\right)^2 + \frac{m}{2}\left(1+\alpha^2\right)\omega^2 x^2 + 2\alpha\omega m\dot\xi(x-\xi) = 0, \qquad (5.28)$$

respectively.

We can solve (5.28) with the result that

$$S(x,t) = \frac{1}{2}m\omega\sqrt{1+\alpha^2}\,[x-\xi(t)]^2 \cot\left(\sqrt{1+\alpha^2}\omega t\right) + mx\dot\xi + g(t), \qquad (5.29)$$

where $g(t)$ is a function of time and is defined by

$$g(t) = m\int^t \left(2\alpha\omega\xi\dot\xi + \frac{1}{2}m\omega\sqrt{1+\alpha^2}\xi^2 - \frac{1}{2}m\dot\xi^2\right)dt. \qquad (5.30)$$

As Eq. (5.5) shows the motion of a particle in a viscous medium with a force of friction linear in velocity can be described by adding a term λS to the Jacobi function of the standard Hamilton-Jacobi equation.

5.3 More About the Hamilton-Jacobi Equation for the Damped Motion

Now we want to point out three interesting facts about the Hamilton-Jacobi equation for damped systems:

(a) The non-uniqueness of the Hamilton-Jacobi equation and thus the principal function.

(b) The separability of the principal function for a system of two interacting particles moving in a viscous medium.

The fact that λS appears linearly in (5.5) is important since the motion of a particle in the viscous medium should not affect the motion of another particle moving in the same medium as long as there is no interaction between the two particles.

(c) The effective mass in classical systems and its connection to the quadratic damping force.

(a) For q-equivalent Hamiltonians, e.g. H_1 and H_2, Eq. (4.32) we have different Hamilton-Jacobi equations. For H_1 the Hamilton-Jacobi equation is given by (5.3) with $V = 0$, but for H_2 the corresponding equation for the principal function S is

$$\mathcal{E}\exp\left[\frac{1}{p_0}\left(\frac{\partial S}{\partial x}\right)\right] + \lambda p_0 x + \frac{\partial S}{\partial t} = 0. \qquad (5.31)$$

This equation is separable, and by writing it as

$$S = -\alpha t + W(\alpha, x), \qquad (5.32)$$

we find the differential equation for W;

$$W(\alpha, x) = p_0 \int^x \ln \left(\frac{\alpha - \lambda p_0 x'}{\varepsilon} \right) dx'. \tag{5.33}$$

Since $(\partial S / \partial \alpha) = \beta$ is a constant, from (5.32) and (5.33) we obtain

$$-\lambda(\beta + t) = \int^x \frac{dx'}{x' - \left(\frac{\alpha}{\lambda p_0} \right)}, \tag{5.34}$$

which is the solution of the equation of motion.

(b) Let $S(\mathbf{r}_1, \mathbf{r}_2, t)$ be the solution of the modified Hamilton-Jacobi for the two particles, then

$$\frac{1}{2m_1} [\nabla_1 S(\mathbf{r}_1, \mathbf{r}_2, t)]^2 + \frac{1}{2m_2} [\nabla_2 S(\mathbf{r}_1, \mathbf{r}_2, t)]^2 + V_1(\mathbf{r}_1) + V_2(\mathbf{r}_2) + \lambda S + \frac{\partial S}{\partial t} = 0. \tag{5.35}$$

This equation can be decomposed into two equations:

$$\frac{1}{2m_i} [\nabla_i S_i (\mathbf{r}_i, t)]^2 + V_i(\mathbf{r}_i) + \lambda S(\mathbf{r}_i, t) + \frac{\partial}{\partial t} S(\mathbf{r}_i, t) = 0, \quad i = 1, 2, \tag{5.36}$$

where

$$S(\mathbf{r}_1, \mathbf{r}_2, t) = S_1 (\mathbf{r}_1, t) + S_2 (\mathbf{r}_2, t). \tag{5.37}$$

Thus each particle moves independently of the other in the medium.

This additivity property of the S function in the corresponding quantum mechanical problem corresponds to the separability of the wave function.

(c) Finally having observed that the Hamiltonian for linear damping can also be interpreted as the Hamiltonian for a system where the mass of the system increases with time, we can inquire whether the same is true about the quadratic damping or not. Let us consider the following Hamiltonian

$$H = \left[\frac{p_r^2}{2m^*(r)} + \frac{p_\theta^2}{2m^*(r)r^2} \right] - \frac{1}{m} \int^r m^* (r') F (r') dr', \tag{5.38}$$

where $m^*(r)$ or the effective mass is given by

$$m^*(r) = m \left(1 - A e^{-2\gamma r} \right), \quad A < 1. \tag{5.39}$$

This Hamiltonian gives us the equations of motion

$$m^* \ddot{r} = -\frac{1}{2} \frac{dm^*}{dr} \dot{r}^2 + \frac{r p_\theta^2}{m^{*2}(r) r^4} \left(m^* + \frac{r}{2} \frac{dm^*}{dr} \right) + \frac{m^*}{m} F(r), \tag{5.40}$$

and

$$\dot{\theta} = \frac{p_\theta}{m^*(r) r^2}, \quad p_\theta = \text{constant}. \tag{5.41}$$

Equation (5.40) shows that the radial motion of the particle is subject to a dissipative force proportional to $e^{-2\gamma r}\dot{r}^2$, i.e. this force becomes important at short distances. The Hamilton-Jacobi equation for the Hamiltonian (5.38) is found from H;

$$\frac{1}{2m^*(r)}\left(\nabla S\right)^2 + V(r) + \frac{\partial S}{\partial t} = 0, \tag{5.42}$$

where

$$V(r) = -\int^r \frac{m^*(r')}{m} F(r')\,dr'. \tag{5.43}$$

Later we will see that Eq. (5.42) is the classical limit of the Schrödinger equation for certain types of velocity-dependent potentials.

Bibliography

[1] H. Goldstein, *Classical Mechanics*, Second Edition (Addison-Wesley, Reading, Mass. 1980) Chapter 10.

[2] L.A. Pars, *A Treatise on Analytical Dynamics*, (John Wiley & Sons, New York, 1965).

[3] J.L. McCauley, *Classical Mechanics*, (Cambrdge University Press, 1997) Chapter 16.

[4] M. Razavy, Z. Phys. B26, 201 (1977).

[5] M. Razavy, Can. J. Phys. 56, 311 (1978).

[6] L. Herrera, L. Nuñez, A Patiño, and H. Rago, Am. J. Phys. 54, 273 (1986).

[7] M. Razavy, Can. J. Phys. 56, 1372 (1978).

Chapter 6

Motion of a Charged Particle in an External Electromagnetic Field in the Presence of Damping

When A particle of charge e moves in an external electromagnetic field and at the same time is subject to a conservative force $(-\nabla V(\mathbf{r}))$, we can either add the Lorentz force $e\left(\mathbf{E} + \frac{1}{c}\mathbf{v} \wedge \mathbf{B}\right)$ to the conservative force in the equation of motion or alternatively use the principle of minimal coupling and replace the canonical momentum $\mathbf{p}$ by $\left(\mathbf{p} - \frac{e}{c}\mathbf{A}\right)$ in the Hamiltonian $H(\mathbf{r}, \mathbf{p})$ and add a term $e\phi$ to H. Here $\mathbf{A}(\mathbf{r}, t)$ and $\phi(\mathbf{r}, t)$ are the vector and the scalar electromagnetic potentials.

The physical meaning of the concept of minimal coupling which plays an essential role in the quantum theory of fields is carefully explained by Wentzel [1]. When the Hamiltonian H represents the energy of the particle, the two formulations lead to the same result [2]. However for the systems where H is not the energy of the system, and also for the dissipative systems in general, the minimum coupling rule does not give us the correct equation of motion. Let us illustrate this point by the following examples:

(a) In the case of a charged particle moving in an external electromagnetic field $(\mathbf{E}(\mathbf{r}, t), \mathbf{B}(\mathbf{r}, t))$, and at the same time experiencing a linear damping force $\lambda \dot{\mathbf{r}}$ the equation of motion is

$$m\ddot{\mathbf{r}} + m\lambda\dot{\mathbf{r}} = \left(e\mathbf{E} + \frac{e}{c}\dot{\mathbf{r}} \wedge \mathbf{B}\right). \tag{6.1}$$

We have already seen that the Hamiltonian

$$H = \frac{\mathbf{p}^2}{2m} \exp(-\lambda t), \tag{6.2}$$

generates the equation of motion

$$m\ddot{\mathbf{r}} + m\lambda\dot{\mathbf{r}} = 0. \tag{6.3}$$

Now by applying the minimal coupling rule to H, Eq. (6.2), we find

$$H = \frac{1}{2m}\left(\mathbf{p} - \frac{e}{c}\mathbf{A}\right)^2 \exp(-\lambda t) + e\phi. \tag{6.4}$$

But this Hamiltonian will not give us the equation of motion (6.1). The Hamiltonian which is the generator of the equation of motion (6.1) is [3]

$$H = \frac{1}{2m}\left[\mathbf{p} - \frac{e}{c}\mathbf{A}_\lambda(\mathbf{r}, t)\right]^2 \exp(-\lambda t) + e\phi_\lambda(\mathbf{r}, t), \tag{6.5}$$

where

$$\mathbf{A}_\lambda(\mathbf{r}, t) = \mathbf{A}(\mathbf{r}, t)\, e^{\lambda t} - \lambda \int^t \mathbf{A}(\mathbf{r}, t')\, e^{\lambda t'}\, dt', \tag{6.6}$$

and

$$\phi_\lambda(\mathbf{r}, t) = \phi(\mathbf{r}, t) e^{\lambda t}. \tag{6.7}$$

This Hamiltonian with the generalized form of minimal coupling will lead to the equation of motion (6.1). Associated with this definition of the electromagnetic potentials we have the corresponding gauge transformation

$$\phi'_\lambda(\mathbf{r}, t) \rightarrow \phi_\lambda(\mathbf{r}, t) - \frac{1}{c}\frac{\partial \chi_\lambda(\mathbf{r}, t)}{\partial t} e^{\lambda t}, \tag{6.8}$$

and

$$\mathbf{A}'_\lambda(\mathbf{r}, t) \rightarrow \mathbf{A}_\lambda(\mathbf{r}, t) + \nabla\chi_\lambda(\mathbf{r}, t), \tag{6.9}$$

where

$$\chi_\lambda(\mathbf{r}, t) = \chi(\mathbf{r}, t)e^{\lambda t} - \lambda \int^t \chi(\mathbf{r}, t')\, e^{\lambda t'}\, dt'. \tag{6.10}$$

(b) As a second example let us study the time-independent Hamiltonian formulation for the motion of a particle of charge e which is moving in a viscous medium and is under the influence of a constant magnetic field $\mathbf{B}$ which is along the z axis. Here the vector potential is given by

$$\mathbf{A} = \frac{1}{2}B\left(\mathbf{j}x - \mathbf{i}y\right), \tag{6.11}$$

where $\mathbf{i}$, $\mathbf{j}$ and $\mathbf{k}$ are the unit vectors along the x, y and z axes respectively. The x and y components of the position of the particle which we assume is moving in the $z = $ constant plane are given by

$$\ddot{x} + \lambda\dot{x} - \Omega\dot{y} = 0, \tag{6.12}$$

and

$$\ddot{y} + \lambda\dot{y} + \Omega\dot{x} = 0, \tag{6.13}$$

where ω is the cyclotron frequency and is given by

$$\Omega = \frac{eB}{mc}. \tag{6.14}$$

We want to construct a time-independent Hamiltonian and this Hamiltonian will be a constant of motion, a result which follows from the Hamilton's canonical equations.

Writing $v_x = \dot{x}$ and $v_y = \dot{y}$, from Eqs. (6.12) and (6.13) we have

$$\frac{dv_x}{dv_y} = \frac{\lambda v_x - \Omega v_y}{\Omega v_x + \lambda v_y}. \tag{6.15}$$

We can integrate this equation to yield

$$\left(v_x^2 + v_y^2\right) \exp\left[\frac{2\lambda}{\Omega} \tan^{-1}\left(\frac{v_x}{v_y}\right)\right] = \text{constant}. \tag{6.16}$$

Next we can replace v_x and v_y in (6.16) by

$$v_x = \frac{\partial H}{\partial p_x}, \quad v_y = \frac{\partial H}{\partial p_y}, \tag{6.17}$$

to get

$$\left[\left(\frac{\partial H}{\partial p_x}\right)^2 + \left(\frac{\partial H}{\partial p_y}\right)^2\right] \exp\left[\frac{2\lambda}{\Omega} \tan^{-1}\left(\frac{\frac{\partial H}{\partial p_x}}{\frac{\partial H}{\partial p_y}}\right)\right] = C(H), \tag{6.18}$$

where C is an arbitrary function of H. When $\lambda = 0$ and $C(H) = \frac{2}{m}H$, Eq. (6.18) can be solved and we find the general form of H to be

$$H = \frac{1}{2m}\left[p_x - f(x,y)\right]^2 + \frac{1}{2m}\left[p_y - g(x,y)\right]^2, \tag{6.19}$$

where $f(x,y)$ and $g(x,y)$ are arbitrary functions of x and y. For the special case of

$$f(x,y) = -m\Omega By, \quad \text{and} \quad g(x,y) = m\Omega Bx, \tag{6.20}$$

we have a Hamiltonian which can be found from the free particle Hamiltonian by minimal coupling. But when $\lambda \neq 0$ the solution of (6.18) for $C(H) = \frac{2}{m}H$ is not known.

The classical equations (6.12) and (6.13) must be modified if we want that in the corresponding quantum description the particle relaxes to a well-defined equilibrium state [4]. Rather than using Eqs. (6.12) and (6.13) one can consider the following classical equations:

$$v_x = \dot{x} + \frac{\lambda}{2}x, \tag{6.21}$$

$$v_y = \dot{y} + \frac{\lambda}{2}y, \tag{6.22}$$

$$\ddot{x} = \Omega\dot{y} - \lambda\dot{x} - \omega_0^2 x + \frac{1}{2}\lambda\Omega y, \tag{6.23}$$

and

$$\ddot{y} = -\Omega\dot{x} - \lambda\dot{y} - \omega_0^2 y - \frac{1}{2}\lambda\Omega x. \tag{6.24}$$

These equations can be derived from the time-dependent Hamiltonian

$$H = \frac{1}{2m}\left(\mathbf{p} - \frac{e}{2c}\mathbf{B} \wedge \mathbf{r}\, e^{\lambda t}\right)^2 e^{-\lambda t} + \frac{1}{2}m\omega_0^2\mathbf{r}^2 e^{\lambda t}, \tag{6.25}$$

in which the minimal coupling rule is violated. For a constant $\mathbf{B}$ the minimal coupling rule implies that $\mathbf{p}$ should be replaced by

$$\mathbf{p} \to \mathbf{p} - \frac{e}{2c}\mathbf{B} \wedge \mathbf{r}, \tag{6.26}$$

rather than $\mathbf{p} - \frac{e}{2c}\mathbf{B} \wedge \mathbf{r}e^{\lambda t}$ as is assumed in the Hamiltonian (6.25).

(c) The second method which can be used to construct a Hamiltonian for the equations of motion (6.12) and (6.13) is to utilize the complex coordinate z. Thus if we define z and α by

$$z = x + iy, \quad \alpha = \lambda + i\Omega, \tag{6.27}$$

then we can combine (6.12) and (6.13) into a single equation

$$\ddot{z} + \alpha\dot{z} = 0. \tag{6.28}$$

By integrating (6.28) with respect to time t we find

$$v_z + \alpha z = \text{constant} = C(H). \tag{6.29}$$

For different choices of $C(H)$ we get different Hamiltonians. However if we want a Hamiltonian to have the dimension of energy and be a quadratic function of momentum, we choose

$$C(H) = \sqrt{\frac{2}{m}H}, \tag{6.30}$$

and substitute this and $v_z = \frac{\partial H}{\partial p_z}$ in (6.29) to find

$$\sqrt{\frac{2}{m}H} = \frac{\partial H}{\partial p_z} + \alpha z. \tag{6.31}$$

This equation can be integrated to yield an implicit equation for H

$$p_z = \pm\sqrt{2mH} + m\alpha z \ln\left[\frac{\pm\sqrt{2mH}}{m\alpha z} - 1\right]. \tag{6.32}$$

Another simple Hamiltonian for (6.28) is found by taking $C(H) = \beta H$ where β is a real constant. In this case the result of integration of (6.29) is

$$H = A \exp\left(\beta p_z\right) + \left(\frac{\alpha}{\beta}\right) z. \tag{6.33}$$

If we write $p_z = p_x + i p_y$ and take the real and imaginary parts of H, we find the new Hamiltonian to be [5]

$$H_R = A \exp\left(\beta p_x\right) \cos\left(\beta p_y\right) + \frac{1}{\beta}\left(\lambda x - \Omega y\right). \tag{6.34}$$

The Lagrangian corresponding to (6.34) is given by

$$L_R = \frac{1}{2\beta}\dot{x}\ln\left(\dot{x}^2 + \dot{y}^2\right) - \frac{1}{\beta}\dot{y}\tan^{-1}\left(\frac{\dot{y}}{\dot{x}}\right) - \frac{\dot{x}}{\beta} + \frac{1}{\beta}\left(\Omega y - \lambda x\right), \tag{6.35}$$

and this Lagrangian yields the following equations of motion:

$$\left(\frac{1}{\dot{x}^2 + \dot{y}^2}\right)\left[\dot{y}\left(\ddot{x} + \lambda\dot{x} - \Omega\dot{y}\right) - \dot{x}\left(\ddot{y} + \lambda\dot{y} + \Omega\dot{x}\right)\right] = 0, \tag{6.36}$$

and

$$\left(\frac{1}{\dot{x}^2 + \dot{y}^2}\right)\left[\dot{x}\left(\ddot{x} + \lambda\dot{x} - \Omega\dot{y}\right) + \dot{y}\left(\ddot{y} + \lambda\dot{y} + \Omega\dot{x}\right)\right] = 0, \tag{6.37}$$

and these are equivalent to (6.12) and (6.13) provided that $\left(\frac{1}{\dot{x}^2 + \dot{y}^2}\right) \neq 0$. The imaginary part of H gives us Eqs. (6.12) and (6.13), but with $\lambda \to \Omega$ and $\Omega \to -\lambda$. A microscopic model of the interaction between a charged particle in a uniform magnetic field and coupled to a heat bath will be discussed later in Chapter 11 [6].

Bibliography

[1] G. Wentzel in *Preludes in Theoretical Physics*, edited by A. De-Shalit, H. Feshbach and L. van Hove, (North-Holland, Amsterdam, 1966) p. 199.

[2] H. Goldstein, *Classical Mechanics*, Second Edition, (Addison-Wesley, Reading, 1980).

[3] A. Pimpale and M. Razavy, Phys. Rev. A36, 2739 (1987).

[4] V.V. Dodonov and O.V. Man'ko, Theor. Math. Phys. vol. 65, 1033 (1986).

[5] V. Dodonov, V.I. Man'ko and V.D. Skarzhinsky, Nuovo Cimento, 69B, 185 (1982).

[6] X.L. Li, G.W. Ford and R.F. O'Connell, Phys. Rev. A41, 5287 (1990).

Chapter 7

Noether and Non-Noether Symmetries and Conservation Laws

The Lagrangian formulation of dissipative systems enable us to determine the constants of motion, if they exist, with the help of an important theorem due to Noether [1]-[4]. When the dynamics of a dissipative system cannot be formulated in terms of the variational principle, there are other methods which can be used to obtain the constants of motion directly from the equations of motion [5] [6]. To begin with, we will follow a very simple approach based on the transformation properties of the generalized coordinates and time in the Lagrangian formulation to obtain the conserved quantities, and then examine other possibilities [7] [8].

Consider the following set of transformations for a system of N degrees of freedom:

$$X_i = X_i\left(x_j, \dot{x}_j, t; \epsilon\right), \quad i, j = 1, 2, \cdots N, \tag{7.1}$$

and

$$T = T\left(x_j, \dot{x}_j, t; \epsilon\right), \quad j = 1, 2, \cdots N. \tag{7.2}$$

These transformations have the property that

$$X_i \rightarrow x_i, \quad \text{as} \quad \epsilon \rightarrow 0, \tag{7.3}$$

and

$$T \rightarrow t, \quad \text{as} \quad \epsilon \rightarrow 0. \tag{7.4}$$

The Lagrangian for the system is given by $L\left(x_i, \dot{x}_i, t\right)$ from which the equation of motion for the damped system can be derived via the Euler-Lagrange equation.

77

If we replace $x_i, \dot{x}_i$ and t in the Lagrangian by $X_i, \dot{X}_i$ and T, then we find the new Lagrangian $L\left(X_i, \frac{dX_i}{dT}, T; \epsilon\right)$, where

$$\frac{dX_i}{dT} = \left(\frac{dX_i}{dt}\right)\left(\frac{dT}{dt}\right)^{-1} = \frac{\dot{X}_i\left(x_i, \dot{x}_i, \ddot{x}_i, t; \epsilon\right)}{\dot{T}\left(x_i, \dot{x}_i, \ddot{x}_i, t; \epsilon\right)}. \tag{7.5}$$

Now let us calculate

$$\left[\frac{\partial}{\partial \epsilon}\left\{L\left(X_i, \frac{dX_i}{dT}, T; \epsilon\right) \dot{T}\left(x_i, \dot{x}_i, \ddot{x}_i, t; \epsilon\right)\right\}\right]_{\epsilon=0}. \tag{7.6}$$

If we write this expression which is a function of $x_i, \dot{x}_i$ and t as $dF(x_i, \dot{x}_i, t)/dt$ then the Noether theorem states that the quantity

$$L\Lambda + \sum_{j=1}^{N} \frac{\partial L\left(x_i, \dot{x}_i, t\right)}{\partial x_j}\left(\Delta_j - \dot{x}_j\Lambda\right) - F, \tag{7.7}$$

is a constant of motion. In this relation

$$\Lambda = \left[\frac{\partial T\left(x_i, \dot{x}_i, t; \epsilon\right)}{\partial \epsilon}\right]_{\epsilon=0}, \tag{7.8}$$

and

$$\Delta_j = \left[\frac{\partial X_j\left(x_i, \dot{x}_i, t; \epsilon\right)}{\partial \epsilon}\right]_{\epsilon=0}. \tag{7.9}$$

In order to prove that (7.7) is indeed a constant we write $\dot{F}$ which is defined by (7.6) as the limit of

$$\left(\frac{\partial L}{\partial T}\right)\left(\frac{\partial T\left(x_j, \dot{x}_j, t; \epsilon\right)}{\partial \epsilon}\right)\left(\frac{dT\left(x_j, \dot{x}_j, t; \epsilon\right)}{dt}\right)$$

$$+ \sum_k \left(\frac{\partial L\left(X_j, \frac{dX_j}{dT}, T\right)}{\partial X_k}\right)\left(\frac{\partial X_k\left(x_j, \dot{x}_j, t; \epsilon\right)}{\partial \epsilon}\right)\left(\frac{dT\left(x_j, \dot{x}_j, t; \epsilon\right)}{dt}\right)$$

$$+ \sum_k \left(\frac{\partial L\left(X_j, \frac{dX_j}{dT}, T\right)}{\partial \left(\frac{dX_k}{dT}\right)}\right)\left[\frac{\partial}{\partial \epsilon}\left(\frac{dX_k\left(x_j, \dot{x}_j, \ddot{x}_j, t; \epsilon\right)}{dT}\right)\right]\left(\frac{dT\left(x_j, \dot{x}_j, t; \epsilon\right)}{dt}\right)$$

$$+ L\left(X_j, \dot{X}_j, T\right)\frac{\partial}{\partial \epsilon}\left(\frac{dT\left(x_j, \dot{x}_j, t; \epsilon\right)}{dt}\right), \tag{7.10}$$

as ϵ tends to zero.

For calculating the derivatives in (7.10) we first expand T and X_j in powers of ϵ;

$$T = t + \Lambda\epsilon + \cdots, \tag{7.11}$$

and

$$X_j = x_j + \Delta_j\epsilon + \cdots. \tag{7.12}$$

Using these expansions we obtain the following relations:

$$\dot{T} = 1 + \dot{\Lambda}\epsilon + \cdots, \tag{7.13}$$

$$\dot{X}_j = \dot{x}_j + \dot{\Delta}_j\epsilon + \cdots, \tag{7.14}$$

$$\left(\frac{dX_j}{dT}\right)_{\epsilon=0} = \left(\frac{dX_j}{dt}\right)\left(\frac{dT}{dt}\right)^{-1} = \dot{x}_j, \tag{7.15}$$

$$\left[\frac{\partial T\left(x_j, \dot{x}_j, t; \epsilon\right)}{\partial \epsilon}\right]_{\epsilon=0} = \Lambda, \tag{7.16}$$

$$\left[\frac{\partial X_j\left(x_j, \dot{x}_j, t; \epsilon\right)}{\partial \epsilon}\right]_{\epsilon=0} = \Delta_j, \tag{7.17}$$

$$\left[\frac{\partial}{\partial \epsilon}\left(\frac{dT\left(x_j, \dot{x}_j, t; \epsilon\right)}{dt}\right)\right]_{\epsilon=0} = \dot{\Lambda}, \tag{7.18}$$

$$\left[\frac{\partial}{\partial \epsilon}\left(\frac{dX_j\left(x_j, \dot{x}_j, t; \epsilon\right)}{dt}\right)\right]_{\epsilon=0} = \dot{\Delta}_j, \tag{7.19}$$

and finally

$$\left[\frac{\partial}{\partial \epsilon}\left(\frac{dX_j\left(x_j, \dot{x}_j, t; \epsilon\right)}{dT}\right)\right]_{\epsilon=0} = \dot{\Delta}_j - \dot{x}_j\dot{\Lambda}. \tag{7.20}$$

We note also that the partial derivatives of $L\left(X_j, \frac{dX_j}{dT}, T\right)$ with respect to T, X_j and $\frac{dX_j}{dT}$ in the limit of $\epsilon \to 0$ tend to the corresponding derivatives of $L\left(x_j, \dot{x}_j, t\right)$ with respect to t, x_j and $\dot{x}_j$. Also in the limit of $\epsilon \to 0$, $L\left(X_j, \frac{dX_j}{dT}, T\right)$ goes over to $L\left(x_j, \dot{x}_j, t\right)$. From Eqs. (7.13)-(7.20) and these limits we find the following relation;

$$\frac{\partial L\left(x_i, \dot{x}_i, t\right)}{\partial t}\Lambda + \sum_j \frac{\partial L\left(x_i, \dot{x}_i, t\right)}{\partial x_j}\Delta_j$$

$$+ \sum_j \frac{\partial L\left(x_i, \dot{x}_i, t\right)}{\partial \dot{x}_j}\left(\dot{\Delta}_j - \dot{x}_j\dot{\Lambda}_j\right) + L\left(x_i, \dot{x}_i, t\right)\dot{\Lambda} = \dot{F}. \tag{7.21}$$

Since now we can write all the arguments in terms of x_i, $\dot{x}_i$ and t, we will suppress the arguments in L appearing in differential relations.

Next we find the total derivatives of $\frac{dL}{dt}$, $\frac{d}{dt}\left(\frac{\partial L}{\partial \dot{x}_j}\Delta_j\right)$ and $\frac{d}{dt}\left(\frac{\partial L}{\partial \dot{x}_j}\dot{x}_j\Lambda\right)$, and from these we calculate $\frac{\partial L}{\partial t}$, $\frac{\partial L}{\partial \dot{x}_j}\dot{\Delta}_j$ and $\frac{\partial L}{\partial \dot{x}_j}x_j\dot{\Lambda}$ respectively. We substitute for these partial derivatives in the above equation for $\dot{F}$ and after simplifying it we obtain

$$\sum_j \left\{\frac{\partial L}{\partial x_j} - \frac{d}{dt}\left(\frac{\partial L}{\partial \dot{x}_j}\right)\right\}\left(\Delta_j - \dot{x}_j\Lambda\right)$$

$$+ \frac{d}{dt}\left[L\Lambda + \sum_j \frac{\partial L}{\partial \dot{x}_j}\left(\Delta_j - \dot{x}_j\Lambda\right) - F\right] = 0. \tag{7.22}$$

The curly bracket in (7.22) is zero since L satisfies the Euler-Lagrange equation, therefore the total time derivative in (7.22) is also zero. This shows that the quantity in the square bracket in (7.22) is constant in agreement with (7.7).

Now let us study some specific examples of the application of the Noether theorem.

(1) In cases where the Lagrangian L is not explicitly dependent on time, i.e. $\frac{\partial L}{\partial t} = 0$, we have a constant of motion. This is the case of invariance of the Lagrangian under time translation. We choose X_j and T to be

$$X_j = x_j, \quad j = 1, 2, \cdots N, \quad \text{and} \quad T = t + \epsilon, \tag{7.23}$$

then $\frac{dX_j}{dT} = \dot{x}_j, \Lambda = 1$, and $\Delta_j = 0$. Also from (7.6) it follows that $\frac{dF}{dt} = 0$ or $F = \text{constant}$. Substituting these values of Δ_j and Λ in (7.7) we find that

$$L - \sum_j x_j \frac{\partial L}{\partial \dot{x}_j} = \text{constant} = -H, \tag{7.24}$$

where H is the Hamiltonian of the system.

For dissipative motions such as the one given by Eqs. (3.76)-(3.77) the Hamiltonian H is not the energy of the system.

(2) If the Lagrangian does not depend on a particular coordinate x_k, then the momentum conjugate to this coordinate p_k is conserved . In this case we choose

$$T = t, \quad \text{and} \quad X_k = x_k + \epsilon, \quad X_j = x_j, \quad j \neq k. \tag{7.25}$$

From Eqs. (7.8) and (7.9) we obtain Δ_k and Λ;

$$\Lambda = 0, \quad \Delta_k = 1, \quad \Delta_j = 0, \quad j \neq k. \tag{7.26}$$

Now from (7.6) it follows that $\dot{F} = 0$, and therefore F is a constant. We can set this constant equal to zero, and hence Noether's theorem, Eq. (7.7), implies that

$$\frac{\partial L}{\partial \dot{x}_k} = p_k = \text{constant}. \tag{7.27}$$

For the one-dimensional motion of a particle in a viscous medium in the absence of any other force except linear damping as we have seen before, the Lagrangian is

$$L = \frac{1}{2} m \dot{x}^2 e^{\lambda t}. \tag{7.28}$$

The transformation (7.25) with $x_k = x$ shows us that according to Noether's theorem

$$p(t) = \frac{\partial L}{\partial \dot{x}} = m \dot{x} e^{\lambda t} = p_0, \tag{7.29}$$

where p_0 is a constant.

The Lagrangian L in (7.28) is not invariant under time translation, Eq. (7.23). However for this motion we can still find a Lagrangian which is not explicitly dependent on time. If we want to transform the time t to another time t^* then the action S, not the Lagrangian must remain invariant.

Let us examine the infinitesimal action Ldt:

$$Ldt = \frac{1}{2}m \left(\frac{dx}{dt}\right)^2 e^{\lambda t} dt, \tag{7.30}$$

and let $t^* = t^*(t)$ be the new time variable, then we want to have

$$dS = L\left(\dot{x}, t\right) dt = L\left(\frac{dt}{dt^*}\right) dt^*, \tag{7.31}$$

invariant or

$$\frac{1}{2}m \left(\frac{dx}{dt^*}\right)^2 \left(\frac{dt^*}{dt}\right)^2 e^{\lambda t} dt = \frac{1}{2}m \left(\frac{dx}{dt^*}\right)^2 dt^*. \tag{7.32}$$

From this relation we find that

$$\frac{dt^*}{dt} = e^{-\lambda t} \quad \text{or} \quad t^* = \frac{1 - e^{-\lambda t}}{\lambda}, \tag{7.33}$$

where the integration constant is chosen in such a way that when $\lambda \to 0$ then $t^* \to t$. Now $L\left(\dot{x}^*\right) dt^*$ is invariant under the new time translation, i.e. $t^* \to t^* + \epsilon$, and this means that the original Lagrangian with the variable t is invariant under the time translation given by [9]

$$t \to t - \frac{1}{\lambda} \ln \left(1 - \lambda \epsilon e^{\lambda t}\right). \tag{7.34}$$

For a number of problems involving dissipative forces, the direct approach of investigating the invariance under a general linear transformation is simpler [10]-[12]. Thus let us consider the case of a one-dimensional motion under the action of a force $f\left(x, \dot{x}, t\right)$ when the coordinate x and the time t are transformed according to

$$X = \alpha x + \beta t + x_0, \tag{7.35}$$

and

$$T = \delta x + \nu t + t_0, \tag{7.36}$$

where $\alpha, \beta, \delta, \nu, x_0$ and t_0 are all constants and

$$\alpha \nu - \beta \delta \neq 0, \tag{7.37}$$

i.e. the transformation is not singular and x and t can be obtained in terms of X and T. Let us denote the time derivatives with respect to T by primes, e.g.

$$X' = \frac{dX}{dT} = \frac{\alpha \dot{x} + \beta}{\delta \dot{x} + \nu}, \tag{7.38}$$

and write the equation of motion using the new variables X and T. By substituting for x, $\dot{x}, \ddot{x}$ and t in terms of X, X', X'' and T in the equation of motion

$m\ddot{x} = f(x, \dot{x}, t)$ and requiring that the transformed equation of motion be the same, i.e.

$$mX'' = f(X, X', T),\tag{7.39}$$

we find that [12]

$$f(x, \dot{x}, t) = \frac{(\delta\dot{x} + \nu)^3}{(\alpha\nu - \beta\delta)}f(X, X', T).\tag{7.40}$$

Now let us suppose that the Lagrangian $L(x, \dot{x}, t)$ remains invariant under this transformation, i.e.

$$L(x, \dot{x}, t) = L(X, X', T),\tag{7.41}$$

then Eq. (3.52) yields

$$\left[\frac{\partial^2 L}{\partial\dot{x}^2}\left(\frac{\partial L}{\partial t} + \dot{x}\frac{\partial L}{\partial x}\right) - 2\frac{\partial L}{\partial\dot{x}}\left(\dot{x}\frac{\partial^2 L}{\partial x\partial\dot{x}} + \frac{\partial^2 L}{\partial\dot{x}\partial t} - \frac{\partial L}{\partial x}\right)\right]\delta = 0.\tag{7.42}$$

This equation can be satisfied apart from some uninteresting Lagrangians only when $\delta = 0$. When this is the case, then (7.40) reduces to

$$f(x, \dot{x}, t) = \frac{\nu^2}{\alpha}f(X, X', T).\tag{7.43}$$

If $g(x, \dot{x}, t)$ is a first integral of motion, i.e. $(dg/dt) = 0$, then g is a solution of the partial differential equation

$$\frac{\partial g}{\partial t} + \dot{x}\frac{\partial g}{\partial x} + \frac{1}{m}f(x, \dot{x}, t)\frac{\partial g}{\partial\dot{x}} = 0.\tag{7.44}$$

Let us see the effect of the transformations (7.35) and (7.36) with $\delta = 0$ on the function $g(x, \dot{x}, t)$. To this end we write $g(x, \dot{x}, t) = G(X, X', T)$ and show that $G(X, X', T)$ is also a solution of (7.44):

$$\begin{aligned}\frac{\partial g}{\partial t} &+ \dot{x}\frac{\partial g}{\partial x} + \frac{1}{m}f(x, \dot{x}, t)\frac{\partial g}{\partial\dot{x}}\\ &= \alpha\frac{\partial G}{\partial X}\dot{x} + \frac{\alpha}{m\nu}f(x, \dot{x}, t)\frac{\partial G}{\partial X'} + \beta\frac{\partial G}{\partial X} + \nu\frac{\partial G}{\partial T}\\ &= \nu\left[\left(\frac{\alpha\dot{x} + \beta}{\nu}\right)\frac{\partial G}{\partial X} + \frac{\alpha}{m\nu^2}f\frac{\partial G}{\partial X'} + \frac{\partial G}{\partial T}\right]\\ &= \nu\left[\frac{\partial G}{\partial X}Q' + \frac{1}{m}f(X, X', T)\frac{\partial G}{\partial X'} + \frac{\partial G}{\partial T}\right] = 0.\end{aligned}\tag{7.45}$$

Before applying these ideas to the problems involving dissipative systems, we observe that there are the following invariances that can be studied using the transformations (7.35) and (7.36):
(1) Time translation invariance:

$$\alpha = \nu = 1, \quad \beta = x_0 = 0, \quad t_0 \text{ arbitrary}.\tag{7.46}$$

(2) Coordinate translation:

$$\alpha = \nu = 1, \quad \beta = t_0 = 0, \quad x_0 \text{ arbitrary.} \tag{7.47}$$

(3) Time-scale transformation:

$$\alpha = 1, \quad \beta = x_0 = t_0 = 0, \quad \nu \text{ arbitrary.} \tag{7.48}$$

(4) Coordinate-scale transformation:

$$\nu = 1, \quad \beta = x_0 = t_0 = 0, \quad \alpha \text{ arbitrary,} \tag{7.49}$$

and finally
(5) Galilean transformation:

$$\alpha = \nu = 1, \quad x_0 = t_0 = 0, \quad \beta \text{ arbitrary.} \tag{7.50}$$

These are not the only possible transformations which lead to the corresponding conservation laws, but they are the important ones in our study of the dissipative systems [12] [13].

As an example of the application of these invariances as applied to the damped motion, let us consider the damped harmonic oscillator

$$f(x, \dot{x}) = -m\lambda\dot{x} - m\omega_0^2 x, \tag{7.51}$$

for which $f(x, \dot{x})$ is a homogeneous function of first degree in x and $\dot{x}$. Here we set $\nu = 1$ but we allow α to be a free coordinate scale factor. If we are looking for a Lagrangian which remains invariant under this particular transformation, thus independent of α, then we search for a Lagrangian which is a function of $z = \dot{x}/x$ and t [12]. In terms of z and t Eqs. (3.52) and (7.44) can be written as

$$\frac{\partial^2 L}{\partial z \partial t} - z^2 \frac{\partial^2 L}{\partial z^2} + \frac{1}{m} f(1, z, t) \frac{\partial^2 L}{\partial z^2} = 0, \tag{7.52}$$

and

$$\frac{\partial g}{\partial t} - \left[z^2 - \frac{1}{m} f(1, z, t) \right] \frac{\partial g}{\partial z} = 0. \tag{7.53}$$

For the velocity dependent force $f(x, \dot{x})$ given by (7.51) these equations reduce to:

$$\frac{\partial^2 L}{\partial z \partial t} - z^2 \frac{\partial^2 L}{\partial z^2} - (\lambda z + \omega_0^2) \frac{\partial^2 L}{\partial z^2} = 0, \tag{7.54}$$

and

$$\frac{\partial g}{\partial t} - (z^2 + \lambda z + \omega_0^2) \frac{\partial g}{\partial z} = 0. \tag{7.55}$$

The particular solutions of (7.54) and (7.55) are given by [12]:

$$L = \frac{\lambda + 2z}{2\omega} \left[\tan^{-1} \left(\frac{2z + \lambda}{2\omega} \right) + \omega t \right] - \frac{1}{2} \ln \left[1 + \left(\frac{2z + \lambda}{2\omega} \right)^2 \right], \tag{7.56}$$

and

$$g = \tan^{-1}\left(\frac{2z + \lambda}{2\omega}\right) + \omega t, \qquad (7.57)$$

respectively, where $\omega = \left(\omega_0^2 - \frac{\lambda^2}{4}\right)^{\frac{1}{2}}$.

The Lagrangian L, Eq. (7.56), and the Lagrangian L_2 given by (3.111) differ from each other by a total time derivative.

7.1 Non-Noether Symmetries and Conserved Quantities

The foregoing discussion of the symmetries was based on the existence of Lagrangians for conservative as well as non-conservative dynamical systems. Now we want to study the symmetries which can be directly derived from the equations of motion which we assume to be of the second order [5] [14].

We write these equations as

$$\frac{d^2 x_j}{dt^2} = f_j\left(x_i, \dot{x}_i, t\right), \quad j = 1, 2, \cdots n, \qquad (7.58)$$

where we have set the mass of the particle(s) equal to unity. In terms of the generalized dissipative forces discussed in Chapter 2, we can derive Eq. (7.58) from the Euler-Lagrange equation

$$\frac{d}{dt}\left(\frac{\partial L}{\partial \dot{x}_j}\right) - \frac{\partial L}{\partial x_j} = Q_j. \qquad (7.59)$$

As before we introduce the infinitesimal time and coordinate transformations (7.11) and (7.12) with Λ and Δ_j being functions of the coordinates x_j and time t, and the small parameter ϵ. If (7.58) is invariant under the transformation (7.11) and (7.12) then by direct substitution for X_i and T in $\frac{d^2 X_j}{dT^2} = f_j\left(X_i, \frac{dX_i}{dT}, T\right)$ and subsequent expansion in powers of ϵ, we find that Λ and Δ_j must satisfy the differential equation

$$\Pi_j \equiv \frac{d^2 \Delta_j}{dt^2} - \dot{x}_j \frac{d^2 \Lambda}{dt^2} - 2 f_j\left(x_i, \dot{x}_i, t\right)\frac{d\Lambda}{dt} - \hat{\mathcal{D}}(f_j) = 0. \qquad (7.60)$$

Here $\hat{\mathcal{D}}$ is the differential operator

$$\hat{\mathcal{D}} = \Lambda \frac{\partial}{\partial t} + \sum_j\left[\Delta_j \frac{\partial}{\partial x_j} + \left(\dot{\Delta}_j - \dot{x}_j \dot{\Lambda}\right)\frac{\partial}{\partial \dot{x}_j}\right], \qquad (7.61)$$

which operates on $f_j\left(x_i, \dot{x}_i, t\right)$. In Eq. (7.61) a dot indicates the total time derivative along the trajectory, i.e.

$$\frac{d}{dt} = \frac{\partial}{\partial t} + \sum_j\left(\dot{x}_j \frac{\partial}{\partial x_j} + f_j \frac{\partial}{\partial \dot{x}_j}\right). \qquad (7.62)$$

Let us define D by

$$D = \det \left[\frac{\partial^2 L}{\partial \dot{x}_j \partial \dot{x}_k} \right], \tag{7.63}$$

and let M_{jk} be the cofactor of $\frac{\partial^2 L}{\partial \dot{x}_j \partial \dot{x}_k}$ in the matrix formed by the second order derivatives. Writing Eq. (7.59) in the expanded form, viz,

$$\sum_k \frac{\partial^2 L}{\partial \dot{x}_j \partial \dot{x}_k} \ddot{x}_k = \frac{\partial L}{\partial x_j} - \frac{\partial^2 L}{\partial \dot{x}_j \partial t} - \sum_k \dot{x}_k \frac{\partial^2 L}{\partial \dot{x}_j \partial x_k} + Q_j, \tag{7.64}$$

we can show that

$$\sum_k \left[\frac{\partial f_k}{\partial \dot{x}_k} - \sum_j \frac{\partial}{\partial \dot{x}_k} \left(\frac{M_{kj}}{D} Q_j \right) \right] + \frac{d}{dt} (\ln D) = 0. \tag{7.65}$$

Now we state the conservation law in the following way: Let $g = g(x_j, \dot{x}_j, t)$ satisfy the equation

$$\frac{dg}{dt} = \sum_{k,j} \frac{\partial}{\partial \dot{x}_j} \left(\frac{M_{jk}}{D} Q_k \right), \tag{7.66}$$

and Λ and Δ be solutions of (7.60), then equations (7.58) admit a conserved quantity

$$\Phi = 2 \sum_j \left(\frac{\partial \Delta_j}{\partial x_j} - \dot{x}_j \frac{\partial \Lambda}{\partial x_j} \right) - N\dot{\Lambda} + \hat{\mathcal{D}}(\ln D) - \hat{\mathcal{D}}(g). \tag{7.67}$$

In order to show that Φ is a constant of motion, we try to relate it to $\sum_j \frac{\partial \Pi_j}{\partial \dot{x}_j}$. To this end we differentiate (7.60) with respect to $\dot{x}_j$ and then using (7.66) we substitute for

$$\sum_{k,j} \left[\frac{\partial}{\partial \dot{x}_j} \left(\frac{M_{jk}}{D} Q_k \right) \right] \tag{7.68}$$

in the resulting expression to find

$$\sum_j \frac{\partial \Pi_j}{\partial \dot{x}_j} = \frac{d}{dt} \left[2 \sum_j \left(\frac{\partial \Delta_j}{\partial x_j} - \dot{x}_j \frac{\partial \Lambda}{\partial x_j} \right) - N\dot{\Lambda} + \hat{\mathcal{D}}(\ln D) - \hat{\mathcal{D}}(g) \right] = \frac{d\Phi}{dt} = 0, \tag{7.69}$$

where for one-dimensional motion $N = 1$. Thus Φ is a conserved quantity.

As an example let us consider the simple case of a dissipative force which is quadratic in velocity, $\ddot{x} + \gamma \dot{x}^2 = 0$. For this damping force Eq. (7.60) becomes

$$\begin{aligned}
\Delta_{tt} &- 2\gamma \dot{x} \Delta_{tx} + \gamma^2 \dot{x}^2 \Delta_{xx} + \gamma^2 \dot{x}^2 \Delta_x + \gamma \dot{x} \Lambda_{tt} \\
&- 2\gamma^2 \dot{x}^2 \Lambda_{tx} + \gamma^3 \dot{x}^3 \Lambda_{xx} + \gamma^3 \dot{x}^3 \Lambda_x - 2\gamma^2 \dot{x}^2 \Lambda_t + 2\gamma^3 \dot{x}^3 \Lambda_x \\
&= -2\gamma \dot{x} \Delta_t + 2\gamma^2 \dot{x}^2 \Delta_x - 2\gamma^2 \dot{x}^2 \Lambda_t + 2\gamma^3 \dot{x}^3 \Lambda_x.
\end{aligned} \tag{7.70}$$

Here subscripts denote partial derivatives. Since Δ and Λ and their partial derivatives do not depend on $\dot{x}$, therefore the coefficients of different powers of $\dot{x}$ on the two sides of (7.70) should match. Thus we find four partial differential equations for Δ and Λ. The most general solution of these equations are:

$$\Lambda = a_0 + a_1 t + a_2 e^{\gamma x} + a_3 t e^{\gamma x}, \tag{7.71}$$

and

$$\Delta = b_0 + b_1 e^{\gamma x} + b_2 t e^{-\gamma x}. \tag{7.72}$$

The nontrivial Φ s obtained from these results are given by [14]

$$\Lambda = e^{\gamma x}, \quad \Delta = 0, \quad \Phi = -2\gamma x e^{\gamma x}, \tag{7.73}$$

and

$$\Lambda = -t e^{\gamma x}, \quad \Delta = e^{\gamma x}, \quad \Phi = -3\left(1 - \gamma t \dot{x}\right) e^{\gamma x}. \tag{7.74}$$

7.2   Noether's Theorem for a Scalar Field

We have already seen that by applying the discrete version of the Noether's theorem to the Lagrangian L, we can find certain conserved quantities for a dissipative system. In this section we want to study the field theoretical description of this theorem which can be used to study the invariances and the conservation laws of the wave equation [15] [16].

Let $\mathcal{L}\left[\psi, \frac{\partial\psi}{\partial x_j}, \frac{\partial\psi}{\partial t}, x_j, t\right]$ be the Lagrangian density which is the generator of the wave equation for a dissipative system. Then we make an infinitesimal change of coordinates x_j s and the time t by writing

$$x_j \to X_j(x_j), \quad \text{and} \quad t \to T(t), \tag{7.75}$$

and at the same time change the wave function

$$\psi(x_j, t) \to \Psi\left[\psi(x_j, t), x_j, t\right] = \psi(x_j, t) + \delta\psi[\psi(x_j, t), x_j, t], \tag{7.76}$$

where $\delta\psi(x_j, t) = \Psi\left[\psi(x_j, t), x_j, t\right] - \psi(x_j, t)$ measures the effect of both the changes in x_j s and t and also in $\psi(x_j, t)$. This transformation changes the Lagrangian density to $\mathcal{L}'$ where

$$\mathcal{L}\left[\psi, \frac{\partial\psi}{\partial x_j}, \frac{\partial\psi}{\partial t}, x_j, t\right] \to \mathcal{L}'\left[\Psi\left(X_j, T\right), \frac{\partial\Psi}{\partial X_j}, \frac{\partial\Psi}{\partial T}, X_j, T\right]. \tag{7.77}$$

The fact that the action $\mathcal{S}$ should remain invariant, i.e.

$$\mathcal{S}'\left[\Psi\right] = \mathcal{S}\left[\psi\right], \tag{7.78}$$

implies that the new Lagrangian density $\mathcal{L}'$, as a function of the new variables X_j and T, must be given by

$$\mathcal{L}'\left[\Psi, X_j, T\right] = \left(\frac{\partial t}{\partial T}\right) J \mathcal{L}\left[\psi, x_j, t\right], \tag{7.79}$$

where

$$J = \left(\frac{\partial(x_1, x_2, x_3)}{\partial(X_1, X_2, X_3)}\right), \tag{7.80}$$

is the Jacobian of the transformation. As in the case of the infinitesimal transformation of the Lagrangian for particles, Eq. (7.6), here we have the infinitesimal change of the Lagrangian which is given by

$$\mathcal{L}'\left[\Psi, X_j, \frac{\partial \Psi}{\partial T}, \frac{\partial \Psi}{\partial X_j}, T\right] - \mathcal{L}\left[\psi, \frac{\partial \psi}{\partial t}, \frac{\partial \psi}{\partial x_j}, x_j, t\right] = \frac{d(\Delta \Lambda^0)}{dt} + \nabla \cdot (\Delta \boldsymbol{\Lambda}). \tag{7.81}$$

That is for the case of a field the total time derivative is replaced by a 4-divergence of a 4-vector $\left(\Delta \Lambda^0, \Delta \boldsymbol{\Lambda}\right)$.

Assuming that the transformation is infinitesimal

$$X_j(x_j) = x_j + \delta x_j, \quad \text{and} \quad T = t + \delta t, \tag{7.82}$$

and

$$\Psi(x_j, t) = \psi(x_j, t) + \delta\psi(x_j, t), \tag{7.83}$$

then the variations of dt and d^3r are expressible as

$$\delta(dt) = dt \frac{d\delta t}{dt}, \quad \text{and} \quad \delta(d^3r) = d^3r \nabla \cdot (\delta \mathbf{r}). \tag{7.84}$$

Thus we obtain the following results:

$$\delta\psi = \delta_0\psi + \delta t\dot{\psi} + \delta \mathbf{r} \cdot \nabla\psi, \tag{7.85}$$

$$\delta\dot{\psi} = \delta_0\dot{\psi} + \delta t\ddot{\psi} + \delta \mathbf{r} \cdot \nabla\dot{\psi}, \tag{7.86}$$

and

$$\delta(\nabla\psi) = \delta_0\nabla\psi + \delta\nabla\dot{\psi} + (\delta \mathbf{r} \cdot \nabla)\nabla\psi, \tag{7.87}$$

where dots denote partial derivatives with respect to t.
The variation of action can be written down as

$$\begin{aligned}
\delta S &= \int \mathcal{L}'\left(\Psi, \frac{\partial \Psi}{\partial T}, \frac{\partial \Psi}{\partial X_j}, X_j, T\right) d^3X\,dT \\
&\quad - \int \mathcal{L}\left(\psi, \frac{\partial \psi}{\partial t}, \frac{\partial \psi}{\partial x_j}, x_j, t\right) d^3x\,dt.
\end{aligned} \tag{7.88}$$

Now by substituting from Eqs. (7.82)-(7.83) in (7.88) we obtain

$$
\begin{aligned}
\delta \mathcal{S} &= \int \left[\{ dt d^3 r + (\delta dt)\, d^3 r + (\delta d^3 r)\, dt \} \{ (\mathcal{L} + \delta \mathcal{L}) - \mathcal{L} \} \right] \\
&= \int \left[\left\{ \frac{d(\delta t)}{dt} + \nabla \cdot \delta \mathbf{r} \right\} \mathcal{L} + \frac{d\mathcal{L}}{dt} \delta t + \delta \mathbf{r} \cdot \nabla \mathcal{L} + \delta_0 \mathcal{L} \right] dt d^3 r \\
&= \int \left[\frac{d\rho}{dt} + \nabla \cdot \mathbf{j} + \left(\frac{\partial \mathcal{L}}{\partial \psi} - \frac{d}{dt} \frac{\partial \mathcal{L}}{\partial \dot{\psi}} - \nabla \cdot \frac{\partial \mathcal{L}}{\partial (\nabla \psi)} \right) \right. \\
&\quad \times \left. \left(\delta \psi - \delta t \dot{\psi} - \delta \mathbf{r} \cdot \nabla \psi \right) \right] d^3 r dt, \tag{7.89}
\end{aligned}
$$

where the quantities $\delta \rho$ and $\delta \mathbf{j}$ are defined by

$$
\delta \rho = \mathcal{L} \delta t + \frac{\partial \mathcal{L}}{\partial \dot{\psi}} \delta_0 \psi, \tag{7.90}
$$

and

$$
\delta \mathbf{j} = \mathcal{L} \delta \mathbf{r} + \frac{\partial \mathcal{L}}{\partial (\nabla \psi)} \delta_0 \psi. \tag{7.91}
$$

Thus from (7.89) it follows that

$$
\frac{d\rho}{dt} + \nabla \cdot \mathbf{j} + \left[\frac{\partial \mathcal{L}}{\partial \psi} - \frac{d}{dt} \left(\frac{\partial \mathcal{L}}{\partial \dot{\psi}} \right) - \nabla \cdot \frac{\partial \mathcal{L}}{\partial (\nabla \psi)} \right] \left(\delta \psi - \delta t \dot{\psi} - \delta \mathbf{r} \cdot \nabla \psi \right) = 0. \tag{7.92}
$$

For a stationary path the sum of the terms in the square bracket vanishes (Euler-Lagrange equation) and Eq. (7.92) becomes the equation of continuity

$$
\frac{d\rho}{dt} + \nabla \cdot \mathbf{j} = 0. \tag{7.93}
$$

Next we define the momentum density $\pi(\mathbf{r}, t)$ by

$$
\pi(\mathbf{r}, t) = \frac{\delta \mathcal{L}}{\delta \dot{\psi}}, \tag{7.94}
$$

where $\frac{\delta \mathcal{L}}{\delta \dot{\psi}}$ is the functional derivative of $\mathcal{L}$. Then the Hamiltonian density for this ψ field is given by

$$
\mathcal{H} = \pi(\mathbf{r}, t) \dot{\psi}(\mathbf{r}, t) - \mathcal{L}, \tag{7.95}
$$

and the action can be written as

$$
S = \int dt \int \left[\pi(\mathbf{r}, t) \dot{\psi}(\mathbf{r}, t) - \mathcal{L} \right]_{\dot{\psi} = \dot{\psi}(\pi)} d^3 r. \tag{7.96}
$$

Now the functional derivative of S with respect to variation $\delta_0 \psi(\mathbf{r}, t)$ with no variation of $\mathbf{r}$ and t must vanish. Since $\pi(\mathbf{r}, t)$ and $\psi(\mathbf{r}, t)$ are independent functions, the functional derivatives of S yield the canonical equations:

$$
\dot{\psi}(\mathbf{r}, t) = \frac{\delta \mathcal{H}}{\delta \pi}, \quad \text{and} \quad \dot{\pi}(\mathbf{r}, t) = -\frac{\delta \mathcal{H}}{\delta \psi}. \tag{7.97}
$$

In §13.13 we will discuss the application of the Noether theorem to obtain conservation laws for linear and nonlinear wave equations for dissipative systems.

Bibliography

[1] E. Noether, Nachr. Ges. Wiss. Göttingen, 235 (1918).

[2] E.T. Whittaker, *A Treatise on the Analytical Dynamics of Particles and Rigid Bodies*, Fourth Edition (Cambridge University Press, London 1965) pp. 54-69.

[3] L.D. Landau and E.M. Lifshitz, *Mechanics*, (Addison-Wesley, Reading, 1960) Chapter 2.

[4] H. Goldstein, *Classical Mechanics*, Second Edition (Addison-Wesley Reading, 1980) Chapter 8.

[5] S. Hojman, J. Phys. A25, L291 (1992).

[6] M. Lutzky, J. Phys. A28, L637 (1995).

[7] E.A. Desloge and R.I. Karsh, Am. J. Phys. 45, 336 (1977).

[8] E.A. Desloge, *Classical Mechanics*, vol. 2 (John Wiley & Sons, New York, 1982) p. 581.

[9] H-J Wagner, Z. Physik, B 95, 261 (1994).

[10] H.H. Denman, J. Math. Phys. 6, 1611 (1965).

[11] H.H. Denman, Am. J. Phys. 36, 516 (1968).

[12] W. Sarlet, J. Math. Phys. 19, 1094 (1978).

[13] P. Havas, Acta Phys. Austriaca, 38, 145 (1973).

[14] J-L. Fu and L-Q. Chen, Phys. Lett A317, 255 (2003).

[15] See for instancs, N.A. Doughty, *Lagrangian Interaction*, (Addison-Wesley, Reading, 1990) p. 214.

[16] D.E. Soper, *Classical Field Theory*, (John Wiley & Sons, New York, 1976) pp. 101-108.

Chapter 8

Dissipative Forces Derived from Many-Body Problems

Consider a large system S of N interacting particles which is isolated from the rest of the universe and therefore has a constant total energy. Now let us divide this system into two parts S_1 and S_2 and study the development of S_1 in time. Because of the coupling between S_1 and S_2 the two parts will exchange energy with each other. For instance initially S_1 and S_2 will be losing and will be gaining energy respectively. When both S_1 and S_2 have finite degrees of freedom then this energy flow from S_1 to S_2 follows by a flow in the opposite direction, i.e. from S_2 to S_1, and therefore there is no fixed direction for the flow of energy for all times. Furthermore, unlike the case of some of the phenomenological frictional forces, the motion of S_1 as well as S_2 will be invariant under the time reversal transformation $t \rightarrow -t$, $\mathbf{r}_i \rightarrow \mathbf{r}_i$ and $\dot{\mathbf{r}}_i = -\dot{\mathbf{r}}_i$, where i refers to the i-th particle of the system S ($i = 1, 2, \cdots N$). However if S_1 contains a few and S_2 has an infinite number of degrees of freedom, and if the initial conditions are such that the direction of the energy flow is from S_1 to S_2, then S_1 will be losing energy at all times and thus it is a dissipative system. In this chapter we study a number of exactly solvable classical systems, where a particle called the "central particle" loses energy to a large number of oscillators usually referred to as a "heat bath".

8.1 The Schrödinger Chain

This many-body problem was originally formulated and solved by Schrödinger before his discovery of wave mechanics [1] [2]. The remarkable feature of this chain is that the decay law of the central particle is non-exponential and is proportional to $t^{-\frac{1}{2}}$.

   The classical as well as the quantal motion of a finite long chain has also been investigated in order to examine the recurrence time and the quantum spectrum of such a system [3] [4] but in this chapter we study an infinite chain.

   Here the large system S consists of a linear chain of infinite mass points each coupled to its nearest neighbor by elastic springs. Let us denote the displacement of the j-th particle in the chain from its equilibrium position by ξ_j. then we can write the equation of motion of the j-th particle as

$$m\left(\frac{d^2\xi_j}{dt^2}\right) = \frac{1}{4}m\nu^2\left(\xi_{j+1} + \xi_{i-1} - 2\xi_j\right), \quad j = 0, \pm 1, \pm 2 \cdots, \tag{8.1}$$

where $\frac{1}{4}m\nu^2$ is the spring constant.
For the initial conditions we assume that the central particle, $j = 0$, which we choose as the decaying part S_1 is displaced by a distance A and then released with zero initial velocity while other particles are at rest in their equilibrium positions

$$\xi_0(t = 0) = A, \quad \xi_j(t = 0) = 0, \quad j \neq 0, \tag{8.2}$$

$$\dot{\xi}_j(t = 0) = 0, \quad j = 0, \pm 1, \pm 2 \cdots. \tag{8.3}$$

   We can solve the system of equations (8.1) by the following method [5] [6].
Let $G(z, t)$ be the generating function defined by

$$G(z, t) = \sum_{j=-\infty}^{+\infty} \xi_j(t)z^{2j}. \tag{8.4}$$

We multiply (8.1) by z^{2j} and sum over all j s using Eq. (8.4). This yields the differential equation for the generating function

$$\frac{d^2}{dt^2}G(z, t) - \frac{1}{4}\nu^2\left(\frac{1}{z} - z\right)^2 G(z, t) = 0. \tag{8.5}$$

The function $G(z, t)$ is subject to the following initial conditions which can be derived from Eqs. (8.2)-(8.4):

$$G(z, t = 0) = A, \quad \left(\frac{dG(z, t)}{dt}\right)_{t=0} = 0. \tag{8.6}$$

The solution of (8.5) which satisfies (8.6) is given by

$$G(z, t = 0) = A\cosh\left[\frac{1}{2}\nu t\left(\frac{1}{z} - z\right)\right]. \tag{8.7}$$

Equation (8.7) shows that $G(z, t)$ is the generating function for the Bessel function of even order [7]

$$\cosh\left[\frac{1}{2}\nu t\left(\frac{1}{z} - z\right)\right] = \sum_{j=-\infty}^{+\infty} J_{2j}(\nu)z^{2j},$$ (8.8)

and therefore

$$\xi_j(t) = AJ_{2j}(\nu)$$ (8.9)

is the solution of (8.1) with the initial conditions (8.2) and (8.3).

The motion of the central particle, $k = 0$, can also be obtained from the Hamiltonian

$$H_0 = \frac{1}{2m}p_0^2 + \frac{1}{4}m\nu^2 x_0^2 + x_0\left\{\frac{1}{4}m\nu^2 A\left(J_{-1}(\nu t) + J_1(\nu t)\right)\right\}.$$ (8.10)

Similar single particle Hamiltonian can be written down for other particles in the system. However, classically we can obtain other q-equivalent Hamiltonians for the motion of the central particle which is damped. For instance we can write

$$H = \frac{p_0^2(t)}{2m\nu t} + \frac{1}{2}m\nu^3 x_0^2(t)t.$$ (8.11)

This Hamiltonian gives us the equation of motion for $x_0(t)$ which is the same as the equation for the Bessel function, i.e. $x_0(t) = AJ_0(\nu t)$, but note that the relation between H and the energy of the system E is given by

$$E = H\nu t.$$ (8.12)

Thus the Hamiltonian does not represent the energy, and we cannot proceed with the quantization of this motion using (8.11).

8.2 A Particle Coupled to a Chain

A model similar to the Schrödinger chain is that of a massive particle, M, coupled to a semi-infinite chain of oscillators (Rubin's model) [8] [9]. In this model the total Hamiltonian is given by

$$H = \frac{p^2}{2M} + V(x) + \sum_{n=1}^{\infty}\left[\frac{p_n^2}{2m} + \frac{1}{2}m\omega_0^2\left(x_{n+1} - x_n\right)^2\right] + \frac{1}{2}m\omega_0^2\left(x - x_1\right)^2.$$ (8.13)

We can diagonalize this Hamiltonian using the Fourier transform method. Let us denote the highest frequency mode of the oscillators by ω_R, and introduce $X(k)$ and $P(k)$ by

$$x_n = \sqrt{\frac{2}{\pi}}\int_0^{\pi}\sin(kn)X(k)dk,$$ (8.14)

and

$$p_n = \sqrt{\frac{2}{\pi}} \int_0^\pi \sin(kn) P(k) dk. \tag{8.15}$$

By substituting (8.14) and (8.15) in (8.13) and simplifying the result we obtain

$$H = \frac{p^2}{2M} + V(x) + \frac{1}{2} m\omega_0^2 x^2 + \int_0^\pi \left(\frac{P^2(k)}{2m} + \frac{m}{2}\omega^2(k)X^2(k) - c(k)X(k)x \right) dk, \tag{8.16}$$

where the eigenfrequencies $\omega(k)$ and the coupling $c(k)$ are related to ω_R by

$$\omega(k) = \omega_R \sin\left(\frac{k}{2}\right), \quad \text{and} \quad c(k) = \sqrt{\frac{2}{\pi}} \left(\frac{m\omega_R^2}{4} \right) \sin(k). \tag{8.17}$$

Next we write the equations of motion for the particles forming the chain, and also for the massive particle, M.

Using the initial conditions

$$x_n(0) = 0, \quad \text{and} \quad p_n(0) = 0, \quad n = 1, 2, 3 \cdots, \tag{8.18}$$

and then eliminating the degrees of freedom of the particles in the chain, i.e. $X(k)$ and $P(k)$ (see §3.4 for the details), we obtain the equation of motion for the particle M,

$$M\frac{d^2 x(t)}{dt^2} + \frac{\partial V}{\partial x} + \frac{1}{2} m\omega^2 x^2(t) - \int_0^t G(t - t') x(t') dt' = 0, \tag{8.19}$$

where the kernel $G(T)$ is given by

$$G(T) = \begin{cases} \frac{m\omega_R}{2\pi} \int_0^\infty \frac{\sin k}{\sin(\frac{k}{2})} \sin\left[\omega_R \sin\left(\frac{k}{2}\right) T\right] dk & \text{for } T > 0 \\ 0 & \text{for } T < 0 \end{cases}. \tag{8.20}$$

In Fig. (8.1) the kernel $G(T)$ as a function of T is shown. This kernel is an oscillating function of time with decreasing amplitude similar to the Bessel function $J_1(\omega_R t)$.

8.3   Dynamics of a Non-Uniform Chain

As a classical model exhibiting the exponential decay law we want to discuss the special case of a non-uniform chain which is again exactly solvable [10]. Denoting the mass of the j-th particle in the chain by m_j and the spring constant connecting the particles j and $j+1$ by K_j, the equation of motion for the j-th particle can be written as

$$m_j \left(\frac{d^2 \xi_j(t)}{dt^2} \right) = K_j \left[\xi_{j+1}(t) - \xi_j(t) \right] + K_{j-1} \left[\xi_{j-1}(t) - \xi_j(t) \right], \quad j = 1, 2, \cdots N. \tag{8.21}$$

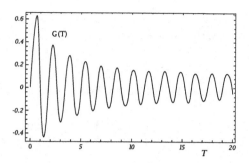

Figure 8.1: The kernel $G(T)$, Eq. (8.20), is shown as a function of T.

Next we introduce the variables η_j by

$$\eta_j(t) = \sqrt{m_j}\xi_j(t), \qquad (8.22)$$

and the constants $\omega_1, \omega_2, \cdots \omega_{2N-2}$ by the following relations [11]

$$\omega_{2j-1}^2 = \frac{K_j}{m_j}, \quad \omega_{2j}^2 = \frac{K_j}{m_{j+1}}, \qquad (8.23)$$

then the equations of motion become

$$\left(\frac{d^2\eta_j(t)}{dt^2}\right) = (\omega_{2j-1}\omega_{2j})\,\eta_{j+1}(t) + (\omega_{2j-3}\omega_{2j-2})\,\eta_{j-1}(t)$$
$$- \left(\omega_{2j-1}^2 + \omega_{2j-2}^2\right)\eta_j(t). \qquad (8.24)$$

Now we introduce a new set of variables $z_1(t), z_2(t), \cdots z_{N-1}(t)$ by

$$\left(\frac{dz_j(t)}{dt}\right) = \omega_{2j}\eta_{j+1}(t) - \omega_{2j-1}\eta_j(t), \qquad (8.25)$$

and write (8.24) as

$$\left(\frac{d\eta_j(t)}{dt}\right) = \omega_{2j-1}z_j(t) - \omega_{2j-2}z_{j-1}(t). \qquad (8.26)$$

We can combine (8.25) and (8.26) to form a single set of equations:

$$\left(\frac{dX_j(t)}{dt}\right) = \omega_j X_{j+1}(t) - \omega_{j-1}X_{j-1}(t), \qquad (8.27)$$

where $X_1(t), X_2(t) \cdots X_{2N-1}(t)$ are defined by

$$X_{2j-1}(t) = \eta_j(t), \quad X_{2j}(t) = z_j(t). \qquad (8.28)$$

Equation (8.27) is a linear differential-difference equation that we want to solve for a semi-infinite chain, i.e. in the limit of $N \to \infty$.

The set of equations (8.27) are quite general and are valid for any set of K_j s and m_j s. For an exactly solvable case we choose

$$\omega_j = \alpha j. \tag{8.29}$$

where α is a positive constant. Then from (8.23) we obtain

$$\frac{m_{j+1}}{m_j} = \left(\frac{\omega_{2j-1}}{\omega_{2j}}\right)^2 = \left(\frac{2j-1}{2j}\right)^2, \quad j = 1, 2, \cdots. \tag{8.30}$$

For this special case we have

$$\omega_{2j-1}^2 = \frac{K_j}{m_j} = \alpha^2(2j-1)^2, \quad \omega_{2j}^2 = \frac{K_{j-1}}{m_j} = 4\alpha^2(j-1)^2, \tag{8.31}$$

and therefore

$$\frac{d^2\xi_j}{dt^2} = \alpha^2(2j-1)^2(\xi_{j+1} - \xi_j) + \alpha^2(2j-2)^2(\xi_{j-1} - \xi_j). \tag{8.32}$$

For the initial conditions we assume that the particle $j = 1$ is displaced by one unit length and zero initial speed, while all of the other particles are at rest with no initial displacement, i.e.

$$\xi_1(0) = 1, \quad \left(\frac{d\xi_j}{dt}\right)_{t=0} = 0, \quad j = 1, 2, .. \quad \xi_j(0) = 0, \quad j \neq 1. \tag{8.33}$$

Thus we can reduce (8.32) to a first order differential equation for $X_j(t)$;

$$\frac{dX_j(t)}{dt} = \alpha\left[jX_{j+1}(t) - (j-1)X_{j-1}(t)\right], \tag{8.34}$$

which is subject to the initial conditions

$$X_1(0) = \sqrt{m_1}, \quad X_j(0) = 0, \quad j \neq 1. \tag{8.35}$$

Now we use the method of generating function to find the solution of (8.34) with the initial conditions (8.35). To this end we define $G(t, z)$ by

$$G(t, z) = \sum_{j=1}^{\infty} X_j(t) z^{j-1}. \tag{8.36}$$

By differentiating (8.36) we find that

$$z\frac{\partial}{\partial z}[zG(t, z)] = \sum_{j=1}^{\infty}(j-1)X_{j-1}(t)z^{j-1}, \tag{8.37}$$

and

$$\frac{\partial}{\partial z}G(t, z) = \sum_{j=1}^{\infty}jX_{j+1}(t)z^{j-1}. \tag{8.38}$$

Next we multiply (8.34) by z^{j-1} and sum over all j s; we obtain the following first order linear partial differential equation for $G(t, z)$

$$\frac{\partial}{\partial t} G(t, z) = \alpha \left[\frac{\partial G(t, z)}{\partial z} - z \frac{\partial}{\partial z} (zG(t, z)) \right].$$

(8.39)

The value of $G(t, z)$ at $t = 0$ is independent of z, this follows from the initial conditions (8.35) and Eq. (8.36)

$$G(t = 0, z) = \sqrt{m_1}.$$

(8.40)

We can solve (8.39) with the initial value (8.40) and the result is

$$G(t, z) = \frac{\sqrt{m_1}}{\cosh(\alpha t)} [1 + z \tanh(\alpha t)]^{-1}.$$

(8.41)

Now if we expand (8.41) in powers of z and compare it to (8.36) we find $X_j(t)$;

$$X_j(t) = \frac{(-1)^{j-1} \sqrt{m_1} [\tanh(\alpha t)]^{j-1}}{\cosh(\alpha t)}.$$

(8.42)

Thus the displacement of the j-th particle in this model is given by

$$\xi_j(t) = \sqrt{\frac{m_1}{m_j}} \frac{[\tanh(\alpha t)]^{2j-2}}{\cosh(\alpha t)},$$

(8.43)

where

$$\sqrt{\frac{m_1}{m_j}} = \frac{2 \times 4 \times 6... \times (2j - 2)}{1 \times 3 \times 5... \times (2j - 3)}.$$

(8.44)

As Eq. (8.43) shows the motion of all of the particles will be damped, and the j-th particle has its maximum displacement at the time t_j where

$$t_j = \left(\frac{1}{\alpha} \right) \cosh^{-1}(\sqrt{2j - 1}),$$

(8.45)

and its maximum displacement is

$$\max \xi_j = \sqrt{\frac{m_1}{m_j}} \left(\frac{1}{\sqrt{2j - 1}} \right) \left(\frac{2j - 2}{2j - 1} \right)^{j-1}.$$

(8.46)

We can find an approximate uncoupled equation for the solution of the j-th particle by replacing for $\xi_{j+1}(t)$ and $\xi_{j-1}(t)$ in terms of the derivatives of $\xi_j(t)$ in Eq. (8.21) to get

$$\ddot{\xi}_j(t) + \alpha \left[2j(2j - 1) + 2(j - 1)(2j - 3) \right] \dot{\xi}_j(t)$$
$$+ \quad \alpha^2 \left[(2j - 1)^2 + 4(j - 1)^2 \right] \xi_j(t) = 0,$$

(8.47)

or in the simpler form of

$$\ddot{\xi}_j(t) + \alpha\gamma(j)\dot{\xi}_j(t) - \frac{1}{m_j}f_j(\xi) = 0, \tag{8.48}$$

with $\gamma(j)$ and $f_j(\xi)$ defined by comparing the two equations (8.47) and (8.48). This last equation shows that the motion of each particle is damped and that the damping constant is proportional to α. We note that while the original equation (8.21) is invariant under time-reversal transformation, the approximate form (8.47) is not. This approximate differential equation (8.47) for $\xi_j(t)$ is valid for $t > \frac{1}{\alpha}$.

8.4   Mechanical System Coupled to a Heat Bath

A heat bath is a collection of harmonic oscillators all of them coupled to the central system and this collection has the following properties:
(1) The spectrum of the oscillator frequencies must be dense.
(2) In almost all cases studied up to now a linear coupling is assumed between the central particle (or system) and the heat bath.
(3) The coupling constant must be a smooth function of the frequencies of the oscillators.

In the cases where a particle is coupled to a field, such as the interaction of an electron with the electromagnetic field (e.g. van Kampen's model) all of these conditions are satisfied. Furthermore we assume that the coupling of a small system to a field is a local coupling, that is the coupling depends on the value of the field at a single point, but occasionally the coupling can be nonlocal in space or in time (e.g. the version of the Wigner-Weisskopf model discussed in §18.2).

Another model of a damped system which has been studied extensively consists of a central particle with the coordinate Q and momentum P, (system S_1), coupled to a large number of harmonic oscillators and these oscillators form the heat bath S_2 (Ullesrma's model) [12]-[15].

Here the coupling between the two is linear and the Hamiltonian for the entire system, S, is given by

$$H = \frac{P^2}{2M} + V(Q) + \sum_{n=1}^{N} \frac{1}{2}\left(\frac{p_n^2}{m_n} + m_n\omega_n^2 q_n^2\right) + \sum_{n=1}^{N} \epsilon_n q_n Q. \tag{8.49}$$

This Hamiltonian has two defects:
(a) For the special case of $V(x) = 0$ there is no lower bound on the energy [15] and
(b) The Hamiltonian is not invariant under spatial transformation. We can remedy these defects by replacing the interaction term $\sum_n \epsilon_n q_n Q$ by

$\frac{1}{2} \sum_n (q_n - Q)^2$. This modification will not affect the main conclusion of the work regarding the dissipative nature of the motion of the central particle. For the sake of simplicity we set all masses appearing in (8.49) equal to one, i.e.

$$M = m_1 = m_2 = \cdots = m_N = 1. \tag{8.50}$$

As for the initial conditions, we assume that at $t = 0$ the oscillators are all at rest at their equilibrium position. These conditions guarantee that the flow of energy is from the central body or S_1 to the heat bath.

$$q_n(0) = p_n(0) = 0, \quad n = 1, 2, \cdots N. \tag{8.51}$$

From (8.49) we find the equations of motion for q_n and p_n

$$\dot{q}_n = p_n. \tag{8.52}$$

and

$$\dot{p}_n = - \left(\omega_n^2 q_n + \epsilon_n Q \right) = \ddot{q}_n. \tag{8.53}$$

The formal solution of (8.53) is given by

$$
\begin{aligned}
q_n(t) \; = \; & q_n(0) \cos (\omega_n t) + \frac{1}{\omega_n} \dot{q}_n(0) \sin (\omega_n t) \\
& - \frac{\epsilon_n}{\omega_n} \int_0^t \sin [\omega_n (t - t')] Q(t') \, dt'.
\end{aligned} \tag{8.54}
$$

Using the initial conditions (8.51) we can simplify (8.54)

$$q_n(t) = -\frac{\epsilon_n}{\omega_n} \int_0^t \sin [\omega_n (t - t')] Q(t') \, dt'. \tag{8.55}$$

Similarly for the motion of the central oscillator we have

$$\dot{Q}(t) = P(t), \quad \text{and} \quad \dot{P}(t) = -\frac{\partial V}{\partial Q} - \sum_{n=1}^N \epsilon_n q_n. \tag{8.56}$$

Thus the equation of motion for this oscillator can be obtained from Eqs. (8.55) and (8.56)

$$\frac{d^2 Q(t)}{dt^2} + \frac{\partial V}{\partial Q} - \sum_{n=1}^N \left(\frac{\epsilon_n^2}{\omega_n} \right) \int_0^t \sin [\omega_n (t - t')] Q(t') \, dt' = 0. \tag{8.57}$$

This is an integro-differential equation for Q which can be written as

$$\frac{dP(t)}{dt} + \frac{\partial V}{\partial Q} - \int_0^t K(t - t') Q(t') \, dt' = 0, \tag{8.58}$$

and

$$\frac{dQ(t)}{dt} = P(t), \tag{8.59}$$

where the kernel $K(t - t')$ is given by

$$K(t - t') = \sum_{n=1}^{N} \left(\frac{\epsilon_n^2}{\omega_n}\right) \int_0^t \sin\left[\omega_n(t - t')\right] dt'. \tag{8.60}$$

We also note that in the time-reversed motion $t \to -t$, $t' \to -t'$ we have an equation similar to the integro-differential equation for $Q(t)$, i.e.

$$\frac{d^2 Q(-t)}{dt^2} + \frac{\partial V}{\partial Q} - \int_0^{-t} K(t - t') Q(-t') dt' = 0. \tag{8.61}$$

An important relation in this case is the time derivative of the equal-time Poisson bracket of $P(t)$ and $Q(t)$;

$$\frac{d}{dt}\{P(t), Q(t)\} = \left\{\frac{dP(t)}{dt}, Q(t)\right\} = \int_0^t K(t - t')\{Q(t), Q(t')\} dt'. \tag{8.62}$$

Since the Poisson bracket $\{Q(t), Q(t')\}$ is not zero for all values of t and t', therefore $\frac{d}{dt}\{P, Q\}\}$ is not zero. In other words the equal-time Poisson bracket of $P(t)$ and $Q(t)$ is not equal to (-1) but is given by

$$\{P(t), Q(t)\} = -1 + \int_0^t dt' \int_0^{t'} K(t' - t'')\{Q(t'), Q(t'')\} dt''. \tag{8.63}$$

In general the coupling between the bath of oscillators and the central particle is nonlinear, i.e. instead of the interaction $\sum_n \epsilon_n q_n(t) Q$, we have

$$H_I = \sum_n \varepsilon_n q_n(t) \Phi(Q), \tag{8.64}$$

where $\Phi(Q)$ is a given function of Q. Now the equation of motion for the central particle takes the form

$$\frac{dP}{dt} + \frac{\partial V}{\partial x} + \sum_n \varepsilon_n q_n \frac{\partial \Phi(Q)}{\partial Q} = 0, \tag{8.65}$$

or by eliminating $q_n(t)$ as before we have

$$\frac{dP}{dt} + \frac{\partial V}{\partial x} - \left(\frac{\partial \Phi(Q)}{\partial Q}\right) \int_0^t K(t - t') \Phi[Q(t')] dt' = 0. \tag{8.66}$$

In this case we find the time derivative of the Poisson bracket is given by

$$\frac{d}{dt}\{P(t), Q(t)\} = \int_0^t K(t - t')\{\Phi[Q(t')], Q(t)\} dt' \tag{8.67}$$

Thus in general the equal-time Poisson bracket of $P(t)$ and $Q(t)$ will depend on time as well as on the variables $P(t)$ and $Q(t)$. This result has important consequences in the problem of quantization of a damped system.

The expression for the Poisson bracket Eq. (8.63) can be used to calculate the effect of dissipation by the heat bath as a perturbation on the motion. Thus to the zeroth order we ignore the effect of the coupling of the central particle to the system of oscillators, and we calculate the bracket $\{Q(t'), Q(t'')\}$ from the equations of motion

$$\frac{dP^{(0)}}{dt} + \frac{\partial V(Q)}{\partial Q} = 0, \quad P^{(0)} = \dot{Q}^{(0)}. \tag{8.68}$$

Then the bracket $\{P^{(1)}(t), Q^{(1)}(t)\}$ to the first order is obtained by substituting $\{Q^{(0)}(t'), Q^{(0)}(t'')\}$ on the right hand side of Eq. (8.63) and carrying out the integrals. Even in this first order we can see whether the bracket $\{P(t), Q(t)\}$ depends on time or depends on the initial values P_0 and Q_0 as well as t.

As an example let us consider the special case where $K(t - t')$ is of the form

$$K(t - t') = 2\lambda \frac{d}{dt'} \delta(t' - t) \tag{8.69}$$

(see below) and when $V(Q) = 0$. Then we have

$$\frac{dP}{dt} + \lambda \dot{Q} = 0, \quad P = \dot{Q}, \tag{8.70}$$

and therefore

$$Q(t) = \frac{P_0}{\lambda} \left(1 - e^{-\lambda t}\right) + Q_0, \tag{8.71}$$

where Q_0 and P_0 are the initial values of $Q(t)$ and $P(t)$. For this case we can calculate the Poisson bracket

$$\{P(t), Q(t)\}_{P_0, Q_0} = -e^{-\lambda t}, \tag{8.72}$$

or use the perturbation form mentioned above to first calculate $Q^{(0)} = P_0 t + Q_0$ and use it to determine $\{Q^{(0)}(t'), Q^{(0)}(t'')\}$. Substituting this bracket in Eq. (8.63) we find

$$\{P^{(1)}(t), Q^{(1)}(t)\}_{P_0, Q_0} = -1 + \lambda t. \tag{8.73}$$

In this case as well as the problem of damped harmonic oscillator, i.e. $V(Q) = \frac{1}{2}m\omega_0^2 x^2$ the bracket depends on time. However for the nonlinear forces $\{P(t), Q(t)\}$ will depend on Q_0 and P_0. For instance if $V(Q) = \frac{A}{Q^2}$, then in the absence of damping

$$\left[Q^{(0)}(t)\right]^2 = Q_0^2 + 2P_0 Q_0 t + \left(P_0^2 + \frac{2A}{Q_0^2}\right) t^2, \tag{8.74}$$

and $\{Q^{(0)}(t'), Q^{(0)}(t'')\}_{Q_0, P_0}$ and consequently $\{P^{(1)}(t), Q^{(1)}(t)\}$ will depend on Q_0, P_0 and t. In the latter case we know that in the quantum formulation of the problem the commutator will not be a c-number.

For the special case of

$$\omega_n = \frac{n\pi c}{L}, \quad n = 1, 2, \cdots \infty, \tag{8.75}$$

and for one of the following choices of ϵ_n

$$\epsilon_n(\pm) = \frac{\lambda a \omega_n \Omega}{(\omega_n^2 \pm \alpha^2)^{\frac{1}{2}}}, \tag{8.76}$$

we can find $K(t - t')$ analytically. The ω_n s are the characteristic frequencies of a wave confined in an enclosure of length L, and α and Ω are constants having the dimension of frequency and λ is a dimensionless constant. Using the summation formulae [16]

$$\sum_{n=1}^{\infty} \frac{n \sin(nx)}{(n^2 + A^2)} = \frac{\pi}{2} \frac{\sinh[A(\pi - x)]}{\sinh(A\pi)}, \quad 0 < x < 2\pi, \tag{8.77}$$

and

$$\sum_{n=1}^{\infty} \frac{n \sin(nx)}{(n^2 - A^2)} = \left(\frac{\pi}{2}\right) \frac{\sin\{A[(2m+1)\pi - x]\}}{\sin(A\pi)},$$
$$0 < x < 2\pi, \quad A \text{ not an integer}, \tag{8.78}$$

we find that for $\epsilon_n(+)$ we have

$$\begin{aligned} K(t - t') &= 0, \quad t - t' = 0, \\ K(t - t') &= \left(\frac{\lambda^2 a^2 \Omega^2 L}{2c}\right) \frac{\sinh\left[\frac{\alpha L}{c} - \alpha(t - t')\right]}{\sinh\left(\frac{L\alpha}{c}\right)}, \\ 0 &< c(t - t') < 2L. \end{aligned}$$
$$\tag{8.79}$$

Similarly for $\epsilon_n(-)$ we obtain

$$\begin{aligned} K(t - t') &= 0, \quad t - t' = 0, \\ K(t - t') &= \left(\frac{\lambda^2 a^2 \Omega^2 L}{2c}\right) \frac{\sin\left[\frac{\alpha L}{c} - \alpha(t - t')\right]}{\sin\left(\frac{L\alpha}{c}\right)}, \\ 0 &< c(t - t') < 2L. \end{aligned}$$
$$\tag{8.80}$$

There are two special cases that we can simplify the problem of the motion of the central particle:

For the special case when the central particle is harmonically bound, the integro-differential equation will be linear in $Q(t)$ and there are two independent solutions for the equation (8.58) for a given $K(t - t')$. The first one $Q_1(t)$ is obtained by imposing the initial conditions

$$Q(0) = 1, \quad \dot{Q}(0) = 0, \tag{8.81}$$

and the second one which we denote by $Q_2(t)$ satisfies the conditions

$$Q(0) = 0, \quad \dot{Q}(0) = 1. \tag{8.82}$$

Writing the potential in the form of $V(Q) = \frac{1}{2} m \Omega^2 Q^2$, we can convert the integro-differential equation (8.57) to a differential equation of fourth order. We first note that $K(t - t')$ is discontinuous at $t - t'$, and we take K at $t - t'' = 0$ to be $\frac{1}{2} C$, where $C = \frac{\lambda^2 \alpha^2 \Omega^2 L}{2c}$. The integro-differential equation for Q in this case is

$$\ddot{Q}(t) + \Omega^2 Q(t) = \int_0^t K(t - t') Q(t') dt'. \tag{8.83}$$

By differentiating (8.83) twice and noting that $K(0) = \frac{1}{2} C$, we get

$$\frac{d^4 Q(t)}{dt^4} + (\Omega^2 + \alpha^2) \frac{d^2 Q(t)}{dt^2} - \frac{1}{2} C \frac{dQ(t)}{dt} + \left[\alpha C \cot\left(\frac{\alpha L}{c} \right) + \alpha^2 \Omega^2 \right] Q(t) = 0. \tag{8.84}$$

If we impose the initial conditions of

$$Q(0) = 1, \quad \text{and} \quad \left(\frac{dQ(t)}{dt} \right)_{t=0} = 0, \tag{8.85}$$

on Eq. (8.83), then (8.84) must satisfy the initial conditions

$$Q(0) = 1, \quad \dot{Q}(0) = 0, \quad \ddot{Q}(0) = -\Omega^2, \quad \text{and} \quad \left(\frac{d^3 Q(t)}{dt^3} \right)_{t=0} = \frac{1}{2} C. \tag{8.86}$$

The condition for the nonexistence of self-accelerating solutions is that (see §10.1) [12]

$$\Omega^2 - \sum_{n=1}^{\infty} \frac{\epsilon_n^2}{\omega_n} \geq 0, \tag{8.87}$$

and we assume that this relation holds. The characteristic roots of the differential equation (8.84) are given by the quartic equation

$$r^4 + (\Omega^2 + \alpha^2) r^2 - \frac{1}{2C} r + \left[\alpha C \cot\left(\frac{\alpha L}{c} \right) + \alpha^2 \Omega^2 \right] = 0. \tag{8.88}$$

This equation has four complex roots which we write as

$$r'_\pm = -\lambda \pm i\beta, \quad \text{and} \quad r''_\pm = -\lambda \pm i\nu. \tag{8.89}$$

Thus we can write Q as

$$Q(t) = e^{-\lambda t} [D_1 \sin(\beta t) + D_2 \cos(\beta t)] + e^{\lambda t} [D_3 \sin(\nu t) + D_4 \cos(\nu t)], \quad t \leq \frac{L}{c}, \tag{8.90}$$

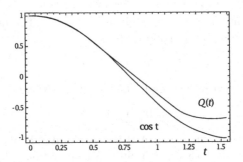

Figure 8.2: The damped oscillation of the central particle $Q(t)$ when it is coupled to the heat bath, and the coupling is small $C = 0.2$. The undamped motion $\cos t$ is also shown.

where $D_1 \cdots D_4$ are four arbitrary constants which can be determined from the initial conditions (8.86). In Fig. (8.2) the result of the integration of differential equation (8.84) with the boundary conditions (8.86) is shown for small damping. For comparison the undamped solution, $\cos t$, is also displayed.

For the general form of the potential $V(Q)$ we can choose ϵ_n and ω_n in such a way that [17]

$$\sum_{n=0}^{\infty} \left(\frac{\epsilon_n^2}{\omega_n^2} \right) \cos \left[\omega_n \left(t - t' \right) \right] = 2\lambda \delta \left(t - t' \right), \tag{8.91}$$

therefore

$$\sum_{n=0}^{\infty} \left(\frac{\epsilon_n^2}{\omega_n} \right) \sin \left[\omega_n \left(t - t' \right) \right] = 2\lambda \frac{d}{dt'} \delta \left(t - t' \right). \tag{8.92}$$

Then the equation of motion is reduced to

$$\frac{d^2 Q(t)}{dt^2} + \lambda \frac{dQ(t)}{dt} + \frac{\partial V(Q)}{\partial Q} = 0, \tag{8.93}$$

in which the damping force is proportional to the velocity (see also §17.5).

8.5   Euclidean Lagrangian

When the central particle moves in a potential $V(Q)$, and this potential forms a barrier to the motion of the particle, then only by the mechanism of tunnelling the particle can pass through the barrier. In the region where the classical momentum becomes imaginary, we can use an imaginary time formulation for the motion of the particle [18] [19].

Denoting the imaginary time by τ, where $\tau = it$, we write the Lagrangian for the interaction between the central particle and the heat bath as

$$L = -\left[\frac{1}{2}\left(\frac{dQ}{d\tau}\right)^2 + V(Q) + \sum_0^\infty \frac{1}{2}\left\{\left(\frac{dq_n}{d\tau}\right)^2 + \omega_n^2 q_n^2\right\} + \sum_{n=0}^\infty \epsilon_n q_n\right]. \quad (8.94)$$

The equations of motion derived from this Lagrangian are:

$$\frac{d^2Q}{d\tau^2} - \frac{\partial V(Q)}{\partial Q} - \sum_{n=0}^\infty \epsilon_n q_n = 0, \quad (8.95)$$

and

$$\frac{d^2q_n}{d\tau^2} - \omega_n^2 q_n - \epsilon_n Q = 0. \quad (8.96)$$

Again we solve (8.96) for $q_n(\tau)$, but now we assume that

$$q_n(-\infty) = q_n(\infty) = 0, \quad (8.97)$$

and substitute the result in (8.95) to get

$$\frac{d^2Q}{d\tau^2} - \frac{\partial V(Q)}{\partial Q} + \int_{-\infty}^\infty \tilde{K}(\tau - \tau')Q(\tau')\,d\tau', \quad (8.98)$$

where

$$\tilde{K}(\tau - \tau') = -\sum_{n=0}^\infty \frac{\epsilon_n^2}{2\omega_n}\exp\left(-\omega_n|\tau - \tau'|\right), \quad (8.99)$$

and the dots refer to derivatives with respect to τ.
When $V(Q)$ is a double well potential, then Eq. (8.98) can only be solved numerically.

Bibliography

[1] E. Schrödinger, Ann. Phys. (Leipzig) 44, 916 (1914).

[2] H. Levine, *Unidirectional Wave Motions*, (North-Holland, Amsterdam, 1978) p. 301.

[3] H.J. Kreuzer, *Nonequilibrium Thermodynamics and its Statistical Foundations*, (Oxford University Press, London, 1981) p. 341.

[4] D.H. Zanette, Am. J. Phys. 62, 404 (1994).

[5] R.E. Bellman and K.L. Cooke, *Differential-Difference Equations*, (Academic Press, New York, N.Y. 1963).

[6] M. Razavy, Can. J. Phys. 57, 1731 (1979).

[7] I.S. Gradshteyn and I.M. Ryzhik, *Tables of Integrals, Series, and Products*, (Academic Press, New York, 1965) p. 973.

[8] R.J. Rubin, Phys. Rev. 131, 964 (1963).

[9] U. Weiss, *Quantum Dissipative Systems*, Second Edition (World Scientific, Singapore, 1999) p. 27.

[10] M. Razavy, Can J. Phys. 58, 1019 (1980).

[11] F.J. Dyson, Phys. Rev. 92, 1331 (1953).

[12] N.G. van Kampen, Dans. mat.-fys. Medd. 26, Nr. 15 (1951).

[13] P. Ullersma, Physica 32, 27 (1966).

[14] M. Razavy, Phys. Rev. 41, 1211 (1990).

[15] G.W. Ford, J.T. Lewis and R.F. O'Connell, Phys. Rev. A37, 4419 (1988).

[16] E.R. Hanson, *A Table of Series and Products*, (Prentice-Hall, Inc. Englewood Cliffs, 1975) p. 222.

[17] A. Pimpale and M. Razavy, Phys. Rev. A 36, 2739 (1987).

[18] A.O. Caldeira and A. Leggett, Ann. Phys. (NY) 149, 374 (1983).

[19] M. Razavy, *Quantum Theory of Tunneling*, (World Scientific, Singapore 2003) Chapter 12.

Chapter 9

A Particle Coupled to a Field

In this section we will study a number of examples of mechanical systems coupled to vector or scalar fields. When a system is coupled to a quantum field, the vacuum fluctuations are always present and should be a part of the equation of motion. The effect of the vacuum noise is to ensure that in the Langevin equation there is a time-dependent driving force in addition to the damping force.

9.1 Harmonically Bound Radiating Electron

We have already seen that a classical radiating electron is an interesting example of a damped system. Now we want to derive a non-relativistic form of this dissipative system from a large and conservative system S consisting of an electron, S_1 interacting with an external electromagnetic field S_2 [1] [2]. Here the field is the transverse electromagnetic field described by the vector potential $\mathbf{A}$ with the constraint $\nabla \cdot \mathbf{A} = 0$ (van Kampen model). The Hamiltonian for the entire system is

$$H = \frac{1}{2m_0} \left(\mathbf{P} - e\tilde{\mathbf{A}} \right)^2 + V(\mathbf{R}) + \frac{1}{8\pi} \int \left[\mathbf{E}^2 + (\text{curl}\mathbf{A})^2 \right] d^3r. \qquad (9.1)$$

In this expression we have set the velocity of light $c = 1$, and m_0 is the bare mass of the electron and e is its charge, and the coupling term, $e\tilde{\mathbf{A}}$, is defined by

$$e\tilde{\mathbf{A}}(\mathbf{R},t) = \int \mathbf{A}(\mathbf{r},t)\rho(|\mathbf{r}-\mathbf{R}|)d^3r = \int \mathbf{A}(\mathbf{r}+\mathbf{R},t)\rho(|\mathbf{r}|)d^3r. \qquad (9.2)$$

Here $\rho(r)$ is the charge distribution for the electron and hence $\int \rho(r)d^3r = e$.

If the motion of the electron is confined to a region around the origin, then we set $\mathbf{R} = 0$ and we have

$$e\tilde{\mathbf{A}}(t) = \int \mathbf{A}(\mathbf{r},t)\rho(|\mathbf{r}|)d^3r. \qquad (9.3)$$

In the dipole approximation we can expand $\mathbf{A}(\mathbf{r})$ in terms of a complete set of functions obtained for waves confined inside a large sphere of radius L, and thus we find

$$\mathbf{A}(\mathbf{r},t) = \text{Transverse part of } \left\{ \sum_{n=1}^{\infty} \sqrt{\frac{3}{L}} \mathbf{q}_n(t) \frac{\sin(k_n r)}{r} \right\}, \qquad (9.4)$$

where $k_n = \frac{n\pi}{L}$. Note that there are three directions of polarization corresponding to the three components of $\mathbf{q}_n$. Similarly we expand $\mathbf{E}(\mathbf{r})$ as

$$\mathbf{E}(\mathbf{r},t) = -\text{Transverse part of } \left\{ \sum_{n=1}^{\infty} \sqrt{\frac{3}{L}} \mathbf{p}_n(t) \frac{\sin(k_n r)}{r} \right\}, \qquad (9.5)$$

where $\mathbf{p}_n(t)$ is the canonical conjugate of $\mathbf{q}_n(t)$. This follows from the fact that $\left(-\frac{\mathbf{E}}{4\pi}\right)$ is the canonical conjugate of $\mathbf{A}$. Then $e\tilde{\mathbf{A}}$ can be obtained from [1]

$$e\tilde{\mathbf{A}}(t) = \sum \mathbf{q}_n(t)\sqrt{\frac{4}{3L}} \int \sin(k_n r)\,\rho(r)4\pi r dr = \sum \epsilon_n \mathbf{q}_n(t), \qquad (9.6)$$

where

$$\epsilon_n = \delta_n \frac{n\pi c}{L}\sqrt{\frac{4e^2}{3L}}, \qquad (9.7)$$

and δ_n is the form factor of the electron ($\delta_n = 1$ for a point particle). Substituting (9.4), (9.5) and (9.6) in (9.1) and carrying out the integrals we find

$$H = \frac{1}{2m_0}(\mathbf{P})^2 + V(\mathbf{R}) - \frac{1}{m_0}\mathbf{P}\cdot\sum_n \epsilon_n \mathbf{q}_n + \frac{1}{2m_0}\left(\sum_n \epsilon_n \mathbf{q}_n\right)^2$$

$$+ \frac{1}{2}\sum_n (\mathbf{p}_n^2 + k_n^2 \mathbf{q}_n^2). \qquad (9.8)$$

This Hamiltonian is similar to the one describing the coupling between a particle and a bath consisting of harmonic oscillators, the only difference is that here the momentum of the particle is coupled to the amplitude of the field.

For the case of a harmonically bound electron, viz, when $V(\mathbf{R}) = \frac{1}{2}K\mathbf{R}^2$

the Hamiltonian is quadratic in the canonical variables and can be diagonalized by orthogonal transformation.

Now let us consider the following canonical transformations

$$\mathbf{p}_n = \mathbf{p}'_n, \quad \mathbf{q}_n = \mathbf{q}'_n + \frac{\epsilon_n}{mk_n^2}\mathbf{P}', \quad \mathbf{P}_n = \mathbf{P}'_n, \tag{9.9}$$

and

$$\mathbf{R}_n = \mathbf{R}'_n + \sum_n \frac{\epsilon_n}{mk_n^2}\mathbf{p}'_n, \tag{9.10}$$

where

$$m = m_0 + \sum_n \frac{\epsilon_n^2}{k_n^2}, \tag{9.11}$$

then the Hamiltonian becomes

$$H = \frac{1}{2m_0}\left(\mathbf{P}'\right)^2 + \frac{1}{2}K\left[\mathbf{R}' + \sum_n \left(\frac{\epsilon_n}{mk_n^2}\right)\mathbf{p}'_n\right]^2 + \frac{1}{2m_0}\left(\sum_n \epsilon_n \mathbf{q}'_n\right)^2$$
$$+ \frac{1}{2}\sum_n \left(\mathbf{p}'^2_n + k_n^2\mathbf{q}'^2_n\right). \tag{9.12}$$

This is one of the several forms of Hamiltonians quadratic in both momenta and coordinates that we can find for realistic physical systems.

9.2 An Oscillator Coupled to a String of Finite Length

Another problem which can be reduced to a quadratic Hamiltonian is that of an oscillator coupled to a finite or infinite string (Sollfrey's model) [3]-[5]. First consider a string of length $2L$ fixed at both ends and right at its mid-point, $x = 0$, is coupled to an oscillator of natural frequency ν_0. We choose the units so that the density of the string and also its tension be equal to unity.

The equations of motion of this system are:

$$\left(\frac{\partial^2}{\partial x^2} - \frac{\partial^2}{\partial t^2}\right)y(x,t) = e\delta(x)\left[y(0,t) - q(t)\right], \tag{9.13}$$

$$\left(\frac{d^2}{dt^2} + \nu_0^2\right)q(t) = \frac{e}{m}\left[y(0,t) - q(t)\right]. \tag{9.14}$$

We note that only the even solutions of (9.13) will be coupled to the oscillator, and we will only consider these solutions.

Let us take the solutions of the coupled set (9.13) and (9.14) to be of the form

$$y(x,t) = y(x,\omega)\exp(i\omega t), \quad \text{and} \quad q(t) = q(\omega)\exp(i\omega t). \tag{9.15}$$

By substituting these in Eqs. (9.13) and (9.14) and imposing the boundary conditions

$$y(\pm L, t) = 0, \tag{9.16}$$

we find the following eigenfunctions

$$y(x, \omega) = \frac{\sin\left[\omega_n(L - |x|)\right]}{\sqrt{N_\omega} \sin\left(\nu_n L\right)}, \quad \nu_n = \left(n - \frac{1}{2}\right)\left(\frac{\pi}{L}\right), \tag{9.17}$$

and

$$q(\omega) = \frac{-e \sin\left(\omega_n L\right)}{\sqrt{m N_\omega} \sin\left(\nu_n L\right)\left(\omega_n^2 - \nu_0'^2\right)}, \tag{9.18}$$

where the eigenvalues ω are the roots of the transcendental equation

$$2\omega\left(\omega^2 - \nu_0'^2\right)\cos(\omega L) + e\left(\omega^2 - \nu_0^2\right)\sin(\omega L) = 0, \tag{9.19}$$

and where

$$\nu_0'^2 = \nu_0^2 + \frac{e}{m}. \tag{9.20}$$

The eigenfunctions $y(x, \omega)$ are normalized in such a way that in the presence of the coupling

$$\int_{-L}^{L} y_i(x, \omega)y_j(x, \omega)dx + q_i(\omega)q_j(\omega) = \delta_{ij}, \tag{9.21}$$

whereas in the absence of coupling this relation is

$$\int_{-L}^{L} y_i(x)y_j(x)dx + q_i q_j = \delta_{ij}. \tag{9.22}$$

In the latter case we have

$$y_n(x) = \frac{\cos\left(\nu_n x\right)}{\sqrt{L}}, \quad q_n = \frac{\delta_{\nu_n, \nu_0}}{\sqrt{m}}. \tag{9.23}$$

From (9.17), (9.18) and (9.21) we can calculate the normalization constant N_{ω_n} which turns out to be

$$N_{\omega_n} = L + \left\{\frac{e \sin^2(\omega_n L)\left[\omega_n^4 + \left(\nu_0'^2 - 3\nu_0^2\right)\omega_n^2 + \nu_0^2 \nu_0'^2\right]}{2\omega_n^2\left(\omega_n^2 - \nu_0'^2\right)^2}\right\}. \tag{9.24}$$

The equations of motion (9.13)-(9.14) can be derived from the Hamiltonian

$$H = \frac{1}{2}\left[\left(\frac{p^2}{m} + m\nu_0^2 q^2\right) + \int_{-L}^{L}\left(\pi^2(x, t) + \left(\frac{\partial y(x, t)}{\partial x}\right)^2\right)dx\right]$$
$$+ \frac{e}{2}\left(y(0, t) - q(t)\right)^2, \tag{9.25}$$

where $\pi(x,t)$ is the momentum density conjugate to $y(x,t)$, i.e. y and π satisfy the Poisson bracket

$$\{\pi(x,t), y(x',t)\} = -\delta(x - x').$$ (9.26)

We expand both $y(x,t)$ and $\pi(x,t)$ in terms of Fourier series:

$$y(x,t) = \sum_{\nu=1}^{\infty} y_\nu(x) q'_\nu(t), \quad q(t) = \frac{q'_0(t)}{\sqrt{m}},$$ (9.27)

and

$$\pi(x,t) = \sum_{\nu=1}^{\infty} y_\nu(x) p'_\nu(t), \quad p(t) = \sqrt{m} p'(t).$$ (9.28)

By substituting these expansions in the Hamiltonian (9.25), we find

$$H = \frac{1}{2} \sum_{\nu}^{\infty} \left(p_\nu'^2 + \nu^2 q_\nu'^2 \right) + \frac{e}{2} \left[\sum_{\nu=1}^{\infty} \frac{q'_\nu}{\sqrt{L}} - \frac{q'_0}{\sqrt{m}} \right]^2.$$ (9.29)

The Hamiltonian H can be diagonalized. This is achieved with the help of the orthogonal matrix $T_{\omega\nu}$, and this matrix satisfies the following conditions:

$$\omega^2 \delta_{\omega\omega'} = \sum_{\nu=0}^{\infty} \nu^2 T_{\omega\nu} T_{\omega'\nu} + e \left[\sum_{\nu=1}^{\infty} \frac{T_{\omega\nu}}{\sqrt{L}} - \frac{T_{\omega\nu_0}}{\sqrt{m}} \right] \times \left[\sum_{\nu=1}^{\infty} \frac{T_{\omega'\nu}}{\sqrt{L}} - \frac{T_{\omega'\nu_0}}{\sqrt{m}} \right],$$ (9.30)

and

$$\sum_{\omega=0}^{\infty} T_{\omega\nu} T_{\omega\nu'} = \delta_{\nu\nu'}.$$ (9.31)

To find the matrix elements of T we observe that they are the Fourier transform components of the coupled system eigenfunctions with respect to the uncoupled system eigenfunctions. By evaluating the corresponding integrals we find that

$$T_{\omega_n \nu_k} = \frac{e \left(\omega_n^2 - \nu_0^2 \right) \sin(\omega_n L)}{\sqrt{L N_{\omega_n}} \left(\omega_n^2 - \nu_0'^2 \right) \left(\omega_n^2 - \nu_k^2 \right) \sin(\nu_n L)},$$ (9.32)

and

$$T_{\omega_n \nu_0} = -\frac{e \sin(\omega_n L)}{\sqrt{m N_{\omega_n}} \left(\omega_n^2 - \nu_0'^2 \right) \sin(\nu_n L)}.$$ (9.33)

A model similar to the one we discussed in this section has been proposed by Unruh and Zurek [6]. In Unruh's model the system is described by the Lagrangian

$$L = L_{F+P} + L_I,$$ (9.34)

where

$$L_{F+P} = \int \left[\left(\frac{\partial y(x,t)}{\partial t} \right)^2 - \left(\frac{\partial y(x,t)}{\partial x} \right)^2 + \delta(x) \left\{ \left(\frac{dq(t)}{dt} \right)^2 - \nu_0^2 q^2 \right\} \right] dx,$$ (9.35)

is the Lagrangian for the field and the particle and

$$L_I = -e \int q(t)\delta(x) \left(\frac{\partial y(x,t)}{\partial t} \right) dx, \tag{9.36}$$

is the Lagrangian for the coupling between the two. The Lagrangian L generates the equations of motion similar to (9.13) and (9.14)

$$\frac{\partial^2 y(x,t)}{\partial t^2} - \frac{\partial^2 y(x,t)}{\partial x^2} = e \frac{dq(t)}{dt}\delta(x). \tag{9.37}$$

and

$$\frac{d^2 q(t)}{dt^2} + \nu_0^2 q(t) = -e \frac{\partial y(0,t)}{\partial t}, \tag{9.38}$$

Unlike the model of Sollfrey [4] this Lagrangian as well as the resulting equation of motion are not invariant under time reversal transformation. Thus the interaction Lagrangian L_I has a distinctive arrow of time.

9.3   An Oscillator Coupled to an Infinite String

In the case of an infinite string the equations of motion (9.13) and (9.14) remain unchanged, however the boundary conditions (9.16) must be lifted. We solve this problem by means of the Fourier integrals [4] [5]. Thus we write

$$y(x,t) = \frac{1}{\sqrt{\pi}} \int_0^\infty q'(\nu,t) \cos(\nu x) d\nu, \tag{9.39}$$

$$q'(\nu,t) = \frac{1}{\sqrt{\pi}} \int_{-\infty}^\infty y(x,t) \cos(\nu x) dx, \tag{9.40}$$

with a similar expansion for $\pi(x,t)$. The Hamiltonian for this case is analogous to Eq. (9.29)

$$\begin{aligned}
H &= \frac{1}{2} \int_0^\infty \left[p'^2(\nu) + \nu^2 q'^2(\nu) \right] d\nu + \left(p_0'^2 + \nu_0^2 q_0'^2 \right) \\
&+ \frac{e}{2} \left[\int_0^\infty \frac{q'(\nu)}{\sqrt{\pi}} d\nu - \frac{q_0'}{\sqrt{m}} \right]^2.
\end{aligned} \tag{9.41}$$

We can diagonalize H by means of the orthogonal transformation $T(\omega,\nu)$ and $T(\omega,\nu_0)$ where

$$T(\omega,\nu) = \frac{\omega}{\sqrt{F(\omega)}} \left[\left(\omega^2 - \nu_0'^2 \right) \delta(\omega - \nu) + e \left(\omega^2 - \nu_0^2 \right) \mathcal{P} \frac{1}{\pi(\omega^2 - \nu^2)} \right], \tag{9.42}$$

and

$$T(\omega, \nu_0) = \frac{-e\,\omega}{\sqrt{\pi m F(\omega)}},$$

(9.43)

where

$$F(\omega) = \omega^2 \left(\omega^2 - \nu_0'^2\right)^2 + \frac{e^2}{4}\left(\omega^2 - \nu_0^2\right)^2.$$

(9.44)

The matrix elements $T(\omega, \nu)$ and $T(\omega, \nu_0)$ have the following properties:

$$
\begin{aligned}
\omega^2 \delta(\omega - \omega') &= \int_0^\infty \nu^2 T(\omega, \nu) T(\omega', \nu)\, d\nu + \nu_0^2 T(\omega, \nu_0) T(\omega', \nu_0) \\
&+ e\left[\int_0^\infty T(\omega, \nu)\frac{d\nu}{\sqrt{\pi}} - \frac{T(\omega, \nu_0)}{\sqrt{m}}\right] \\
&\times \left[\int_0^\infty T(\omega', \nu)\frac{d\nu}{\sqrt{\pi}} - \frac{T(\omega', \nu_0)}{\sqrt{m}}\right],
\end{aligned}
$$

(9.45)

$$\int_0^\infty T(\omega, \nu) T(\omega, \nu')\, d\omega = \delta(\nu - \nu'),$$

(9.46)

$$\int_0^\infty T(\omega, \nu) T(\omega, \nu_0)\, d\omega = 0,$$

(9.47)

and

$$\int_0^\infty T^2(\omega, \nu_0)\, d\omega = 1.$$

(9.48)

In the next chapter the motion of the central particle, in this case a harmonic oscillator, will be discussed in detail.

A slightly different way of considering the coupling between a harmonic oscillator and an infinite string is the one suggested by Yurke [7] [8].

Here the equations are similar to the equations (9.13)-(9.14), but we now write them in the form

$$\rho\frac{\partial^2 y(x,t)}{\partial t^2} - T\frac{\partial^2 y(x,t)}{\partial x^2} = 0 \quad x > 0,$$

(9.49)

$$m\frac{\partial^2 y(0,t)}{\partial t^2} - T\left(\frac{\partial y(x,t)}{\partial x}\right)_{x=0^+} + m\Omega_0^2 y(0,t) = 0, \quad x = 0.$$

(9.50)

In these equations ρ is the density of the string, m is the mass of the oscillator and Ω_0 is its natural frequency. The infinite string is under the tension T.

Now the D'Alembert general solution of (9.49) is

$$y(x,t) = y_{in}\left(\frac{x}{v} + t\right) + y_{out}\left(-\frac{x}{v} + t\right),$$

(9.51)

where v is the velocity of propagation of the transverse wave in the string and is given by

$$v = \sqrt{\frac{T}{\rho}}.$$

(9.52)

By substituting (9.51) in (9.50) we find

$$m\frac{\partial^2 y(0,t)}{\partial t^2} + m\lambda\frac{\partial y(0,t)}{\partial t} + m\Omega_0^2 y(0,t) = 2m\lambda\frac{\partial y_{in}(0,t)}{\partial t}, \tag{9.53}$$

where

$$\lambda = \frac{\sqrt{\rho T}}{m}, \tag{9.54}$$

which is the damping constant has the dimension of time^{-1}. In the absence of any incoming wave $\frac{\partial y(0,t)_{in}}{\partial t}$ is zero and Eq. (9.53) represents the equation of motion of a damped harmonic oscillator. Equations of motion (9.49)-(9.50) can be derived from the Lagrangian density

$$
\begin{aligned}
\mathcal{L} \;=\; & \left\{ \frac{1}{2}\rho\left(\frac{\partial y(x,t)}{\partial t}\right)^2 - \frac{1}{2}T\left(\frac{\partial y(x,t)}{\partial x}\right)^2 \right\}\theta(x) \\
& + \; \delta(x)\left[m\left(\frac{\partial y(0,t)}{\partial t}\right)^2 - m\Omega_0^2 y^2(0,t) \right],
\end{aligned}
\tag{9.55}
$$

where $\theta(x)$ is the step function

$$\theta(x) = \begin{cases} 1 & \text{for } x > 0 \\ 0 & \text{for } x < 0 \end{cases}, \tag{9.56}$$

and where we have used the integral [10]

$$\int_0^x f(x)\delta(x)dx = \frac{1}{2}f(0). \tag{9.57}$$

The momentum density for the transverse motion of the string is found from the Lagrangian density $\mathcal{L}$

$$\pi(x,t) = \frac{\partial\mathcal{L}}{\partial\left(\frac{\partial y(x,t)}{\partial t}\right)} = [2m\delta(x) + \theta(x)\rho]\left(\frac{\partial y(x,t)}{\partial t}\right). \tag{9.58}$$

Using this expression for momentum density we can determine the Hamiltonian density of the system $\mathcal{H}$

$$
\begin{aligned}
\mathcal{H} \;=\; & \frac{1}{2}\left[\rho\left(\frac{\partial y(x,t)}{\partial t}\right)^2 + T\left(\frac{\partial y(x,t)}{\partial x}\right)^2 \right]\theta(x) \\
& + \; \delta(x)\left[m\left(\frac{\partial y(0,t)}{\partial t}\right)^2 + m\Omega_0^2 y^2(0,t) \right],
\end{aligned}
\tag{9.59}
$$

where $\frac{\partial y}{\partial t}$ should be replaced by $\pi(x,t)$ using Eq. (9.58).

The Lamb model [3] [9] is one of the earliest models in which the motion of a particle which is subject to the potential $V(x)$ is coupled to a field $y(x,t)$. The Lagrangian for this model is similar to Yurke's model and is given by

$$L = \frac{1}{2}m\dot{q}^2(t) - V(q) + \frac{1}{2}\int_0^\infty \left[\rho\left(\frac{\partial y(x,t)}{\partial t}\right)^2 - T\left(\frac{\partial y(x,t)}{\partial x}\right)^2\right] dx. \quad (9.60)$$

Since there is no coupling between $y(x,t)$ and $q(t)$ in the Lagrangian, one imposes the following condition:

$$q(t) = y(0,t). \quad (9.61)$$

Thus the equation of motion of the particle is given by

$$m\ddot{q} + \frac{dV(q)}{dq} = f(t), \quad (9.62)$$

where $f(t)$ is the force exerted by the string on the particle and has to be determined from (9.61).

The equation for the wave motion of the string is given by the inhomogeneous wave equation

$$\frac{\partial^2 y(x,t)}{\partial t^2} - c^2\frac{\partial y(x,t)}{\partial x^2} = -\left(\frac{f(t)}{\rho}\right)\delta(x), \quad (9.63)$$

where $c = \sqrt{\frac{T}{\rho}}$ is the speed of the wave. The retarded solution of (9.63) can be written as

$$y(x,t) = y_g(x,t) - \frac{1}{2\rho c}\int_{-\infty}^{t-\frac{|x|}{c}} f(t')\,dt', \quad (9.64)$$

where $y_g(x,t)$ is the solution of the homogeneous wave equation. By taking $x = 0$ in Eq. (9.64) and then differentiating with respect to t we find $f(t)$

$$f(t) = 2\rho c\left[\frac{\partial y_g(0,t)}{\partial t} - \frac{\partial y(0,t)}{\partial t}\right]. \quad (9.65)$$

From Eqs. (9.61), (9.62) and (9.65) we obtain

$$m\ddot{q} + m\lambda\dot{q} + \frac{dV(q)}{dq} = F(t), \quad (9.66)$$

where λ has the same form as in Yurke's model, Eq. (9.54), and the particle is subject to the additional time-dependent force

$$F(t) = 2\sqrt{\rho T}\left(\frac{\partial y_g(0,t)}{\partial t}\right). \quad (9.67)$$

Bibliography

[1] H.G. van Kampen, Dans. mat.-fys. Medd. 26, Nr.15 (1951).

[2] P. Ullersma, Physica 32, 27 (1966).

[3] H. Lamb, Proc. London, Math. Soc. 32, 208 (1900).

[4] W. Sollfrey and G. Goertzel, Phys. Rev. 83, 1038 (1951).

[5] W. Sollfrey, Ph.D. Dissertation, New York University (1950).

[6] W.G. Unruh and W.H. Zurek, Phys. Rev. D40, 1071 (1989).

[7] B. Yurke and O. Yurke, MSC report # 4240 (1980).

[8] B. Yurke, Am. J. Phys. 54, 1133 (1986).

[9] G.W. Ford, J.T. Lewis and R.F. O'Connell, J. Stat. Phys. 53, 439 (1988).

[10] See for instance B. Friedman, *Principles and Techniques of Applied Mathematics*, (John Wiley & Sons, New York, 1957) p. 154.

Chapter 10

Damped Motion of the Central Particle

In the previous chapter we showed that the motion of a particle coupled to a field can be described by a Hamiltonian which is similar to that of an oscillator coupled to a heat bath. Now we want to study the motion of the particle (or the central oscillator) when all of the oscillators forming the bath are initially at the equilibrium position with zero velocity. All of the Hamiltonians which we have considered are quadratic functions of momenta and coordinates and can be diagonalized exactly by canonical transformations [1]-[3]. For the specific case of an oscillator either coupled to a finite or to an infinite string (Sollfrey's model) we discussed the technique of diagonalization in the last chapter. Here we consider a general linear coupling between the central particle and the heat bath or the field, where the coupling can depend on a number of parameters.

10.1 Diagonalization of the Hamiltonian

Let us write a typical quadratic Hamiltonian as (see Eq. (8.49))

$$H = \frac{1}{2}\left(P^2 + \Omega_0^2 Q^2\right) + \sum_{n=1}^{N} \frac{1}{2}\left(p_n^2 + \omega_n^2 q_n^2\right) + \sum_{n=1}^{N} \epsilon_n q_n Q, \qquad (10.1)$$

where P and Q are the momenta and the coordinate of the central oscillator. We have written H when there are N oscillators in the bath, but later we take the limit as $N \to \infty$. Now we want to find the time evolution of both Q

and P, and to this end we first diagonalize the Hamiltonian H. The canonical transformations that we need are of the forms

$$Q = \sum_{\nu=0}^{N} X_{0\nu} q'_\nu, \quad P = \sum_{\nu} X_{0\nu} p'_\nu, \tag{10.2}$$

and

$$q_n = \sum_{\nu=0}^{N} X_{n\nu} q'_\nu, \quad p_n = \sum_{\nu} X_{n\nu} p'_\nu. \tag{10.3}$$

By substituting these in Eq. (10.1) and using the orthogonality of the transformations we find

$$H = \frac{1}{2} \sum_{\nu=0}^{N} \left(p'^2_\nu + s^2_\nu q'^2_\nu \right), \tag{10.4}$$

where s_ν s are the eigenfrequencies of the system that we would like to determine. To this end from the Hamiltonian (10.1) we derive the equations of motion for Q and q_n

$$\frac{d^2 Q(t)}{dt^2} + \Omega_0^2 Q(t) = \sum_{n=1}^{N} \epsilon_n q_n(t), \tag{10.5}$$

and

$$\frac{d^2 q_n(t)}{dt^2} + \omega_n^2 q_n(t) = \epsilon_n Q(t), \quad n = 1, 2, \cdots N. \tag{10.6}$$

By substituting

$$Q(t) = Q(s) e^{ist} \quad \text{and} \quad q_n(t) = q(s) e^{ist}, \tag{10.7}$$

in Eqs. (10.5) and (10.6) we get the set of homogeneous equations

$$\left(\Omega_0^2 - s^2 \right) Q(s) = \sum_{n=1}^{N} \epsilon_n q_n(s) \tag{10.8}$$

and

$$\left(\omega_n^2 - s^2 \right) q_n(s) = \epsilon_n Q(s). \tag{10.9}$$

Equations (10.8) and (10.9) will have nontrivial solutions if and only if

$$\left(\Omega_0^2 - s^2 \right) = \sum_{n=1}^{N} \frac{\epsilon_n^2}{\omega_n^2 - s^2}, \tag{10.10}$$

where s_ν s the roots of (10.10) are the normal mode frequencies. These s_ν s have the following properties [4]:
(1) The number of s_ν s is one more than N, i.e. $\nu = 0, 1, \cdots N$.
(2) Between any two successive s_ν s there is just one ω_n. Moreover $s_N^2 > \omega_n^2$ and $s_0^2 < \omega_1^2$.

(3) The smallest eigenvalue s_1^2 can be positive or negative depending on the sign of the quantity

$$\Omega_0^2 - \sum_{n=1}^{N} \frac{\epsilon_n^2}{\omega_n^2}. \tag{10.11}$$

If all s_ν^2 s are positive then there is no self-accelerating solution. Therefore we assume that the condition

$$\Omega_0^2 - \sum_{n=1}^{N} \frac{\epsilon_n^2}{\omega_n^2} \geq 0 \tag{10.12}$$

is satisfied.

Now we want to determine the orthogonal transformation which diagonalizes the Hamiltonian (10.1). From (10.4) and the Hamilton canonical equations we find $q_\nu'(t)$ and $p_\nu'(t)$ to be

$$q_\nu'(t) = q_\nu'(0) \cos(s_\nu t) + p_\nu'(0) \left[\frac{\sin(s_\nu t)}{s_\nu} \right], \tag{10.13}$$

$$p_\nu'(t) = -q_\nu'(0) s_\nu \sin(s_\nu t) + p_\nu'(0) \cos(s_\nu t). \tag{10.14}$$

Next we invert the transformation (10.2) and (10.3);

$$q_\nu'(t) = X_{0\nu} Q(t) + \sum_{n=1}^{N} X_{n\nu} q_n(t), \tag{10.15}$$

$$p_\nu'(t) = X_{0\nu} P(t) + \sum_{n=1}^{N} X_{n\nu} p_n(t). \tag{10.16}$$

Then from Eqs. (10.2)-(10.3), (10.13)-(10.14) and (10.15)-(10.16) we find $Q(t)$ and $q_m(t)$;

$$Q(t) = \frac{dA(t)}{dt} Q(0) + A(t)P(0) + \sum_{n=1}^{N} \left\{ \frac{dA_n(t)}{dt} q_n(0) + A_n(t) p_n(0) \right\}, \tag{10.17}$$

$$q_j(t) = \frac{dA_j(t)}{dt} Q(0) + A_j(t) P(0) + \sum_{n=1}^{N} \left\{ \frac{dA_{nj}(t)}{dt} q_n(0) + A_{nj}(t) p_n(0) \right\}, \tag{10.18}$$

$$P(t) = \frac{dQ(t)}{dt}, \quad p_n(t) = \frac{dq_n(t)}{dt}. \tag{10.19}$$

In these equations $A(t)$, $A_n(t)$ and $A_{nj}(t)$ are used to denote the following sums:

$$A(t) = \sum_{\nu=0}^{N} X_{0\nu}^2 \left[\frac{\sin(s_\nu t)}{s_\nu} \right], \tag{10.20}$$

$$A_n(t) = \sum_{\nu=0}^{N} X_{0\nu} X_{n\nu} \left[\frac{\sin(s_\nu t)}{s_\nu} \right], \tag{10.21}$$

and

$$A_{nj}(t) = \sum_{\nu=0}^{N} X_{n\nu} X_{j\nu} \left[\frac{\sin(s_\nu t)}{s_\nu} \right]. \tag{10.22}$$

The initial values for $A(t)$, $A_n(t)$ and $A_{nj}(t)$ and their derivatives can be found from the definitions of these quantities and the orthogonality conditions for $X_{0\nu}$ and $X_{n\nu}$. These initial values are:

$$A(0) = 0, \quad \left(\frac{dA(t)}{dt} \right)_{t=0} = 1, \quad \left(\frac{d^2 A(t)}{dt^2} \right)_{t=0} = 0, \tag{10.23}$$

$$A_n(0) = 0, \quad \left(\frac{dA_n(t)}{dt} \right)_{t=0} = 0, \quad \left(\frac{d^2 A_n(t)}{dt^2} \right)_{t=0} = 0, \tag{10.24}$$

and

$$A_{nj}(0) = 0, \quad \left(\frac{dA_{nj}(t)}{dt} \right)_{t=0} = \delta_{nj}, \quad \left(\frac{d^2 A_{nj}(t)}{dt^2} \right)_{t=0} = 0, \tag{10.25}$$

and these agree with Eqs. (10.17) and (10.18).

Our next task is to determine the coefficients $X_{0\nu}$ and $X_{n\nu}$ of the transformation. From the eigenvector equation

$$X_{n\nu} = \frac{\epsilon_n}{s_\nu^2 - \omega_n^2} X_{0\nu}, \tag{10.26}$$

we find the normalization condition of the eigenvectors as

$$\sum_{n=1}^{N} X_{n\nu}^2 + X_{0\nu}^2 = \left[1 + \sum_{n=1}^{N} \frac{\epsilon_n^2}{(s_\nu^2 - \omega_n^2)^2} \right] X_{0\nu}^2 = 1, \tag{10.27}$$

or if we define $G(z)$ by

$$G(z) = z - \Omega_0^2 - \sum_{n=1}^{N} \frac{\epsilon_n^2}{(z - \omega_n^2)}, \tag{10.28}$$

we can write the eigenvalue equation as

$$G(z) = 0, \quad z_\nu = s_\nu^2. \tag{10.29}$$

From the definition of $G(z)$, Eq. (10.28), we can see that the frequencies of the bath oscillators ω_n, $n = 1, 2, \cdots N$ are the poles of $G(z)$ whereas the normal mode frequencies s_ν, $\nu = 0, 1, 2, \cdots N$ are the zeros of $G(z)$. Therefore we can write $G(z)$ as

$$G(z) = \frac{\Pi_{\nu=0}^{N} (z - s_\nu^2)}{\Pi_{n=1}^{N} (z - \omega_n^2)}. \tag{10.30}$$

The normalization condition for the eigenvectors can be expressed as

$$X_{0\nu}^2 = \left[\left(\frac{dG(z)}{dz} \right)_{z=s_\nu^2} \right]^{-1}. \tag{10.31}$$

This last relation shows that $X_{0\nu}^2$ is the residue of the pole of $[G(z)]^{-1}$ at $z = s_\nu^2$.

Having found the complete solution for the quadratic Hamiltonian (10.1), we want to study the dissipative motion of the central particle with the canonical coordinates $Q(t)$ and $P(t)$. For this purpose we assume that all of the oscillators, $q_n(t)$, are initially at rest with zero velocity, i.e.

$$q_n(0) = 0, \quad p_n(0) = 0. \tag{10.32}$$

Using these conditions we can simplify Eq. (10.17),

$$Q(t) = \frac{dA(t)}{dt} Q(0) + A(t) P(0). \tag{10.33}$$

Thus once $A(t)$ is known, the position and momentum of the central particle can be completely determined. For finite N, we can solve Eq. (10.20) for $A(t)$ since the eigenvalues s_ν and the eigenvectors $X_{0\nu}, X_{n\nu}$ are all known. However it is simpler to take the limit of $N \to \infty$ for our model (this is definitely the case for an oscillator coupled to a string). Neglecting the detailed structure of the spectrum and assuming that it is continuous, we can replace the summation over ν by integration. We will do this in the following way (see also §18.3):

$$\frac{dA(t)}{dt} = \sum_{\nu=0}^\infty X_{0\nu}^2 \cos{(s_\nu t)} = \sum_{\nu=0}^\infty \cos{(s_\nu t)} \left[\left(\frac{dG(z)}{dz} \right)_{z=s_\nu^2} \right]^{-1}, \tag{10.34}$$

where ν now runs from zero to infinity. By applying Cauchy's theorem we can write (10.34) as a contour integral

$$\frac{dA(t)}{dt} = \frac{1}{2\pi i} \oint_C \frac{\cos{(\sqrt{z}t)}}{G(z)} dz, \tag{10.35}$$

where the contour C encircles the positive part of the real axis in the z-plane. In the limit where s_ν s are infinite, we can approximate $G(z)$ by

$$G(z) = z - \Omega_0^2 - \int_0^\infty \left(\frac{\gamma(\omega)}{z - \omega^2} \right) d\omega, \tag{10.36}$$

where γ is defined by

$$\gamma(\omega) d\omega = \sum_{\omega < \omega_n < \omega + \Delta\omega} \epsilon_n^2. \tag{10.37}$$

In this limit the condition for non-existence of self-accelerating solution, Eq. (10.12) becomes

$$\int_0^\infty \frac{\gamma(\omega)}{\omega^2} d\omega \leq \Omega_0^2. \tag{10.38}$$

As the approximate equation for $G(z)$, Eq. (10.36), shows, this function has a cut along the positive real axis, therefore we can write the contour integral as

$$\frac{1}{2\pi i} \oint_C \cos\left(\sqrt{x}t\right) \left\{ \frac{1}{G_-(x)} - \frac{1}{G_+(x)} \right\} dx, \tag{10.39}$$

where $G_\pm$ is defined by

$$G_\pm(x) = x - \Omega_0^2 + \mathcal{P} \int_0^\infty \frac{\gamma(\omega)}{\omega^2 - x} d\omega \pm i\pi \frac{\gamma(\sqrt{x})}{2\sqrt{x}}. \tag{10.40}$$

In Eq. (10.40) $\mathcal{P}$ denotes the principal value of the integral. Changing the variable x to s where $s = \sqrt{x}$ we have

$$\frac{dA(t)}{dt} = \int_0^\infty \frac{\gamma(s)\cos(st)ds}{\{s^2 - \Omega_0^2 + \mathcal{P} \int_0^\infty \frac{\gamma(\omega)d\omega}{\omega^2 - s^2}\}^2 + \left(\frac{\pi^2}{4s^2}\right)\gamma^2(s)}. \tag{10.41}$$

Knowing $\dot{A}(t)$ and hence $A(t)$ enables us to calculate $Q(t)$ and $P(t)$ from Eqs. (10.33) and (10.20), and the result shows that the motion of the central particle is damped. We can simplify the result when the following approximation is valid

$$\frac{\pi^2 s^2 \gamma(s)}{4s^2} \approx \frac{\pi^2 s^2}{4\Omega_0^2} \gamma(\Omega_0), \tag{10.42}$$

i.e. when $\frac{\gamma(s)}{s^2}$ does not change rapidly around $s = \Omega_0$. If this is the case we have

$$\Omega_1^2 = \Omega_0^2 - \mathcal{P} \int_0^\infty \frac{\gamma(\omega)}{(\omega^2 - \Omega_0^2)} d\omega \approx \Omega_0^2 - \int_0^\infty \frac{\gamma(\omega)}{\omega^2} d\omega. \tag{10.43}$$

By substituting Eqs. (10.42) and (10.43) in (10.41) we find

$$\frac{dA(t)}{dt} = \frac{2\lambda}{\pi} \int_0^\infty \frac{s^2 \cos(st)ds}{\left(s^2 - \Omega_1^2\right)^2 + \lambda^2 s^2}, \tag{10.44}$$

where λ denotes the quantity

$$\lambda = \left(\frac{\pi}{2}\right) \frac{\gamma(\Omega_0)}{\Omega_0^2}. \tag{10.45}$$

Under these conditions $\dot{A}(t)$ can be calculated analytically:

$$\frac{dA(t)}{dt} = e^{-\frac{1}{2}\lambda t} \left[\cos(\omega t) - \frac{\lambda}{2\Omega} \sin(\omega t) \right], \quad \omega^2 = \Omega_1^2 - \frac{\lambda^2}{4}. \tag{10.46}$$

From (10.46) it follows that $A(t)$ is the solution of the differential equation

$$\frac{d^2 A(t)}{dt^2} + \lambda \frac{dA(t)}{dt} + \Omega_1^2 A(t) = 0, \tag{10.47}$$

and therefore both $Q(t)$ and $P(t)$ satisfy this differential equation, i.e.

$$\frac{d^2Q(t)}{dt^2} + \lambda\frac{dQ(t)}{dt} + \Omega_1^2 Q(t) = 0, \tag{10.48}$$

$$\frac{d^2P(t)}{dt^2} + \lambda\frac{dP(t)}{dt} + \Omega_1^2 P(t) = 0. \tag{10.49}$$

Here the equation of motion for $P(t)$ is not the same as the phenomenological Eq. (4.29) for $p(t)$, even though Eq.(10.48) for $Q(t)$ agrees with Eq. (4.27). In particular we note that in this derivation we have the canonical momentum $P(t)$ equal to the mechanical momentum $\dot{Q}(t)$ (note that we have set the mass of the central particle equal to unity). Thus we have obtained the same set of equations of motion as the ones generated by Dekker's, by $H(x,p,;m,\theta)$, Eq. (4.79) or other Hamiltonians qp-equivalent to these (§4.5).

Bibliography

[1] H.G. van Kampen, Dans. mat.-fys. Medd. 26, Nr.15 (1951).
[2] P. Ullersma, Physica 32, 27 (1966).
[3] M. Razavy, Phys. Rev. A41, 1211 (1990).
[4] G.W. Ford, J.T. Lewis and R.F. O'Connell, J. Stat. Phys. 53, 439 (1988).

Chapter 11

Classical Microscopic Models of Dissipation and Minimal Coupling Rule

In classical dynamics, for the motion of a charged particle in an external electromagnetic field, one can simply solve the equations of motion by adding the Lorentz force to the other forces acting on the particle. However for quantizing such a classical motion we need either the Lagrangian or the Hamiltonian, and in the presence of damping the canonical momentum is not simply related to the mechanical momentum and this creates problems both in connection with the minimal coupling rule and also in regard to the correct form of the momentum operator.

As we discussed in Chapter 4 this problem arises from the non-uniqueness of the Hamiltonians and the canonical momenta for the classical dissipative systems. A possible way of resolving this difficulty is to consider the full conservative system consisting of the subsystem of interest and the heat bath with which it is interacting. Then we can apply the minimal coupling prescription to the full system and eliminate the bath degrees of freedom to get the equation of motion for a charged damped system interacting with electromagnetic field.

Here we will study two microscopic models of damping in which the central particle of charge e is linearly coupled to a bath of neutral harmonic oscillators. The first is a model similar to the van Kampen model [1] and Ullersma model [2] that we have discussed earlier in Chapter 10. The Hamiltonian for this model is given by [3] [4]

$$H = \left[\frac{1}{2}\mathbf{p}^2 + V(\mathbf{r})\right] + \left[\sum_j \left(\frac{1}{2m_j}\mathbf{p}_j^2 + \frac{1}{2}m_j\omega_j^2\mathbf{x}_j^2\right)\right] +$$

$$+ \left(\mathbf{r} \cdot \sum_j \epsilon_j\mathbf{x}_j + \mathbf{r}^2 \sum_j \frac{\epsilon_j^2}{2m_j\omega_j^2}\right). \tag{11.1}$$

In this relation $\mathbf{p}$ is the canonical momentum of the particle, and $\mathbf{x}_j$ and $\mathbf{p}_j$ are the coordinate and momentum of the j-th oscillator with mass m_j and frequency ω_j, and ϵ_j s are the coupling coefficients. The last term in (11.1) is added so as to cancel the shift in the energy of the central particle in the corresponding quantum mechanical problem [5]-[7].

As we have seen earlier (§8.4) for the case of linear damping we choose ϵ_j, m_j and ω_j in such a way that

$$\sum_j \frac{\epsilon_j^2}{m_j\omega_j^2} \cos\left[\omega_j\left(t - t'\right)\right] = \lambda\delta\left(t - t'\right), \tag{11.2}$$

and this gives us a very simple expression for $K\left(t - t'\right)$.

When the central particle of charge e moves in an external electromagnetic field $(\mathbf{A}, \phi)$, then using the minimal coupling rule to incorporate the effect of this external field we get the equations of motion (note that the mass of the central particle is $m = 1$)

$$\dot{x}_\alpha = p_\alpha - \frac{e}{c}A_\alpha(\mathbf{r}, t), \tag{11.3}$$

$$\dot{p}_\alpha = -\frac{\partial V}{\partial x_\alpha} - e\frac{\partial \phi}{\partial x_\alpha} + \frac{e}{c}\mathbf{p} \cdot \left(\frac{\partial \mathbf{A}}{\partial x_\alpha}\right) - \frac{e^2}{c^2}\mathbf{A} \cdot \left(\frac{\partial \mathbf{A}}{\partial x_\alpha}\right) - \sum_j \left[\epsilon_j x_{j\alpha} + \frac{x_\alpha \epsilon_j^2}{m_j\omega_j^2}\right],$$

$$\tag{11.4}$$

$$m_j\dot{x}_{j\alpha} = p_{j\alpha}, \tag{11.5}$$

and

$$\dot{p}_{j\alpha} = -m_j\omega_j^2 x_{j\alpha} - \epsilon_j x_\alpha, \quad \alpha = 1, 2, 3. \tag{11.6}$$

In Eqs. (11.3)-(11.6) x_α, p_α are the components of $\mathbf{r}$ and $\mathbf{p}$ and $x_{j\alpha}, p_{j\alpha}$ are the components of $\mathbf{x}_j$ and $\mathbf{p}_j$ respectively. By eliminating $p_{j\alpha}$ between (11.5) and (11.6) we find the equation of motion for $x_{j\alpha}$;

$$\ddot{x}_{j\alpha} = -\omega_j^2 x_{j\alpha} - \frac{\epsilon_j}{m_j}x_\alpha. \tag{11.7}$$

This equation with the initial conditions $x_{j\alpha}(0) = 0$, $\dot{x}_{j\alpha}(0) = 0$ has the solution

$$x_{j\alpha}(t) = -\frac{\epsilon_j}{m_j\omega_j}\int_0^t x_\alpha(t')\sin\left[\omega_j\left(t - t'\right)\right]dt'. \tag{11.8}$$

From Eqs. (11.3), (11.4) and (11.8) we get

$$\ddot{x}_\alpha = -\frac{\partial V}{\partial x_\alpha} - e\frac{\partial \phi}{\partial x_\alpha} - \frac{e}{c}\dot{A}_\alpha + \frac{e}{c}\mathbf{p}\cdot\left(\frac{\partial \mathbf{A}}{\partial x_\alpha}\right) - \frac{e^2}{c^2}\mathbf{A}\cdot\left(\frac{\partial \mathbf{A}}{\partial x_\alpha}\right)$$
$$- x_\alpha \sum_j \left[\frac{\epsilon_j^2}{m_j\omega_j^2}\right] + \sum_j \frac{\epsilon_j^2}{m_j\omega_j}\int_0^t \sin\left[\omega_j\left(t-t'\right)\right]x_\alpha\left(t'\right)dt'.$$

$$(11.9)$$

We now use (11.3) to replace $\mathbf{p}$ by $\dot{\mathbf{x}} + \frac{e}{c}\mathbf{A}$ and write the right hand side of (11.9) in terms of the Lorentz force. Integrating the last term by parts and using (11.2) we obtain

$$\ddot{x}_\alpha = -\frac{\partial V}{\partial x_\alpha} + e\left[\mathbf{E} + \frac{1}{c}\left(\dot{\mathbf{r}} \wedge \mathbf{B}\right)\right]_\alpha - \lambda\dot{x}_\alpha - \lambda x_\alpha\delta(t). \qquad (11.10)$$

This result shows that in the case of linear coupling to the heat bath we can recover the phenomenological damping force with the explicit inclusion of the Lorentz force. The last term in (11.10) comes from the particular type of coupling between the bath and the particle that we have used.

In the second model that we want to study, the interaction Hamiltonian

$$H_I = \left(\mathbf{r}\cdot\sum_j \epsilon_j\mathbf{x}_j + \mathbf{r}^2\sum_j \frac{\epsilon_j^2}{2m_j\omega_j^2}\right), \qquad (11.11)$$

is replaced by

$$H_I' = \mathbf{p}\cdot\sum_j \varepsilon_j\mathbf{x}_j. \qquad (11.12)$$

Obviously ϵ_j and ε_j have different dimensions, and for H_I' to be invariant under time-reversal invariance ε_j must change sign when t is changed to $-t$. This type of coupling has been studied by Vyatchnin [8].

Now we follow the same steps that led us to the equation of motion for $\ddot{x}_\alpha$ (11.9). Here we have the canonical equations

$$\dot{x}_\alpha = p_\alpha - \frac{e}{c}A_\alpha\left(\mathbf{r}, t\right) + \sum_j \varepsilon_j x_{j\alpha}, \qquad (11.13)$$

$$\dot{p}_\alpha = -\frac{\partial V}{\partial x_\alpha} - e\frac{\partial \phi}{\partial x_\alpha} + \frac{e}{c}\mathbf{p}\cdot\left(\frac{\partial \mathbf{A}}{\partial x_\alpha}\right) - \frac{e^2}{c^2}\mathbf{A}\cdot\left(\frac{\partial \mathbf{A}}{\partial x_\alpha}\right) + \frac{e}{c}\sum_j \varepsilon\mathbf{x}_j\cdot\left(\frac{\partial \mathbf{A}}{\partial x_\alpha}\right), \quad (11.14)$$

$$m\dot{x}_{j\alpha} = p_{j\alpha}, \qquad (11.15)$$

and

$$\dot{p}_{j\alpha} = -m_j\omega_j^2 x_{j\alpha} - \varepsilon_j\left(p_\alpha - \frac{e}{c}A_\alpha\right). \qquad (11.16)$$

Just as in the case of the first model, using the initial conditions $x_{j\alpha}(0) = \dot{x}_{j\alpha}(0) = 0$ we find the solution of the Eqs. (11.13) and (11.16) to be

$$x_{j\alpha}(t) = -\frac{\varepsilon_j}{\omega_j} \int_0^t \left[p_\alpha(t') - \frac{e}{c} A_\alpha(t') \right] \sin\left[\omega_j(t - t') \right] dt'. \qquad (11.17)$$

From Eqs. (11.13), (11.14) and (11.17) after some algebra it follows that

$$\ddot{x}_\alpha = -\frac{\partial V}{\partial x_\alpha} + e \left[\mathbf{E} + \frac{1}{c}(\dot{\mathbf{r}} \wedge \mathbf{B}) \right]_\alpha - \lambda' \dot{x}_\alpha, \qquad (11.18)$$

where λ' is now defined by a relation similar to (11.2)

$$\sum_j \varepsilon_j^2 \cos\left[\omega_j(t - t') \right] = \lambda' \delta(t - t'). \qquad (11.19)$$

Equation (11.18) is identical with the equation of motion of a charged particle moving in an external electromagnetic field, and there is no additional impulse term that we found in (11.10).

Bibliography

[1] N.G. van Kampen, Dans. mat.-fys. Medd. 26, Nr. 15 (1951).

[2] P. Ullersma, Physica 32, 27 (1966).

[3] X.L. Li, G.W. Ford and R.F. O'Connell, Phys. Rev. A41, 5287 (1990).

[4] G.W. Ford, J.T. Lewis and R.F. O'Connell, Phys. Rev. A37, 4419 (1988).

[5] A.O. Caldeira and A. Leggett, Ann. Phys. (NY) 149, 374 (1983).

[6] G.W. Ford, J.T. Lewis and R.F. O'Connell, J. Stat. Phys. 53, 439 (1988).

[7] A. Pimpale and M. Razavy, Phys. Rev. A36, 2739 (1987).

[8] S.P. Vyatchanin, Dok. Akad. Nauk. SSSR 286, 1379 (1986).

Chapter 12

Quantization of Dissipative Systems

12.1 Early Attempts to Quantize the Damped Oscillator

From the early days of quantum mechanics, the question of the radiation damping and its effect on the stability of quantum states (and or Bohr orbits) was the subject of discussion. For instance we know that in a seminar at E.T.H. on the topic of Bohr atom, Max von Laue criticized the Bohr theory by saying "This is all nonsense! Maxwell's equations are valid under all circumstances. An electron in a circular orbit must emit radiation" [1]. The discovery of the matrix mechanics and later the Schrödinger equation renewed the interest in this problem and one of the first papers on the subject of radiation damping and the quantum mechanics of the damped oscillator was written by Seeger [2]. Seeger considered the problem of quantization of a damped harmonic oscillator using Rayleigh's dissipation function, Eq. (3.3), and the matrix mechanics of Born, Jordan and Heisenberg [3].

Starting with the Hamiltonian operator

$$H = \frac{1}{2m}p^2 + \frac{1}{2}m\Omega_0^2 x^2,$$ (12.1)

and the damping term

$$\mathcal{F} = \frac{m}{2}\lambda\dot{x}^2,$$ (12.2)

in the Heisenberg picture, we find the equations of motion for p and x to be

$$\dot{p} = -m\Omega_0^2 x - m\lambda\dot{x}, \quad \text{and} \quad m\dot{x} = p, \tag{12.3}$$

or

$$\ddot{x} + \lambda\dot{x} + \Omega_0^2 x = 0. \tag{12.4}$$

This equation must be satisfied by the matrix elements

$$x_{jk}(t) = x_{jk}(0)\exp\left(i\omega_{jk}t\right), \tag{12.5}$$

i.e.

$$\left\{-\omega_{jk}^2 + i\lambda\omega_{jk} + \Omega_0^2\right\} x_{jk}(0) = 0. \tag{12.6}$$

Therefore $x_{jk}(0)$ is zero unless

$$\omega_{jk} = \pm a + i\frac{\lambda}{2}, \tag{12.7}$$

where

$$a = \sqrt{\Omega_0^2 - \frac{\lambda^2}{4}} \approx \Omega_0 - \frac{\lambda^2}{8\Omega_0}. \tag{12.8}$$

Here we are assuming that the motion is under-damped and that $2\Omega_0 >> \lambda$. Let us now consider the commutation relation

$$\sum_j \left(p_{nj}x_{jn} - x_{nj}p_{jn}\right) = -i\hbar. \tag{12.9}$$

If

$$\alpha_{jk} = \text{Re}\left[\omega_{jk}\right], \tag{12.10}$$

then

$$\sum_j \alpha_{jk}|x_{jk}|^2 = -\frac{\hbar}{2m}. \tag{12.11}$$

We assume that there is no degeneracy. Then for each n there is at least one n' such that $x_{nn'} \neq 0$, but since $\omega_{nn'} = \pm a + i\frac{\lambda}{2}$, therefore there are at most two values n', say n_1' and n_2' for which $x_{nn'} \neq 0$. From the Planck-Bohr relation we have

$$E_n - E_{n_1'} = \hbar a, \quad E_n - E_{n_2'} = -\hbar a, \tag{12.12}$$

or

$$E_n - E_{n_1'} = -\hbar a, \quad E_n - E_{n_2'} = \hbar a. \tag{12.13}$$

Next from the Hamiltonian (12.1) we calculate H_{nn}

$$H_{nn} = \frac{m}{2}e^{-\lambda t}\left(a^2 + \frac{\lambda^2}{4} + \Omega_0^2\right)\sum_j |x_{nj}|^2, \tag{12.14}$$

which, in view of what we have said earlier, can be written as

$$H_{nn} = \frac{m}{2} e^{-\lambda t} \left(a^2 + \frac{\lambda^2}{4} + \Omega_0^2 \right) \left\{ |x_{nn_2'}|^2 + |x_{nn_1'}|^2 \right\}. \tag{12.15}$$

On the other hand from the commutation relation (12.11) we have

$$\alpha_{nn_1'} \left\{ |x_{nn_2'}|^2 - |x_{nn_1'}|^2 \right\} = -\frac{\hbar}{2m}. \tag{12.16}$$

Just as in the problem of undamped harmonic oscillator we suppose that for the ground state n_2' does not exist, hence

$$\alpha_{n_0 n_0'} |x_{n_0 n_0'}|^2 = \frac{\hbar}{2m}, \tag{12.17}$$

and

$$H_{n_0 n_0} = m \Omega_0^2 e^{-\lambda t} |x_{n_0 n_0'}|^2. \tag{12.18}$$

From (12.17) we obtain

$$|x_{n_0 n_0'}|^2 = \frac{\hbar}{2ma}, \tag{12.19}$$

and substituting this in (12.18) we find

$$H_{n_0 n_0} = \frac{\hbar \Omega_0^2}{2a} e^{-\lambda t} \approx \frac{\hbar \Omega_0}{2} \left(1 + \frac{\lambda^2}{8\Omega_0^2} \right) e^{-\lambda t}. \tag{12.20}$$

For higher levels we have

$$H_{jj} - H_{kk} = \frac{\hbar \Omega_0^2}{a} (j - k) e^{-\lambda t}, \tag{12.21}$$

and for the energy differences [2]

$$E_j - E_k = (j - k) a \hbar \approx \hbar \Omega_0 (j - k) \left(1 - \frac{\lambda^2}{8\Omega_0^2} \right), \tag{12.22}$$

or

$$E_j - E_k = (H_{jj} - H_{kk}) \left(\frac{a^2}{\Omega_0^2} \right) e^{\lambda t} \approx (H_{jj} - H_{kk}) \left(1 - \frac{\lambda^2}{4\Omega_0^2} \right) e^{\lambda t}. \tag{12.23}$$

Thus Seeger concludes that the energies of the states of oscillator when $\lambda \ll 2\Omega_0$ remain constant over a considerable time.

Let us emphasize the following aspects of this formulation:

(a) The commutator $[p, x]$ stays constant in time.

(b) Equations for the matrix elements of x and p are identical.

(c) The Hamiltonian used is directly related to the energy, i.e. it is the energy operator.

(d) The change in the energy differences caused by damping is proportional to

$\frac{\lambda^2}{8\Omega_0^2}$.

This method of quantizing a dissipative system has been criticized by Brittin [4]. Brittin has argued that for frictional forces depending on velocity, the commutation relation and the Heisenberg equations of motion are incompatible. Let us consider a one-dimensional motion under the force of friction Q. In Chapter 3 we have seen that the classical equation of motion can be derived from the Euler-Lagrange equation

$$\frac{d}{dt}\left(\frac{\partial L}{\partial \dot{x}}\right) - \frac{\partial L}{\partial x} = Q(x, \dot{x}). \tag{12.24}$$

By introducing the canonical momentum p by

$$p = \frac{\partial L}{\partial \dot{x}}, \tag{12.25}$$

we can write the Hamiltonian as

$$H(x, p) = p\dot{x} - L. \tag{12.26}$$

The equations of motion can also be obtained from (12.26) and the Hamilton canonical equations, which in this case are:

$$\dot{x} = \frac{\partial H}{\partial p}, \quad \text{and} \quad \dot{p} = -\frac{\partial H}{\partial x} + Q\left[x, \dot{x}(x, p)\right]. \tag{12.27}$$

In classical dynamics we can replace (12.27) by the definition of the time derivative of any dynamical variable $A(x, p, t)$ in terms of the Poisson bracket [5]

$$\frac{dA}{dt} = \{A, H\} + \frac{\partial A}{\partial p}Q(x, p), \tag{12.28}$$

where the Poisson bracket $\{A, H\}$ is defined by

$$\{A, H\} = \frac{\partial A}{\partial x}\frac{\partial H}{\partial p} - \frac{\partial A}{\partial p}\frac{\partial H}{\partial x}. \tag{12.29}$$

The transition to quantum mechanics can be made by associating the Poisson bracket with the commutator using the well-known rule [6]

$$\{A, H\} \rightarrow \frac{1}{i\hbar}[A, H] = \frac{1}{i\hbar}(AH - HA). \tag{12.30}$$

Hence for the time derivative of an operator A we find the Heisenberg equation

$$i\hbar\frac{dA}{dt} = [A, H] + i\hbar\frac{\partial A}{\partial p}Q(x, p). \tag{12.31}$$

Next we differentiate the commutator

$$[x, p] = i\hbar, \tag{12.32}$$

with respect to time to obtain

$$\left[\frac{dx}{dt}, p\right] + \left[x, \frac{dp}{dt}\right] = 0. \tag{12.33}$$

With the help of Eq. (12.31), taking A first to be x and then to be p, we rewrite (12.33) as

$$[[x, H], p] + [x, [p, H] + i\hbar Q] = 0. \tag{12.34}$$

From the last relation it follows that

$$[x, Q] = 0, \tag{12.35}$$

which in turn implies that Q can only be a function of x and not of $\dot{x}$ [4].

Brittin's argument shows the inconsistency between the canonical equations of motion, the Hamiltonians and the commutation relations. For a group of classical time-independent Hamiltonians constructed for the damped systems we can investigate the question of consistency in the following way:
Let us start with a separable Lagrangian, in x and $\dot{x}$ for a particle with unit mass written in the form [7]

$$L(x, \dot{x}) = \dot{x} \int^{\dot{x}} \frac{1}{v^2} G(v) dv - V(x), \tag{12.36}$$

where $G(v)$ is an arbitrary function of v. As before we impose the condition that $(1/v)(dG(v)/dv) \neq 0$. The canonical momentum conjugate to x is given by

$$p = \int^{\dot{x}} \frac{dv}{v^2} G(v) + \frac{G(\dot{x})}{\dot{x}} = \int^{\dot{x}} \frac{dv}{v} \frac{dG(v)}{dv}. \tag{12.37}$$

The Hamiltonian found from (12.36) but expressed in terms of x and p is

$$H(x, p) = G[\dot{x}(p)] + V(x), \tag{12.38}$$

where we have solved (12.37) for $\dot{x} = \dot{x}(p)$. Using the canonical commutation relation $[x, p] = i\hbar$ we can calculate $\frac{i}{\hbar}[H, x]$,

$$\frac{i}{\hbar}[H, x] = \frac{i}{\hbar}[G(\dot{x}), x] = \frac{d}{dp} G(\dot{x}) = \frac{d\dot{x}}{dp} \frac{dG(\dot{x})}{d\dot{x}} = \dot{x}. \tag{12.39}$$

This result follows from Eq. (12.37). Equation (12.39) is the correct Heisenberg equation for the time derivative of the position operator. Next we consider the time derivative of $\dot{x}$

$$\ddot{x} = \frac{i}{\hbar}[H, \dot{x}] = \frac{i}{\hbar}[V, \dot{x}]. \tag{12.40}$$

To calculate the right hand side of (12.40) we first note that from the commutation relation, $[x, p] = i\hbar$, it follows that

$$\frac{i}{\hbar}[x, P(p)] = -\frac{dP(p)}{dp}, \tag{12.41}$$

for any function $P(p)$, therefore

$$\frac{i}{\hbar}[x, \dot{x}] = -\frac{d\dot{x}}{dp} = -\frac{\dot{x}}{\left(\frac{dG(\dot{x})}{d\dot{x}}\right)} = -R(\dot{x}), \tag{12.42}$$

where $R(\dot{x})$ is defined by (12.42).

Next we calculate the commutator

$$\frac{i}{\hbar}[x^n, \dot{x}] = \frac{i}{\hbar}\sum_{j=0}^{n-1} x^j[x, \dot{x}]x^{n-1-j} = -\sum_{j=0}^{n-1} x^j R(\dot{x})x^{n-1-j}. \tag{12.43}$$

In the classical limit the right hand side of (12.43) becomes

$$-nx^{n-1}R(\dot{x}) = -R(\dot{x})\frac{d}{dx}x^n, \tag{12.44}$$

thus for this case we have the rule of association

$$\left[R(\dot{x})\frac{d}{dx}x^n\right]_{Q.M.} \rightarrow \sum_{j=0}^{n-1} x^j R(\dot{x})x^{n-1-j}. \tag{12.45}$$

Now we assume that $V(x)$ can be written as a power series in x, i.e.

$$V(x) = \sum_{n=0}^{\infty} a_n x^n, \tag{12.46}$$

then we can calculate (12.40),

$$\frac{i}{\hbar}[V(x), \dot{x}] = -\left(R(\dot{x})\frac{dV(x)}{dx}\right)_{Q.M.}. \tag{12.47}$$

Using (12.47) we find the Heisenberg equation of motion for x

$$\ddot{x} + \left(R(\dot{x})\frac{dV(x)}{dx}\right)_{Q.M.} = 0. \tag{12.48}$$

What we have shown here is that an equation of motion of the form (12.48) is compatible with the canonical commutation relation. But the mathematical consistency does not imply that the resulting quantum equation is physically meaningful.

Again let us consider the simple example where $R(\dot{x}) = \dot{x}$ and thus the classical equation of motion is

$$\ddot{x} + \dot{x}\left(\frac{dV(x)}{dx}\right) = \ddot{x} + \frac{dV(x)}{dt} = 0. \tag{12.49}$$

For this case the Lagrangian and the canonical momentum are:

$$L = \frac{1}{2}\dot{x}\ln\dot{x}^2 - \dot{x} - V(x), \tag{12.50}$$

and

$$p = \frac{\partial L}{\partial \dot{x}} = \frac{1}{2}\ln\dot{x}^2. \tag{12.51}$$

The Hamiltonian obtained from (12.50) is given by

$$H = \pm e^p + V(x). \tag{12.52}$$

The quantum mechanical solution of this problem when $V(x) = x$ will be discussed later in this section where it will be shown that the result is unphysical.

Similar argument for obtaining the Heisenberg equation compatible with the commutator has been obtained for a Hamiltonian of the form $H(x,p) = R(p)S(x)$ for conservative systems. Apparently the consistency requirement is closely related to the problem of the Hamiltonian operator giving the correct equation of motion in the phase space.

In some cases the problem of consistency of the quantization procedure can be circumvented by changing the Lagrangian and the Hamiltonian for the system. As an example consider the following simple case [8] where the equation of motion is given by

$$m\ddot{x} + m\lambda\dot{x} = 0. \tag{12.53}$$

As we have already seen we can construct different Hamiltonians in addition to the one mentioned earlier i.e. $H = (p^2/2m)e^{-\lambda t}$. Let us take the mass of the particle to be equal to one and write as alternative Hamiltonian the function that we found in Eq. (3.73);

$$H = \mathcal{E}\exp\left(\frac{p}{p_0}\right) + \lambda p_0 x, \tag{12.54}$$

where $\mathcal{E}$ and p_0 are constants. Let p and x be the canonically conjugate operators satisfying the commutation relation $[p, x] = i\hbar$, and let us write the Heisenberg equations of motion as

$$\dot{x} = \frac{i}{\hbar}\left[\hat{H}, x\right] = \frac{\mathcal{E}}{p_0}\exp\left(\frac{p}{p_0}\right), \tag{12.55}$$

and

$$\dot{p} = \frac{i}{\hbar}\left[\hat{H}, p\right] = -\lambda p_0. \tag{12.56}$$

In these equations $\hat{H}$ represents the Hamiltonian operator corresponding to the classical function (12.54). By differentiating (12.55) with respect to t we find

$$\ddot{x} = \frac{\mathcal{E}}{2p_0^2}\left\{\dot{p}\exp\left(\frac{p}{p_0}\right) + \exp\left(\frac{p}{p_0}\right)\dot{p}\right\} = -\lambda\dot{x}, \tag{12.57}$$

where in (12.57) we have substituted for $\dot{p}$ and for $\exp\left(p/p_0\right)$ from Eqs. (12.55) and (12.56). Thus the canonical commutation relation and the equations of motion are compatible and there is no inconsistency. The Hamiltonian (12.54) also generates the correct equation of motion in phase space but it does not have the correct form as $\lambda \to 0$, and it is not the energy of the system. Therefore as a quantum-mechanical operator it is not acceptable. This can easily be seen by writing the "wave equation" for the operator (12.54) in momentum space

$$-i\hbar\lambda p_0 \frac{\partial\psi}{\partial p} + \mathcal{E}\exp\left(\frac{p}{p_0}\right)\psi = E'\psi, \qquad (12.58)$$

and solving it to find

$$\psi = \exp\left[\frac{-i}{\lambda\hbar p_0}\left\{\mathcal{E} - E'p_0\exp\left(\frac{p}{p_0}\right)\right\}\right]. \qquad (12.59)$$

Again we note that ψ does not have a well-defined limit as $\lambda \to 0$.

12.2   Yang-Feldman Method of Quantization

In Chapters 3 and 4 we observed that the classical equations of motion in the coordinate space are the basis for construction of families of q-equivalent Lagrangian and Hamiltonian, and that these Hamiltonians (or Lagrangians) give different quantum mechanical results. We may ask whether it is possible to bypass the canonical formalism and to try to quantize the equation of motion (or its solution) directly. This method of quantization has been studied in quantum theory of fields and is known as the Yang-Feldman method [9] [10].

The starting point here is the classical equation of motion which for a particle of mass $m = 1$ can be written as

$$\ddot{x}_i(t) = f_i(\mathbf{x}, \dot{\mathbf{x}}, t). \qquad (12.60)$$

A formal solution of this classical equation with the initial values $x_i(0)$ and $\dot{x}_i(0)$ can be found with the aid of the Green function $G(t - t')$ defined by the differential equation

$$\frac{d^2}{dt^2}G(t - t') = \delta(t - t'), \qquad (12.61)$$

and has the solution

$$G(t - t') = (t - t')\,\theta(t - t'). \qquad (12.62)$$

From Eqs. (12.60) and (12.61) we have the solution

$$x_i(t) = x_i(0) + \dot{x}_i(0)t + \int_0^t (t - t')\,f_i(\mathbf{x}', \dot{\mathbf{x}}', t')\,dt', \qquad (12.63)$$

which is the solution of the classical problem (12.60) subject to the initial conditions $x_i(0)$ and $\dot{x}_i(0)$. We can take Eq. (12.63) as the starting point for quantization in the Heisenberg picture. To this end we assume that for a given force $f_i(\mathbf{x}', \dot{\mathbf{x}}', t')$ we can construct a Hermitian operator by any one of the standard rules of association of classical function and quantum operators [11] [12]. But even in this method the operator $\hat{p}_i(0) = \hat{\dot{x}}_i(0)$ which satisfies the commutation relation $[\hat{x}_i(0), \hat{p}_j(0)] = i\hbar\delta_{ij}$ can appear in the formal solution of the Heisenberg equation corresponding to (12.63) in different ways.

Let us demonstrate this by a simple example using two different q-equivalent Hamiltonians which we found in (4.32). First we note that if we choose $H_1 = (p^2/2)e^{-\lambda t}$ (we have set $m = 1$), then $p(t) = \dot{x}e^{\lambda t}$ and therefore as an operator

$$\hat{p}(0) = \hat{\dot{x}} = -i\hbar\frac{\partial}{\partial x}, \tag{12.64}$$

where the symbol $\hat{\dot{x}}$ indicates that we are using operators. Thus the Heisenberg equation corresponding to the classical equation (12.63) becomes

$$\hat{x}(t) = \hat{x} - i\hbar t\frac{\partial}{\partial x} - \lambda\int_0^t (t - t')\,\hat{\dot{x}}\,(t')\,dt', \tag{12.65}$$

and this equation can be solved by iteration with the result that

$$\hat{x}(t) = \hat{x} - \frac{i\hbar}{\lambda}\left(1 - e^{-\lambda t}\right)\frac{\partial}{\partial x}. \tag{12.66}$$

On the other hand for $\hat{H}_2 = e^p + \lambda x$ which is the same as H_2 in Eq. (4.32) but with $\mathcal{E} = p_0 = 1$ we have

$$\dot{x} = e^p \quad \text{and} \quad p = \ln\dot{x}. \tag{12.67}$$

Using the operator identity

$$\exp\hat{A}\,\exp\hat{B} = \exp\hat{B}\,\exp\hat{A}\,\exp[\hat{A}, \hat{B}], \tag{12.68}$$

with $\hat{B} = \hat{p}$ and $\hat{A} = \frac{i}{\hbar}\hat{H}_2 t$, we get

$$\begin{aligned}\hat{\dot{x}}(t) &= \exp\left(\frac{i}{\hbar}\hat{H}_2 t\right)e^{\hat{p}}\exp\left(-\frac{i}{\hbar}\hat{H}_2 t\right) = e^{\hat{p}}\exp\left(\frac{i}{\hbar}\left[\hat{H}_2, \hat{p}\right]\right)\\ &= e^{\hat{p}}e^{-\lambda t} = \exp\left(-i\hbar\frac{\partial}{\partial x} - \lambda t\right).\end{aligned} \tag{12.69}$$

By direct integration of (12.69) we find

$$\hat{x}(t) = \hat{x} + \frac{1}{\lambda}\left(1 - e^{-\lambda t}\right)\exp\left(-i\hbar\frac{\partial}{\partial x}\right), \tag{12.70}$$

and this is the solution of equation

$$\hat{x}(t) = \hat{x} + t\exp\left(-i\hbar\frac{\partial}{\partial x}\right) - \lambda\int_0^t (t - t')\,\hat{\dot{x}}\,(t')\,dt', \tag{12.71}$$

which is different from (12.65). We can also show that the commutators $\left[\hat{x}(t), \hat{\dot{x}}(t)\right]$ for the two Hamiltonians are not the same. Thus from Eq. (12.66) and its time derivative we can calculate the commutator $\left[\hat{x}(t), \hat{\dot{x}}(t)\right]$;

$$\left[\hat{x}(t), \hat{\dot{x}}(t)\right] = \left[x - \frac{i\hbar}{\lambda}\left(1 - e^{-\lambda t}\right)\frac{\partial}{\partial x}, \ i\hbar e^{-\lambda t}\frac{\partial}{\partial x}\right] = i\hbar e^{-\lambda t}. \qquad (12.72)$$

Similarly from (12.69) and (12.70) we find

$$\left[\hat{x}(t), \hat{\dot{x}}(t)\right] = \left[x, e^p e^{-\lambda t}\right] = i\hbar e^p e^{-\lambda t}. \qquad (12.73)$$

Making use of the relation $\dot{p} = (i/\hbar)[p, H]$, Eq. (12.56), or its integral $p = \lambda t$ we find that $\left[\hat{x}(t), \hat{\dot{x}}(t)\right] = i\hbar$.

12.3   Heisenberg's Equations of Motion for Dekker's Formulation

The complex Hamiltonian of Dekker (§4.5) in operator form can be regarded as the generator for the Heisenberg equations of motion [13]-[15]. In this formalism we start with the operator H given by

$$\mathsf{H} = H + i\Gamma = -i\omega\pi q - \frac{1}{4}\lambda\hbar. \qquad (12.74)$$

The operators q and π satisfy the commutation relation

$$[q, \pi] = i\hbar, \qquad (12.75)$$

and this follows from the commutation relation $[x, p] = i\hbar$. Thus we have linear canonical transformations (4.66)-(4.67). The time evolution of q and π is given by

$$\dot{q} = -\frac{i}{\hbar}[q, H] + \frac{1}{\hbar}[q, \Gamma]_+ + \mathcal{N}_q(t) \qquad (12.76)$$

and

$$\dot{\pi} = -\frac{i}{\hbar}[\pi, H] + \frac{1}{\hbar}[\pi, \Gamma]_+ + \mathcal{N}_\pi(t), \qquad (12.77)$$

where $[\ ,]$ is the usual commutator, $[\ ,]_+$ is the anti-commutator and $\mathcal{N}_q$ and $\mathcal{N}_\pi$ are the noise operators with the property that [14]

$$\langle\mathcal{N}_q(t)\rangle = \langle\mathcal{N}_\pi(t)\rangle = 0, \qquad (12.78)$$

where the average is taken with respect to noise only.

We can rewrite H in terms of the original coordinate and momentum operators x and p;

$$H = \frac{1}{2}p^2 + \frac{\lambda}{4}(px + xp) + \frac{1}{2}m\left(\omega^2 + \frac{\lambda^2}{4}\right)x^2, \quad \Gamma = -\frac{1}{4}\lambda\hbar. \qquad (12.79)$$

For x and p operators we find the equations of motion

$$\dot{x} = -\frac{i}{\hbar} [x, \mathsf{H}] + \mathcal{N}_x(t) = p + \mathcal{N}_x(t) \tag{12.80}$$

and

$$\dot{p} = -\frac{i}{\hbar} [p, \mathsf{H}] + \mathcal{N}_p(t) = -\lambda p - \left(\omega^2 + \frac{\lambda^2}{4} \right) x + \mathcal{N}_p(t). \tag{12.81}$$

Since both $\langle \mathcal{N}_p(t) \rangle_N$ and $\langle \mathcal{N}_x(t) \rangle_N$ are zero, we have the correct result for the Ehrenfest theorem. The properties of the noise operators in this formulation have been studied in detail by Dekker [14] [15]. This particular formulation of the damped oscillator will be further discussed in §14.3.

12.4 Quantization of the Bateman Hamiltonian

In §4.8 we discussed the Bateman dual Lagrangian and Hamiltonian for a damped-amplified harmonic oscillator. We also found a similar Hamiltonian, Eq. (4.178), resulting from the extended phase space formulation. Now we want to study the quantized version of these systems. First we note that the Bateman Hamiltonian (4.102) does not depend on time explicitly, and therefore is a constant of motion. We can obtain the eigenvalues of this system by introducing the creation and annihilation operators $A, B, A^\dagger$ and $B^\dagger$ and writing the Hamiltonian in terms of these operators.

Let us define these operators by:

$$A = \frac{1}{2\sqrt{m\omega\hbar}} \left[(p_x + p_y) - im\omega(x + y) \right], \tag{12.82}$$

$$B = \frac{1}{2\sqrt{m\omega\hbar}} \left[(p_x - p_y) - im\omega(x - y) \right], \tag{12.83}$$

$$A^\dagger = \frac{1}{2\sqrt{m\omega\hbar}} \left[(p_x + p_y) + im\omega(x + y) \right], \tag{12.84}$$

and

$$B^\dagger = \frac{1}{2\sqrt{m\omega\hbar}} \left[(p_x - p_y) + im\omega(x - y) \right], \tag{12.85}$$

where $\omega^2 = (\omega_0^2 - \lambda^2/4)$. These operators satisfy the commutation relations

$$[A, A^\dagger] = [B, B^\dagger] = 1, \tag{12.86}$$

and

$$[A, B] = [A^\dagger, B^\dagger] = 0. \tag{12.87}$$

By standard factorization method we can write the operator form of the Hamiltonian (4.102) as [15] [16]

$$H = \hbar\omega \left(A^\dagger A - B^\dagger B\right) + \frac{i\Gamma}{2}\left(A^\dagger B^\dagger - AB\right), \qquad (12.88)$$

where

$$\Gamma = \hbar\lambda. \qquad (12.89)$$

We note that the first term in (12.88) represents the difference between the two free oscillator Hamiltonians. Denoting the eigenstates of H by $|n_A, n_B\rangle$, where $n_A, n_B = 0, 1, 2 \cdots$ we observe that in the limit of $\Gamma = \hbar\lambda \to 0$ we obtain the harmonic oscillator states provided that $B|n_A, 0\rangle = 0$, i.e. the B oscillator is kept in the ground state.

Next let us define spin-like operators ϕ_0, ϕ_x, ϕ_y and ϕ_z by the relations [15]

$$\phi_0 = \frac{1}{2}\left(A^\dagger A - B^\dagger B\right), \qquad (12.90)$$

$$\phi_x = \frac{1}{2}\left(A^\dagger B^\dagger + AB\right), \qquad (12.91)$$

$$\phi_y = \frac{i}{2}\left(A^\dagger B^\dagger - AB\right), \qquad (12.92)$$

and

$$\phi_z = \frac{1}{2}\left(A^\dagger A + BB^\dagger\right). \qquad (12.93)$$

These ϕ operators satisfy the commutation relations in a similar way to the spin operators

$$[\phi_x, \phi_y] = i\phi_z, \qquad (12.94)$$

$$[\phi_z, \phi_y] = i\phi_x, \qquad (12.95)$$

and

$$[\phi_x, \phi_z] = i\phi_y. \qquad (12.96)$$

Now we write the Hamiltonian (12.88) in terms of these operators:

$$H = 2\hbar\omega\phi_0 + \Gamma\phi_y. \qquad (12.97)$$

This form of H is not diagonal, therefore let us try to express H in terms of ϕ_0 and ϕ_z. For this we observe that just as in the case of spin operators, we have

$$\phi_0^2 = \frac{1}{4} + \phi_z^2 - \left(\phi_x^2 + \phi_y^2\right), \qquad (12.98)$$

where ϕ_0 commutes with ϕ_x, ϕ_y and ϕ_z. Since we have replaced $A, A^\dagger, B$ and $B^\dagger$ by ϕ s, we also label the eigenstates by j and m, i.e. $|jm\rangle$ where

$$j = \frac{1}{2}(n_A - n_B), \quad \text{and} \quad m = \frac{1}{2}(n_A + n_B). \qquad (12.99)$$

Here m is an eigenstate of $(\phi_z - 1/2)$;

$$\phi_0|jm\rangle = j|jm\rangle, \quad \text{and} \quad \phi_z|jm\rangle = \left(m + \frac{1}{2}\right)|jm\rangle. \tag{12.100}$$

In order to find the eigenvalues of H we note that since ϕ_0 and ϕ_y in (12.97) commute with each other, we can find their simultaneous eigenstates. To this end we introduce the operators $\phi_\pm$ which we define by

$$\phi_\pm = \phi_x \mp \phi_z, \tag{12.101}$$

where ϕ_+ and ϕ_- are the raising and the lowering operators respectively. These two operators satisfy the commutation relation

$$[\phi_x, \phi_\pm] = \pm i\phi_\pm. \tag{12.102}$$

Using Baker-Hausdorff relation [17] [18]

$$e^A B e^{-A} = B + [A, B] + \frac{1}{2!}[A, [A, B]] + \frac{1}{3!}[A, [A, [A, B]]] + \cdots, \tag{12.103}$$

we can express ϕ_y in terms of ϕ_z and ϕ_x;

$$\phi_y = \pm i \exp\left[\mp\frac{\pi}{2}\phi_x\right]\phi_z \exp\left[\pm\frac{\pi}{2}\phi_x\right]. \tag{12.104}$$

Now we define $\psi_{jm}^{(\pm)}$ by

$$\psi_{jm}^{(\pm)} = \exp\left[\mp\frac{\pi}{2}\phi_x\right]|jm\rangle, \tag{12.105}$$

then from (12.104) and (12.105) we can show that $\psi_{jm}^{(\pm)}$ is an eigenfunction of ϕ_y;

$$\phi_y\psi_{jm}^{(\pm)} = \pm i\left(m + \frac{1}{2}\right)\psi_{jm}^{(\pm)}. \tag{12.106}$$

Suppose that m_0 is the smallest value that m can take, then

$$\phi_\pm\psi_{jm_0}^{(\pm)} = 0. \tag{12.107}$$

By multiplying (12.107) from the left by $\phi_\pm$ and then substituting for $\phi_\pm\phi_\mp$ from (12.101), for $\phi_+\phi_-$ we find

$$\left(\phi_x^2 - \phi_z^2 + [\phi_x, \phi_z]\right)\psi_{jm_0}^{(\pm)} = \left(\phi_x^2 - \phi_z^2 + i\phi_y\right)\psi_{jm_0}^{(\pm)} = 0, \tag{12.108}$$

and a similar result for $\phi_-\phi_+$. We substitute for $\phi_x^2 - \phi_z^2$ from Eq. (12.98) in (12.108) so that only ϕ_0 and ϕ_y operate on $\psi_{jm_0}^{(\pm)}$,

$$\left(\frac{1}{4} - \phi_0^2 - \phi_y^2 + i\phi_y\right)\psi_{jm_0}^{(\pm)} = 0, \tag{12.109}$$

and then we use the eigenvalue equations (12.100) and (12.106) in (12.109) with the result that $m_0^2 = j^2$, therefore

$$m = |j|, |j| + \frac{1}{2}, |j| + 1, \cdots \tag{12.110}$$

Thus the eigenvalue equation

$$H\psi_{jm}^{(\pm)} = \left(2\hbar\omega\phi_0 + \Gamma\phi_y\right)\psi_{jm}^{(\pm)} = i\hbar\frac{\partial}{\partial t}\psi_{jm}^{(\pm)} \tag{12.111}$$

can be integrated to yield

$$\psi_{jm}^{(\pm)} = \exp\left[-2i\omega jt \pm \frac{\Gamma}{2}(2m+1)t \mp \frac{\pi}{2}\phi_x\right]|jm\rangle. \tag{12.112}$$

As we mentioned earlier we want the B oscillator to be in the ground state, i.e. $n_B = 0$. Imposing this condition on the eigenvalues we have

$$2j = 2m = n_A = n. \tag{12.113}$$

Thus the eigenvalues of the Hamiltonian H are

$$H_n^{(\pm)} = n\hbar\omega \pm \frac{i\Gamma}{2}(n+1). \tag{12.114}$$

12.5 Fermi's Nonlinear Equation for Quantized Radiation Reaction

One of the earliest attempts to include radiative reaction or damping force in the wave mechanics of charged particles was due to Fermi [19]-[22]. If we go back to the classical description we note that for a radiating electron moving in a conservative potential $V(r)$ the equation of motion is

$$m\frac{d^2\mathbf{r}}{dt^2} + \nabla V(r) = m\tau\frac{d^3\mathbf{r}}{dt^3}, \tag{12.115}$$

which is the three-dimensional form of Eq. (2.17) for the potential $V(r)$. Since the radiation reaction force on the right hand side of (12.115) is very small, we can find an iterative solution of (12.115) by writing

$$m\frac{d^2\mathbf{r}^{(0)}}{dt^2} + (\nabla V(r))_{r^{(0)}} = 0, \tag{12.116}$$

and

$$m\frac{d^2\mathbf{r}^{(1)}}{dt^2} + (\nabla V(r))_{r^{(1)}} = \frac{m\tau}{e}\frac{d^3}{dt^3}\left(e\mathbf{r}^{(0)}\right) = \frac{m\tau}{e}\frac{d^3\mathbf{d}(t)}{dt^3}, \tag{12.117}$$

where $\mathbf{d}(t)$ is the classical dipole moment of the electron. The right hand side of (12.117) is only a function of time, therefore we can write it as

$$\frac{m\tau}{e} \frac{d^3\mathbf{d}(t)}{dt^3} = \frac{m\tau}{e}\left[\nabla\left(\frac{d^3\mathbf{d}(t)}{dt^3}\cdot\mathbf{r}\right)\right]_{\mathbf{r}^{(0)}} = -(\nabla W)_{\mathbf{r}^{(0)}}, \qquad (12.118)$$

where

$$W = -\mathbf{r}\cdot\frac{d^3}{dt^3}\left(\frac{m\tau}{e}\mathbf{d}(t)\right). \qquad (12.119)$$

Thus the electron is subject to a conservative potential $V(r)$ and a time-dependent potential

$$\left\{-\left(\frac{m\tau}{e}\right)\left[\mathbf{r}\cdot\frac{d^3\mathbf{d}(t)}{dt^3}\right]\right\}. \qquad (12.120)$$

In wave mechanics we replace the dipole moment by the dipole generated by the charge distribution given in terms of the wave function [6]

$$\rho(\mathbf{r},t) = e\psi^*(\mathbf{r},t)\psi(\mathbf{r},t). \qquad (12.121)$$

Thus we find

$$\mathbf{d}(t) \to \mathbf{d}\left(\psi^*(\mathbf{r},t)\psi(\mathbf{r},t)\right) = \frac{\int \mathbf{r}\rho(\mathbf{r},t)d^3r}{\int \rho(\mathbf{r},t)d^3r}, \qquad (12.122)$$

and the Schrödinger equation for the combined potential will be

$$i\hbar\frac{\partial\psi(\mathbf{r},t)}{\partial t} = H_0\psi(\mathbf{r},t) + W\left(\psi^*\psi\right)\psi(\mathbf{r},t), \qquad (12.123)$$

where

$$H_0 = -\frac{\hbar^2}{2m}\nabla^2 + V(r). \qquad (12.124)$$

While the most general solution of the nonlinear equation (12.123) is not known, for some special cases we can find exact solutions. First we observe that for any eigenstate of the Hamiltonian H_0, the radiation reaction potential W vanishes. Therefore the stationary states of the atom will not be affected by this damping force. We will see that some of the other nonlinear equations describing the dissipation such as Schrödinger-Langevin equation also have this property [23] [24].

For a two level system we can simplify (12.122) by noting that the two levels have energies E_1 and E_2 and these are given by the solution of the Schrödinger equation

$$H\psi_i(\mathbf{r}) = E_i\psi_i(\mathbf{r}), \quad i = 1, 2. \qquad (12.125)$$

The wave function for this two level system is a linear combination of $\psi_1(\mathbf{r})$ and $\psi_2(\mathbf{r})$;

$$\psi(\mathbf{r},t) = C_1(t)\exp\left(-\frac{iE_1t}{\hbar}\right)\psi_1(\mathbf{r}) + C_2(t)\exp\left(-\frac{iE_2t}{\hbar}\right)\psi_2(\mathbf{r}). \qquad (12.126)$$

Now by substituting (12.125) and (12.126) in (12.123) we find the following equations:

$$\frac{dC_1(t)}{dt} = \frac{1}{2} A C_1(t) C_2(t) C_2^*(t), \tag{12.127}$$

and

$$\frac{dC_2(t)}{dt} = -\frac{1}{2} A C_1^*(t) C_1(t) C_2(t), \tag{12.128}$$

where

$$A = \frac{4e^2}{3c^3\hbar} \left(\frac{E_2 - E_1}{\hbar} \right)^3 \left| \int \psi_1^*(\mathbf{r}) \mathbf{r} \psi_2(\mathbf{r}) d^3r \right|^2. \tag{12.129}$$

By changing the independent functions $C_1(t)$ and $C_2(t)$ in (12.127) and (12.128) to $|C_1(t)|^2$ and $|C_2(t)|^2$ we can rewrite these equations as

$$\frac{d|C_1(t)|^2}{dt} = A |C_1(t)|^2 |C_2(t)|^2, \tag{12.130}$$

and

$$\frac{d|C_2(t)|^2}{dt} = -A |C_1(t)|^2 |C_2(t)|^2. \tag{12.131}$$

From these relations it follows that

$$\frac{d|C_1(t)|^2}{dt} + \frac{d|C_2(t)|^2}{dt} = 0. \tag{12.132}$$

Thus we can set

$$|C_1(t)|^2 + |C_2(t)|^2 = 1. \tag{12.133}$$

These coupled first order differential equations can be integrated with the result that [21]

$$|C_1(t)|^2 = \frac{|C_1(0)|^2 e^{At}}{1 + |C_1(0)|^2 (e^{At} - 1)}, \tag{12.134}$$

and

$$|C_2(t)|^2 = \frac{1 - |C_1(0)|^2}{1 + |C_1(0)|^2 (e^{At} - 1)}. \tag{12.135}$$

If at $t = 0$ the electron is in one of these states, say level 2, i.e. if $|C_2(0)|^2 = 1$, then $|C_1(0)|^2 = 0$, and the initial state does not decay. However if we choose $|C_2(0)|^2 = 1 - \epsilon$, where ϵ is a small positive number, then the initial state will decay, and for $At \gg 1$ this decay is exponential, viz,

$$|C_2(t)|^2 \approx \exp(-At). \tag{12.136}$$

The above formulation by Fermi which results in a nonlinear wave equation containing the radiation reaction potential suffers from the wrong behavior for the spontaneous emission. An exact eigenstate ($\epsilon = 0$) cannot radiate spontaneously and when $\epsilon \neq 0$, it takes a long time for the atom to get de-excited [21].

12.6 Attempts to Quantize Systems with a Dissipative Force Quadratic in Velocity

As we have seen earlier (§3.3) the Lagrangian for the one-dimensional motion of a particle of mass m subject to a damping force $\frac{1}{2}m\gamma\dot{x}^2$ and the conservative force $-\frac{dV}{dx}$ is given by

$$L = \frac{1}{2}m\dot{x}^2 e^{\gamma x} - \int^x e^{\gamma y}\frac{dV}{dy}dy. \tag{12.137}$$

The corresponding Hamiltonian for this system is

$$H = \frac{1}{2m}p^2 e^{-\gamma x} + \int^x e^{\gamma y}\frac{dV}{dy}dy, \tag{12.138}$$

where the canonical momentum p is related to the velocity by

$$p = \frac{\partial L}{\partial \dot{x}} = m\dot{x}e^{\gamma x}. \tag{12.139}$$

The classical system is invariant under the time-reversal transformation $t \to -t$ as well as time translation $t \to t + t_0$, but the energy of the system is not conserved.

In quantizing this system just as the case of linear damping we have the violation of the uncertainty principle but we also encounter the additional problem of ordering the factors of x and p appearing in the Hamiltonian H. To illustrate this point let us consider the case where the potential V in (12.138) is zero and the Hamiltonian has just one term:

$$H = \frac{p^2}{2m}\exp(-\gamma x). \tag{12.140}$$

Using the Dirac rule of association [11] [12] [25] and noting that p^3, $e^{-\gamma x}$, $pe^{-\frac{1}{3}\gamma x}p$ and $e^{-\frac{1}{3}\gamma x}pe^{-\frac{1}{3}\gamma x}$ are all Hermitian, we can write

$$\begin{aligned}\hat{H}_1 &= \hat{O}_1\left[\frac{p^2}{2m}e^{-\gamma x}\right] = \frac{1}{6im\gamma\hbar}\left[p^3, e^{-\gamma x}\right] \\ &= \frac{1}{2m}e^{-\gamma x}\left[p^2 + i\hbar\gamma p - \frac{1}{3}\hbar^2\gamma^2\right],\end{aligned} \tag{12.141}$$

where $\hat{O}$ denotes a Hermitian ordering of the argument. But we can order H in other ways. For instance,

$$\begin{aligned}\hat{H}_2 &= \hat{O}_2\left[\frac{p^2}{2m}e^{-\gamma x}\right] = \frac{i}{2m\gamma\hbar}\left[e^{-\frac{1}{3}\gamma x}pe^{-\frac{1}{3}\gamma x}, pe^{-\frac{1}{3}\gamma x}p\right] \\ &= \frac{1}{2m}e^{-\gamma x}\left[p^2 + i\hbar\gamma p - \frac{2}{9}\hbar^2\gamma^2\right]\end{aligned} \tag{12.142}$$

is another Hermitian Hamiltonian operator for the classical H, Eq. (12.140).
From Eqs. (12.141) and (12.142) we find that

$$\hat{H}_2 = \hat{H}_1 + \frac{1}{18m}\hbar^2\gamma^2 e^{-\gamma x}, \qquad (12.143)$$

and hence the commutator of $\hat{H}_2$ and $\hat{H}_1$ is not zero:

$$\left[\hat{H}_2, \hat{H}_1\right] = \left(\frac{1}{18m^2}\right)(\hbar\gamma)^3 e^{-2\gamma x}\left(-\hbar\gamma + 2ip\right). \qquad (12.144)$$

In this way we can construct an infinite number of operators and these, in general, do not commute with each other. However in the classical limit of $\hbar \to 0$ or when $\gamma \to 0$ all of the Hamiltonians constructed by different rules of association commute. As we discussed earlier the classical H is a constant of motion, but in quantum mechanics, if we choose one of the different Hamiltonians to be a constant of motion, the others will not have this essential property [26].

Other ordering rules such as symmetric rule [12]

$$e^{-\gamma x}p^2 \to \frac{1}{2}\left(p^2 e^{-\gamma x} + e^{-\gamma x}p^2\right), \qquad (12.145)$$

or Weyl's rule [27] or Born-Jordan's rule [12] can also be used for constructing Hermitian operators. In fact one can write analogues of these three forms as special cases of the general expression [28]

$$\hat{O}_a\left[p^2 e^{-\gamma x}\right] = \left[ae^{-\gamma x}p^2 + (1 - 2a)pe^{-\gamma x}p + ap^2 e^{-\gamma x}\right]. \qquad (12.146)$$

Denoting any of these Hermitian Hamiltonians by $\hat{H}_a(x, p)$, we have the Heisenberg equations of motion

$$\dot{p} = \frac{1}{i\hbar}\left[p, \hat{H}_a\right] = \gamma\hat{H}_a, \qquad (12.147)$$

and

$$\dot{x} = \frac{1}{i\hbar}\left[x, \hat{H}_a\right] = \frac{1}{2}\left[e^{-\gamma x}p + pe^{-\gamma x}\right]. \qquad (12.148)$$

The expectation value of these operators amounts to what can be termed as Ehrenfest's theorem, and the infinite set of the Hamiltonians constructed by the Dirac's rule of association, or by other rules all satisfy Ehrenfest's theorem. Therefore the classical limit or the correspondence principle will not be helpful in discriminating among the infinite set of the Hamiltonians.

12.7 Solution of the Wave Equation for Linear and Newtonian Damping Forces

Let us start with the assumption that one of the Hermitian Hamiltonians, say $\hat{H}_1$, is a constant of motion. Then the time-dependent Schrödinger equation

$$-\frac{\hbar^2}{2m}e^{-\gamma x}\left[\frac{\partial^2}{\partial x^2} - \gamma\frac{\partial}{\partial x} + \frac{1}{3}\gamma^2\right]\psi(x,t)$$

$$+ \left(\int^x V(y)e^{\gamma y}dy\right)\psi(x,t) = i\hbar\frac{\partial\psi(x,t)}{\partial t} \qquad (12.149)$$

can be separated into space- and time-dependent parts. If we write

$$\psi(x,t) = \psi(x,\mathcal{E})\exp\left(\frac{i\mathcal{E}t}{\hbar}\right), \qquad (12.150)$$

then $\psi(x,\mathcal{E})$ the solution of

$$\frac{-\hbar^2}{2m}e^{-\gamma x}\left[\frac{d^2}{dx^2} - \gamma\frac{d}{dx} + \frac{1}{3}\gamma^2\right]\psi(x,\mathcal{E})$$

$$+ \left(\int^x V(y)e^{\gamma y}dy\right)\psi(x,\mathcal{E}) = \mathcal{E}\psi(x,\mathcal{E}). \qquad (12.151)$$

This equation is exactly solvable for $V(x) = 0$, $V(x) = Ax$ and for $V(x) = e^{-2\gamma x}$ [29], but these solutions are not physically acceptable.

Another possible way of quantizing this type of motion is by the path integral method which essentially yields similar results [30].

We have noted that the classical Hamiltonian found for a damped motion is not the same as the energy of the particle, but it is a first integral of motion. If the damping is proportional to a power of velocity, say $\dot{x}^\nu$, then we can find an energy first integral in the following way:

Let us write the equation of motion as

$$\left(m\ddot{x} + \frac{dV(x)}{dx} + \alpha\dot{x}^\nu\right) = 0, \qquad (12.152)$$

and as in the case of a conservative system, we find the energy first integral by multiplying (12.152) by $\dot{x}$ and in this way we find the rate of change of energy of the particle plus environment;

$$\frac{dE}{dt} = \dot{x}\left(m\ddot{x} + \frac{dV(x)}{dx} + \alpha\dot{x}^\nu\right) = 0. \qquad (12.153)$$

Thus the total energy which is conserved is given by

$$E = \frac{1}{2}m\dot{x}^2 + V(x) + \Delta E, \qquad (12.154)$$

where

$$\Delta E = \alpha \int_0^x \dot{x}^2 dx. \tag{12.155}$$

Denoting the mechanical (kinematical) momentum by $p = m\dot{x}$, we can write Eq. (12.154) as a Hamiltonian

$$H = H_0 + \Delta H = \frac{p^2}{2m} + V(x) + \Delta H, \tag{12.156}$$

where

$$\Delta H = \alpha \int_0^x \dot{x}^\nu dx. \tag{12.157}$$

Next we formally quantize (12.157) by replacing the mechanical momentum p by the operator $-i\hbar\partial/\partial x$, i.e.

$$H = -\frac{\hbar^2}{2m}\frac{\partial^2}{\partial x^2} + V(x) + \alpha \left(-\frac{i\hbar}{m}\right)^\nu \left(\frac{\partial^{\nu-1}}{\partial x^{\nu-1}}\right). \tag{12.158}$$

This Hamiltonian is, in general, complex and non-Hermitian. The time-dependent Schrödinger equation which is obtained from H and is given by

$$i\hbar\frac{\partial\psi(x,t)}{\partial t} = \left[H_0 + \alpha \left(-\frac{i\hbar}{m}\right)^\nu \left(\frac{\partial^{\nu-1}}{\partial x^{\nu-1}}\right)\right]\psi(x,t), \tag{12.159}$$

can be separated by introducing $\psi(x, E)$ by

$$\psi(x,t) = \psi(x,E)\exp\left(-\frac{iEt}{\hbar}\right), \tag{12.160}$$

and substituting $\psi(x,t)$ in (12.159) with the result that

$$-\frac{\hbar^2}{2m}\frac{d^2\psi(x,E)}{dx^2} + [V(x) - E]\psi(x,E) + \alpha \left(\frac{-i\hbar}{m}\right)^\nu \frac{d^{\nu-1}\psi(x,E)}{dx^{\nu-1}} = 0. \tag{12.161}$$

Now let us consider the specific cases of linear and quadratic damping.
   (1) For linear damping $\nu = 1$ and the wave equation is

$$-\frac{\hbar^2}{2m}\frac{d^2\psi(x,E)}{dx^2} + \left[V(x) - \frac{i\alpha\hbar}{m}\right]\psi(x,E) = E\psi(x,E), \tag{12.162}$$

i.e. the potential $\mathcal{V}(x) = V(x) - i\alpha\hbar/m$ is an optical potential with negative imaginary part (Chapter 18).
   From the time-dependent wave equation (12.159) and its complex conjugate with $\nu = 1$ we get

$$\frac{\partial\rho(x,t)}{\partial t} + \frac{\partial}{\partial x}j_x(x,t) = -\frac{2\alpha}{m}\rho(x,t), \tag{12.163}$$

where $\rho(x, t) = |\psi(x, t)|^2$ is the probability density and

$$j_x(x, t) = \frac{-i\hbar}{2m} \left(\psi^* \frac{\partial \psi}{\partial x} - \psi \frac{\partial \psi^*}{\partial x} \right), \tag{12.164}$$

is the current density. By integrating (12.163) over all x and assuming that as $x \to \pm\infty$, $j_x(x, t)$ vanishes we obtain

$$\frac{d}{dt} \int_{-\infty}^{\infty} \rho(x, t) dx = -\frac{2\alpha}{m} \int_{-\infty}^{\infty} \rho(x, t) dx, \tag{12.165}$$

or integrating (12.165) we find

$$\int_{-\infty}^{\infty} \rho(x, t) dx = \exp\left(-\frac{2\alpha t}{m} \right) \left(\int_{-\infty}^{\infty} \rho(x, 0) dx \right). \tag{12.166}$$

Thus in this method of quantization all of the eigenstates including the ground state decay in time. The nonexistence of a stable ground state is a defect of this model.

(2) If we choose $\nu = 2$, then the equation for $\psi(x, E)$ becomes [31]

$$\frac{\hbar^2}{2m} \frac{d^2\psi(x, E)}{dx^2} + \frac{\alpha\hbar^2}{m^2} \frac{d\psi(x, E)}{dx} + (E - V(x))\psi(x, E) = 0. \tag{12.167}$$

Here all of the coefficients in the differential equation are real and in contrast with linear damping one can get normalizable eigenstates. For instance in the case of a harmonic oscillator potential $V(x) = \frac{1}{2} m\omega_0^2 x^2$, we find the solution of (12.167) for odd and even states to be [31]

$$\psi(x, E_j) = N_{2j+1} \exp\left(-\frac{\alpha x}{m} \right) \exp\left(-\frac{m\omega_0}{2\hbar} x^2 \right) H_{2j+1}\left(\sqrt{\frac{m\omega_0}{\hbar}} x \right), \tag{12.168}$$

and

$$\psi(x, E_j) = N_{2j} \exp\left(-\frac{\alpha x}{m} \right) \exp\left(-\frac{m\omega_0}{2\hbar} x^2 \right) H_{2j}\left(\sqrt{\frac{m\omega_0}{\hbar}} x \right), \tag{12.169}$$

where N_j s are the normalization constants.
The eigenvalues corresponding to these wave functions are given by

$$E_j = \left(j + \frac{1}{2} \right) \hbar\omega + \frac{\alpha^2\hbar^2}{2m^3}, \quad j = 0, 1, 2, \cdots. \tag{12.170}$$

We note that (a) all of the eigenfunctions are stable but distorted slightly by the presence of the dissipative force, and (b) the eigenfunctions, apart from a constant shift, are the same as those of the undamped oscillator.

The stability of the bound states in this case, or generally for any confining potential, $V(x)$, can be understood by an examination of the classical motion.

The term $\alpha \dot{x}^2(t)$ in the energy equation shows a reduction in the energy when $\dot{x}(t) > 0$, but for $\dot{x}(t) < 0$ there will be a gain of energy for the particle. The solution of the classical equation

$$\ddot{x} + \omega_0^2 x + \frac{\alpha}{m}\dot{x}^2 = 0 \tag{12.171}$$

can be found exactly [32];

$$t = t_0 + \int_{x_0}^{x} \frac{dx}{\sqrt{Y(x)}}, \tag{12.172}$$

where

$$Y(x) = \omega_0 A^2 \exp\left(-\frac{2\alpha x}{m}\right) + \frac{m^2 \omega_0^2}{2\alpha^2}\left[1 - \frac{2\alpha x}{m} - \exp\left(-\frac{2\alpha x}{m}\right)\right], \tag{12.173}$$

and like the undamped oscillator it is periodic with a constant amplitude. Again we find that this model is unrealistic for the description of a dissipative motion. However if we replace $\frac{\alpha}{m}\dot{x}^2$ in (12.171) by $\frac{\alpha}{m}\dot{x}|\dot{x}|$, then $x(t)$ will exhibit damped oscillations similar to the linear damping.

12.8   The Classical Limit of the Schrödinger Equation with Velocity-Dependent Forces

There are indications, for example from the phase shift analysis, that the interaction between two nucleons at very short distances is velocity-dependent [33] [34]. Similar conclusions can be reached by considering potentials derived from the meson theory of nuclear forces [35]. In low and intermediate energies the two nucleon interaction can be described by the Schrödinger equation

$$-\frac{\hbar^2}{2m}\nabla^2\psi - \frac{\hbar^2}{2m}\nabla \cdot [g(r)\nabla\psi] + V(r)\psi = i\hbar\frac{\partial\psi}{\partial t}, \tag{12.174}$$

where $\nabla \cdot g(r)\nabla$ is a short-range velocity-dependent force and $V(r)$ represents all other (static) forces.

Similar Schrödinger equation with a space-dependent effective mass, is commonly used used in the band theory of solids [36] except that in this case $\left(\frac{m}{m^*}\right)$ is regarded as a tensor, i.e. the first two terms of (12.174) is replaced by

$$-\frac{\hbar^2}{2m}\sum_{i,j}\frac{\partial}{\partial x_i}\left(\frac{m}{m^*(x_k)}\right)_{i,j}\frac{\partial\psi}{\partial x_j}. \tag{12.175}$$

Now let us write (12.174) as

$$-\frac{\hbar^2}{2m}\nabla \cdot \left[\frac{1}{m^*(r)}\nabla\psi\right] + V(r)\psi = i\hbar\frac{\partial\psi}{\partial t}, \tag{12.176}$$

where $m^*(r) = [m/(1 - g(r))]$. To find the classical limit of (12.174) we take $\psi(\mathbf{r}, t)$ to be

$$\psi(\mathbf{r}, t) = \exp\left(\frac{iS(\mathbf{r}, t)}{\hbar}\right). \tag{12.177}$$

Substituting (12.177) in (12.174) and ignoring terms proportional to $\hbar$ we find

$$\frac{1}{2m^*(r)}(\nabla S)^2 + V(r) + \frac{\partial S}{\partial t} = 0, \tag{12.178}$$

which is the same as (5.42) provided that we choose m^* to be given by (5.39).

12.9 Quadratic Damping as an Externally Applied Force

In the case of Stoke's drag force we have seen that how the damping can be assumed to be external and only dependent on time, and formulate the problem accordingly.

For the Newtonian frictional force we can also use a similar technique. Let us study the classical Hamiltonian

$$H = \frac{p^2}{2m} + V(x) + \frac{1}{2}m\gamma\dot{\eta}^2\left[x - \eta(t)\right], \tag{12.179}$$

where x and p are the coordinate and the momentum of the particle respectively, and $\eta(t)$ is the solution of the classical equation

$$m\ddot{\eta} + \frac{1}{2}m\gamma\dot{\eta}^2 + \frac{dV(\eta)}{d\eta} = 0. \tag{12.180}$$

If we compare (12.180) with the equation of motion derived from H, Eq. (12.179), i.e. with

$$m\ddot{x} + \frac{dV(x)}{dx} + \frac{1}{2}m\gamma\dot{\eta}^2 = 0, \tag{12.181}$$

we find that $x = \eta(t)$ is the particular solution of (12.181). The Schrödinger equation for the quantized form of the Hamiltonian (12.179) is

$$-\frac{\hbar^2}{2m}\frac{\partial^2\psi}{\partial x^2} + \left[V(x) + \frac{1}{2}\gamma m\dot{\eta}^2(t)\left(x - \eta(t)\right)\right]\psi = i\hbar\frac{\partial\psi}{\partial t}. \tag{12.182}$$

An examination of this Schrödinger equation shows that it shares certain features with the classical equation of motion, and some of these features are lost in the Hamiltonian formulation. For instance if we consider the special and simple case where $V(x) = 0$, then from (12.180) it is clear that the equation of motion is invariant under the displacement $\eta(t) \to \eta(t) + x_0$, whereas (12.181), with $V = 0$, is not invariant under the translation of the coordinate $x \rightarrow x + x_0$. On the other hand Eq. (12.182), again with $V = 0$, remains unchanged when we replace $\eta(t)$ by $\eta(t) + x_0$ and x by $x + x_0$.

The solution of Eq. (12.182) with $V = 0$ is given by

$$\psi(x,t) = \exp\left\{ \frac{i}{\hbar} \left[m\dot{\eta}(t)(x - \eta(t)) + \frac{1}{2}\int^t m\dot{\eta}^2(t)dt \right] \right\}, \qquad (12.183)$$

which has the correct limit as $\gamma \rightarrow 0$. Since there is no potential energy the energy of the particle is purely kinematical

$$-\frac{\hbar^2}{2m}\frac{\partial^2\psi}{\partial x^2} = \frac{1}{2}m\dot{\eta}^2(t)\psi, \qquad (12.184)$$

and is a decreasing function of time.

We can also solve the wave equation for other potentials. For instance when $V(x) = \frac{1}{2}m\omega_0^2 x^2$, the solution of (12.182) is expressible as

$$\psi(x,t) = \phi_n(x - \eta(t))$$
$$\times \ \exp\left[\frac{i}{\hbar}\left(m\dot{\eta}(x - \eta(t)) - \epsilon_n t + \int^t \frac{1}{2}\left\{ m\dot{\eta}(t)^2 - m\omega_0^2\eta^2(t) \right\} dt \right) \right],$$
$$(12.185)$$

where ϕ_n s are the normalized harmonic oscillator wave functions and

$$\epsilon_n = \left(n + \frac{1}{2} \right)\hbar\omega_0, \quad n = 0,1,2\cdots, \qquad (12.186)$$

are the harmonic oscillator eigenvalues. This result is similar to what we will find for damping linear in velocity §13.4. In both cases the frequency eigenvalues are not affected by damping.

By substituting (12.185) in (12.182), we find the differential equation satisfied by ϕ_n

$$-\left(\frac{\hbar^2}{2m} \right)\frac{d^2\phi_n}{dy^2} + \frac{1}{2}m\omega_0^2 y^2\phi_n = \epsilon_n\phi_n, \qquad (12.187)$$

where we have replaced the coordinate x by the "coherent state" coordinate

$$y = x - \eta(t). \qquad (12.188)$$

We observe that in the absence of damping, $\gamma \rightarrow 0$, both (12.185) and (12.187) reduce to the coherent wave function for the harmonic oscillator.

From the wave function (12.185) we can calculate the mean energy, E_n, for this system when it is in its n-th state

$$\begin{aligned} E_n &= \int_{-\infty}^{\infty} \psi^*(x,t)\left[-\left(\frac{\hbar^2}{2m} \right)\frac{\partial^2}{\partial x^2} + \frac{1}{2}m\omega_0^2 x^2 \right]\psi(x,t)dx \\ &= \epsilon_n + \frac{m}{2}\left(\dot{\eta}^2(t) + \omega_0^2\eta^2(t) \right). \end{aligned} \qquad (12.189)$$

Using the same wave function we can also determine the expectation values of the coordinate and momentum of the particle

$$\langle x \rangle = \eta(t), \quad \text{and} \quad \langle p \rangle = m\dot{\eta}(t). \tag{12.190}$$

12.10 Motion in a Viscous Field of Force Proportional to an Arbitrary Power of Velocity

In §4.10 we have seen that

$$H = p\left[C - \beta(2-a)x\right]^{\frac{1}{(2-a)}} = pf(x), \tag{12.191}$$

is the Hamiltonian for the motion of a particle in a dissipative force, $\beta\dot{x}^\nu$. We can formally "quantize" (12.191) by first symmetrizing H and then constructing the Hermitian operator

$$H = -i\hbar f(x)\frac{\partial}{\partial x} - \frac{i}{2}\hbar\frac{df(x)}{dx}. \tag{12.192}$$

The "wave equation" for this time-independent Hamiltonian is

$$\frac{d\psi}{dx} + \left[\frac{1}{2}\frac{df(x)}{dx} - \frac{i\epsilon}{\hbar}f(x)\right]\psi = 0. \tag{12.193}$$

This first order differential equation can be integrated to yield

$$\psi(x) = \frac{1}{\sqrt{f(x)}}\exp\left[\frac{i\epsilon}{\hbar}\int^x\frac{dx'}{f(x')}\right]. \tag{12.194}$$

Again we observe that this solution is unphysical.

We can also start with the classical Hamiltonian given by Eq. (4.115) for $\dot{x}^\nu$ and write the Hamiltonian operator as

$$H = \frac{m}{2-\nu}\left[i\hbar\left(\frac{\nu-1}{m}\right)\frac{\partial}{\partial x}\right]^{\frac{\nu-2}{\nu-1}} + \alpha x. \tag{12.195}$$

For an arbitrary value of ν this method of canonical quantization is problematic. However for some special cases such as $\nu = \frac{n}{n+1}$, where n is an integer Geicke has shown that the Schrödinger equation leads to the correct classical limit according to the Ehrenfest theorem [37].

By a method similar to what we have seen in the previous section we can formulate a wave equation for a general damping force $F(\eta, \dot{\eta}, t)$ where η is the solution of the classical equation

$$m\ddot{\eta} + F(\eta, \dot{\eta}, t) + \frac{dV(\eta)}{d\eta} = 0. \tag{12.196}$$

Thus starting with the Scrödinger equation

$$-\left(\frac{\hbar^2}{2m}\right)\frac{\partial^2\psi}{\partial x^2} + V(x)\psi + F\left[\eta(t), \dot{\eta}(t), t\right]\left[x - \eta(t)\right]\psi = i\hbar\frac{\partial\psi}{\partial t}, \qquad (12.197)$$

we note that ψ will have the same functional dependence as (12.182). Again when $V(x) = 0$, we have the solution given by (12.183), but now $\eta(t)$ must satisfy the solution of the Newtonian equation of motion (12.196) with $V = 0$.

12.11   The Classical Limit and the Van Vleck Determinant

The formal connection between the classical equation of motion and the time-dependent Schrödinger equation can be established via Ehrenfest's theorem, and this is what we will use to discuss the time-dependence of the expectation values of x and p. But there is another way of relating the wave function and the classical probability of finding the particle in a given part of space. According to Van Vleck [38] in the limit of $\hbar \to 0$, for a system where H is a constant of motion, we have

$$|\psi(x)|^2\, dx \to \left|\frac{\partial H}{\partial p}\right|^{-1} dx = \frac{dx}{|\dot{x}(\mathcal{E}, x)|}, \qquad (12.198)$$

where the last term is the classical probability for finding the particle in the range x and $x + dx$, and $\mathcal{E}$ is the constant value of H.

In the case of a free particle moving in a viscous medium where the drag force is $\frac{1}{2}m\gamma\dot{x}^2$, we have the classical Hamiltonian given by (12.140) and therefore

$$\dot{x} = \sqrt{\frac{2\mathcal{E}}{m}}\exp\left(-\frac{\gamma x}{2}\right). \qquad (12.199)$$

For the dissipative force $\beta\dot{x}^a$ from (12.191) we have

$$\dot{x} = \frac{\partial H}{\partial p} = f(x). \qquad (12.200)$$

We have also the wave function $\psi(x)$ which is given by (12.194). Thus

$$|\psi(x)|^2 = \frac{1}{|f(x)|} = \frac{1}{|\dot{x}(x)|}. \qquad (12.201)$$

The result shows that this wave function, though unphysical, has the correct classical limit.

Bibliography

[1] See J. Mehta and H. Rechenberg *The Historical Development of Quantum Theory*, vol. 1, part 1 (Springer-Verlag, New York, 1982) p. 201.

[2] R.J. Seegert, Proc. Nat. Acad. Sci. (USA) 18, 303 (1932).

[3] M. Born, and P. Jordan, Z. Phys. 34, 873 (1925).

[4] W.E. Brittin, Phys. Rev. 77, 396 (1950).

[5] H. Goldstein, *Classical Mechanics*, Second Edition (Addison-Wesley Reading, 1980).

[6] L.I. Schiff, *Quantum Mechanics*, Third Edition (McGraw-Hill, New York, 1968).

[7] S. Okubo, Phys. Rev. D22, 919 (1980).

[8] S. Okubo Phys. Rev. A 23, 2776 (1981).

[9] C.N. Yang and D. Feldman, Phys. Rev. 79, 972 (1950).

[10] V. Dodonov, V.I. Man'ko and V.D. Skarzhinsky, in *Quantization and Group Methods in Physics*, Edited by A.A. Koma (Nova Science, Commack, 1988).

[11] H.J. Groenewold, Physica 12, 405 (1946).

[12] J.R. Shewell, Am. J. Phys. 27, 16 (1959).

[13] H. Dekker, Z. Phys. B 12, 295 (1975).

[14] H. Dekker, Phys. Rev. A16, 2126 (1977).

[15] H. Dekker, Phys. Rep. 1, 80 (1981).

[16] H. Feshbach and Y. Tikochinsky, Trans. N.Y. Acad. Sci. 38, 44 (1977).

[17] See for instance A.Z. Capri, *Nonrelativistic Quantum Mechanics*, Third Edition (World Scientific, Sigapore, 2002) p. 205.

[18] S. Gasiorowicz, *Quantum Mechanics*, Second Edition, (John Wiley & Sons, New York, 1996) p. 438.

[19] E. Fermi, Rend. Lincei. 5, 795 (1927).

[20] E. Fermi, *Collected Papers*, Vol. I (University of Chicago Press, 1962) p. 271.

[21] K. Wodkiewicz, in *Foundations of Radiation Theory and Quantum Electrodynamics* edited by A.O. Barut (Plenum, New York, 1980) p. 109.

[22] K. Wodkiewicz and M.O. Scully, Phys. Rev. A42, 5111 (1990).

[23] M.D. Kostin J. Chem Phys. 57, 3589 (1972).

[24] M. Razavy, Lett. Nuovo Cimento 24, 293 (1979).

[25] P.A.M. Dirac, *The Principles of Quantum Mechanics*, Fourth Edition (Oxford University Press, London, 1958).

[26] M. Razavy, Phys. Rev. A36, 482 (1987).

[27] H. Weyl, Z. Phys. 46, 1 (1927).

[28] P. Crehan, J. Phys. A22, 811 (1989).

[29] F. Negro and A. Tartaglia, Phys. Rev. A23, 1591 (1981)

[30] C. Stukens and D.H. Kobe Phys. Rev. A34, 3569 (1986).

[31] J. Geicke, J. Phys. A 22, 1017 (1989).

[32] E. Kamke, *Differentialgleichungen, Lösungsmethoden und Lösungen*, (Chelsea, New York, 1948) p. 551.

[33] M. Razavy, G. Field and J.S. Levinger, Phys. Rev. 125, 269 (1962).

[34] M. Lacombe, B. Loiseau, J.M. Richard, R. Vinh Mau, J. Cote', P. Pires and R. de Tourreil, Phys. Rev. C21, 861 (1980).

[35] A.E.S. Green, T. Sawada and D.S. Saxon, *The Nuclear Independent Particle Model*, (Academic Press, New York, 1968).

[36] See for example, W.A. Harrison, *Solid State Theory*, (McGraw-Hill, New York, 1970) p. 143.

[37] J. Geicke, unpublished (2000).

[38] J.H. Van Vleck, Proc. Nat. Acad. Sci. U.S.A. 14, 178 (1928).

Chapter 13

Quantization of Explicitly Time-Dependent Hamiltonian

We have already seen that for a damping force linear in velocity we can derive the equation of motion from the Kanai-Caldirola Hamiltonian, Eq (4.24),

$$H = \frac{p^2}{2m} \exp(-\lambda t) + V(x) \exp(\lambda t). \tag{13.1}$$

Similarly in the case of a radiating electron when the equation of motion is given by (4.54) we find the explicitly time-dependent Hamiltonian (4.57). In this chapter we will study a number of problems that we encounter when we try to quantize these Hamiltonians. Just as the case of time-independent Hamiltonians for nonconservative systems, here we have many wave equations corresponding to the same classical motion.

13.1 Wave Equation for the Kanai-Caldirola Hamiltonian

The Kanai-Caldirola Hamiltonian Eq. (4.24) for the potential $V(x)$, i.e.

$$H = \frac{p^2}{2m} \exp(-\lambda t) + V(x) \exp(\lambda t), \tag{13.2}$$

leads to the correct equation of motion in configuration space but not in phase space (Chapter 4), and that the Hamiltonian H is not the energy function of the system. The operator $\hat{H}$ corresponding to the classical function H is found by using the conventional quantization rule and replacing the canonical momentum p by $-i\hbar\frac{\partial}{\partial x}$,

$$\hat{H} = -\frac{\hbar^2}{2m}\exp(-\lambda t)\frac{\partial^2}{\partial x^2} + V(x)\exp(\lambda t). \tag{13.3}$$

The Schrödinger equation for (13.3) is given by

$$i\hbar\frac{\partial\psi}{\partial t} = -\frac{\hbar^2}{2m}e^{-\lambda t}\frac{\partial^2\psi}{\partial x^2} + V(x)e^{\lambda t}\psi. \tag{13.4}$$

This equation can be separated and solved only for few special forms of $V(x)$, e.g. when $V(x) = 0$ or for the damped harmonic oscillator [1]-[6]. In the latter case the wave equation can be transformed to the Schrödinger equation for the harmonic oscillator, $V(x) = \frac{1}{2}m\omega_0^2 x^2 = \frac{1}{2}m\left(\omega^2 + \frac{\lambda^2}{4}\right)x^2$, if we change the variable x to [1]

$$y = x\exp\left(\frac{\lambda t}{2}\right), \tag{13.5}$$

and thus change (13.3) to

$$i\hbar\left(\frac{\partial\psi}{\partial t} + \frac{\lambda}{2}y\frac{\partial\psi}{\partial y}\right) = -\frac{\hbar^2}{2m}\frac{\partial^2\psi}{\partial y^2} + \frac{1}{2}m\left(\omega^2 + \frac{\lambda^2}{4}\right)y^2\psi. \tag{13.6}$$

Now we write $\psi(y,t)$ as the product of two factors [2]

$$\psi(y,t) = \phi(y)\exp\left(-\frac{im\lambda}{4\hbar}y^2 - i\omega' t\right), \tag{13.7}$$

and substitute it in (13.6) to find the eigenvalue equation for $\phi(y)$,

$$\frac{-\hbar^2}{2m}\frac{d^2\phi}{dy^2} + \frac{1}{2}m\omega^2 y^2\phi = \hbar\Omega\phi, \tag{13.8}$$

where Ω is related to ω' by

$$\omega' = \Omega + \frac{i}{4}\lambda = \left(n + \frac{1}{2}\right)\omega + \frac{i\lambda}{4}, \tag{13.9}$$

and $\phi(y)$ s are the harmonic oscillator wave functions. Substituting for $\phi(y)$, y and ω' in (13.7) we obtain the eigenfunction $\psi_n(x,t)$. The solution of (13.4) for the n-th excited state of the damped harmonic oscillator in terms of the original variable x can be obtained from Eqs. (13.7) and (13.8);

$$\begin{aligned}\psi_n(x,t) &= N_n\exp\left\{\left[-i\omega\left(n + \frac{1}{2}\right) + \frac{\lambda}{4}\right]t - \frac{m}{\hbar}\left(\frac{i\lambda}{4} + \frac{\omega}{2}\right)x^2 e^{\lambda t}\right\} \\ &\quad\times H_n\left[\left(\frac{m\omega}{\hbar}\right)^{\frac{1}{2}}xe^{\frac{\lambda t}{2}}\right],\end{aligned} \tag{13.10}$$

where

$$N_n = \frac{1}{\sqrt{2^n n!}} \left(\frac{m\omega}{\pi\hbar}\right)^{\frac{1}{4}}, \tag{13.11}$$

and H_n is the Hermite polynomial.

The eigenvalues (13.9) found for the damped oscillator are different from those obtained from the dual Hamiltonian formulation of Bateman Eq. (12.113). They also differ from the set of eigenvalues found from the "loss-energy states" which will be defined below [7] [8].

The loss-energy states are special states with certain symmetry property. To obtain these we first we note that the Hamiltonian (13.4) has the property that

$$\hat{H}\left(t + \frac{2\pi i}{\lambda}\right) = \hat{H}(t), \tag{13.12}$$

therefore let us study the class of solutions of the wave equation for the Hamiltonian (13.3) with $V(x) = \frac{1}{2}m\omega_0^2 x^2$ which remains invariant under the transformation

$$\psi_n(x, t + iT) = \exp\left(\varepsilon_n T\right) \psi_n(x, t), \quad \text{Im } T = 0. \tag{13.13}$$

Such states are called "loss-energy states" [7], and these, in some respect, are similar to Bloch states for a pure imaginary one-dimensional lattice. The simplest first integrals of motion are of the type

$$I(t) = \eta(t)x + \xi(t)p, \tag{13.14}$$

and satisfy the relation

$$\left[i\hbar\frac{\partial}{\partial t} - \hat{H}, I\right] = 0. \tag{13.15}$$

From this commutator and the time-dependent Hamiltonian for the damped harmonic oscillator, Eq. (13.3), with $V(x) = \frac{1}{2}m\omega_0^2 x^2$, we find $\dot{\eta}(t)$ and $\dot{\xi}(t)$,

$$\dot{\eta}(t) = m\omega_0^2 e^{\lambda t}\xi(t), \quad \dot{\xi}(t) = -\frac{1}{m}e^{-\lambda t}\eta(t), \tag{13.16}$$

where $\xi(t)$ is a solution of $\ddot{\xi} + \lambda\dot{\xi} + \omega_0^2\xi = 0$ (see Eq. (5.20)). Next we search for the first integrals of motion with the property that

$$I\left(t + \frac{2i\pi}{\lambda}\right) = \alpha I(t), \tag{13.17}$$

where α is a constant. If we choose $\xi(t)$ to be

$$\xi(t) = \exp\left[\left(-\frac{\lambda}{2} + i\omega\right)t\right], \quad \omega = \sqrt{\omega_0^2 - \frac{\lambda^2}{2}}, \tag{13.18}$$

then we find the following two first integrals of motion [7];

$$I_- = e^{i\omega t}\left[e^{-\frac{\lambda t}{2}}p + m\left(\frac{\lambda}{2} - i\omega\right)e^{\frac{\lambda t}{2}}x\right], \tag{13.19}$$

and

$$I_+ = e^{-i\omega t}\left[e^{-\frac{\lambda t}{2}}p + m\left(\frac{\lambda}{2} + i\omega\right)e^{\frac{\lambda t}{2}}x\right],\tag{13.20}$$

with both I_- and I_+ having the time translation invariance

$$I_\pm\left(t + \frac{2\pi i}{\lambda}\right) = -\exp\left(\pm\frac{2\pi\omega}{\lambda}\right)I_\pm(t).\tag{13.21}$$

Using I_+ and I_- we can construct the first integral $K(t)$ with the property

$$K\left(t + \frac{2\pi i}{\lambda}\right) = K(t),\tag{13.22}$$

where this K is related to the Hamiltonian by

$$K(t) = \frac{1}{4m}\left(I_-I_+ + I_+I_-\right) = H(t) + \frac{\lambda}{4}(xp + px).\tag{13.23}$$

The eigenstates of the operator $K(t)$, are the solutions of the eigenvalue equation

$$K(t)\psi_n(x, t) = \hbar\omega\left(n + \frac{1}{2}\right)\psi_n(x, t),\tag{13.24}$$

and are the loss-energy states

$$\begin{aligned}
\psi_n(x, t) &= \frac{1}{\sqrt{2^n n!}}\left(\frac{m\omega}{\pi\hbar}\right)^{\frac{1}{4}}\\
&\times \exp\left[\frac{1}{2}\left(\frac{\lambda}{2} - i\omega\right)t - in\omega t - \frac{m}{2\hbar}\left(\omega + \frac{i\lambda}{2}\right)e^{\lambda t}x^2\right]\\
&\times H_n\left(\sqrt{\frac{m\omega}{\hbar}}xe^{\frac{\lambda t}{2}}\right),\quad n = 0, 1, 2\cdots.
\end{aligned}\tag{13.25}$$

These eigenfunctions have the symmetry property

$$\psi_n\left(x, t + \frac{2\pi i}{\lambda}\right) = \psi_n(x, t)\exp\left[\pi\left(i + \frac{2\omega}{\lambda}\right)\left(n + \frac{1}{2}\right)\right].\tag{13.26}$$

By comparing (13.13) and (13.26) we have, for the loss-energy states, the eigenvalues

$$\varepsilon_n = \hbar\left(\omega + \frac{i\lambda}{2}\right)\left(n + \frac{1}{2}\right),\quad n = 0, 1, 2\cdots.\tag{13.27}$$

and these are similar, but not identical to the eigenvalues of the damped oscillator H^+ of the Bateman dual Hamiltonian, Eq. (12.114).

Let us examine some of the properties of the solution of the wave equation (13.8) when $\psi_n(x, t)$ is given by (13.25). Here we note that (a) the expectation values $\langle n|x|n\rangle$ and $\langle n|p|n\rangle$ are both equal to zero and (b) $\psi_n(x, t)$ is not an eigenfunction of $\hat{H}$ but the expectation value of $\hat{H}$

$$\langle n|\hat{H}|n\rangle = \left(n + \frac{1}{2}\right)\hbar\frac{4\omega^2 + \lambda^2}{4\omega},\tag{13.28}$$

is a constant in time.

From Eq. (13.6) and its complex conjugate we find that [2]

$$i\hbar \left(\frac{\partial}{\partial t} + \frac{\lambda}{2} y \frac{\partial}{\partial y} \right) \psi^* \psi = -\frac{\hbar^2}{2m} \frac{\partial}{\partial y} \left(\psi^* \frac{\partial \psi}{\partial y} - \psi \frac{\partial \psi^*}{\partial y} \right). \tag{13.29}$$

Now we multiply (13.29) by $(i\hbar)^{-1} \exp\left(\frac{\lambda t}{2}\right)$ and we write the result as

$$\frac{\partial}{\partial t} \left(\psi^* \psi e^{-\frac{\lambda t}{2}} \right) + \frac{\partial}{\partial y} \left[\frac{\hbar}{2im} \left(\psi^* \frac{\partial \psi}{\partial y} - \psi \frac{\partial \psi^*}{\partial y} \right) e^{-\frac{\lambda t}{2}} + \frac{\lambda}{2} y \psi^* \psi e^{-\frac{\lambda t}{2}} \right] = 0. \tag{13.30}$$

This corresponds to the equation of continuity if we define the probability density $\rho(y,t)$ and the probability current $j(y,t)$ by

$$\rho(y,t) = \left(\psi^* \psi e^{-\frac{\lambda t}{2}} \right), \tag{13.31}$$

and

$$j(y,t) = \frac{\hbar}{2im} \left(\psi^* \frac{\partial \psi}{\partial y} - \psi \frac{\partial \psi^*}{\partial y} \right) e^{-\frac{\lambda t}{2}} + \frac{\lambda}{2} y \psi^* \psi e^{-\frac{\lambda t}{2}}, \tag{13.32}$$

respectively.

The probability density $\rho(y,t)$ satisfies the normalization condition

$$\int \rho(y,t) dy = \int \psi^*(y,t) \psi(y,t) e^{-\frac{\lambda t}{2}} dy = \int \psi^*(x,t) \psi(x,t) dx = 1. \tag{13.33}$$

An interesting solution of the Schrödinger equation (13.4) is of the form of a wave packet [5]. Let $\xi(t)$ represent the classical solution of the equation of the damped harmonic oscillator (5.20), then the wave packet

$$\begin{aligned} \psi(x,t) &= \left(\frac{m\omega_0 \left(\omega_0 + \frac{i\lambda}{2} \right) e^{\lambda t}}{\pi \omega^2} \right)^{\frac{1}{4}} \exp\left[-m \left(\frac{\omega_0 + i\frac{\lambda}{2}}{2\hbar} \right) e^{\lambda} (x - \xi(t))^2 \right. \\ &\quad + \left. \frac{i}{\hbar} \left\{ m\dot{\xi}(t) e^{\lambda t} (x - \xi(t)) + \frac{1}{2}\hbar\omega_0 t + \int^t L\left(\xi, \dot{\xi}, t\right) dt \right\} \right] \end{aligned} \tag{13.34}$$

is a solution of (13.4). In this equation $L\left(\xi, \dot{\xi}, t\right)$ represents the classical Lagrangian,

$$L = \frac{1}{2} \left(m\dot{\xi}^2 - m\omega_0^2 \xi^2 \right) e^{\lambda t}. \tag{13.35}$$

The probability density for the wave packet (13.34) is given by

$$|\psi(x,t)|^2 = \left(\frac{m\omega_0 e^{\lambda t}}{\pi} \right)^{\frac{1}{2}} \exp\left[-\frac{m\omega_0}{\hbar} e^{\lambda t} (x - \xi(t))^2 \right], \tag{13.36}$$

and thus the center of the wave packet (13.34) oscillates exactly as that of a damped harmonic oscillator, i.e. $\xi(t)$. From the wave packet (13.34) we can also calculate the expectation values of $\langle x \rangle$, $\langle p_{mech} \rangle$, $\langle p_{mech}^2 \rangle$ and $\langle x^2 \rangle$, where p_{mech} denotes the mechanical (or kinematical) momentum. These are given by

$$\langle x \rangle = \langle \psi(x,t)|x|\psi(x,t) \rangle = 0, \tag{13.37}$$

$$\langle p_{mech} \rangle = \langle \psi(x,t)| - i\hbar \frac{\partial}{\partial x} |\psi(x,t) \rangle = 0, \tag{13.38}$$

$$\langle x^2 \rangle = \langle \psi(x,t)|x^2|\psi(x,t) \rangle = \left(\frac{\hbar \omega^2}{2m\omega_0^3} \right) e^{-\lambda t}, \tag{13.39}$$

$$\langle p_{mech}^2 \rangle = \langle \psi(x,t)| - \hbar^2 \frac{\partial^2}{\partial x^2} |\psi(x,t) \rangle = \left(\frac{\hbar \omega^2}{2\omega_0} \right) e^{-\lambda t}. \tag{13.40}$$

The total energy of the wave packet is obtained from the last two relations

$$\frac{1}{2m} \langle p_{mech}^2 \rangle + \frac{1}{2} \omega_0^2 \langle x^2 \rangle = \frac{\hbar \omega^2}{\omega_0} e^{-\lambda t}, \tag{13.41}$$

and this total energy tends to zero as $t \to \infty$. The canonical momentum space wave function corresponding to the wave packet Eq. (13.34) is found from the Fourier transform of $\psi(x,t)$;

$$\phi(p,t) = \left(\frac{m\omega_0}{\pi} e^{\lambda t} \right)^{\frac{1}{4}}$$

$$\times \; \exp \left[- \frac{e^{-\lambda t}}{2m\hbar \left(\omega_0 + i\frac{\lambda}{2} \right)} (p - \eta(t))^2 - \frac{i}{\hbar} \left\{ p\xi(t) - \int^t L dt + \frac{1}{2} \omega_0 t \right\} \right],$$

$$\tag{13.42}$$

where $\eta(t) = m\dot{\xi}(t) e^{\lambda t}$. Again the probability density in momentum space is given by

$$|\phi(p,t)|^2 = \left(\frac{m\omega_0}{\pi} e^{\lambda t} \right)^{\frac{1}{2}} \exp \left[- \frac{m\omega_0}{\hbar} e^{\lambda t} \left(p - m\dot{\xi}(t) e^{\lambda t} \right) \right], \tag{13.43}$$

and from this last relation it follows that the center of this wave packet moves as $m\dot{\xi}(t) e^{\lambda t}$. We note that according to Eqs. (13.39) and (13.40) the expectation values of $\langle p^2 \rangle$ and $\langle x^2 \rangle$ both tend to zero as $t \to \infty$, and the wave function in the limit of $t \to \infty$ collapses to a δ-function. Thus after a long time both the position and the mechanical momentum of the particle can be measured precisely, and in this limit the uncertainty principle will be violated. We can reach this conclusion directly by starting with the commutation relation which is

$$[x, p] = i\hbar. \tag{13.44}$$

Now if we write this commutation relation in terms of the mechanical or kinematical momentum operator $m\dot{x}$, we have

$$[x, m\dot{x} e^{\lambda t}] = i\hbar, \tag{13.45}$$

or

$$[x, \dot{x}] = \frac{i\hbar}{m} e^{-\lambda t}. \tag{13.46}$$

From the commutation relation (13.46) we can derive the uncertainty relation exactly as it is done for conservative systems. Thus if Δx and $\Delta \dot{x}$ represent the uncertainties in the position and the velocity of the particle then

$$\Delta x \Delta \dot{x} \geq \frac{\hbar}{2m} e^{-\lambda t}. \tag{13.47}$$

This expression shows that as $t \to \infty$ one can, in principle, measure the position and the velocity of the particle with arbitrary accuracy, and this is not an acceptable result.

By studying the propagation of the wave packet associated with the motion of the particle Caldirola [9] has shown that the presence of friction reduces the product of the uncertainties $\Delta x \Delta \dot{x}$ shown in (13.47) compared to the corresponding values when $\lambda = 0$, and he has concluded that the Heisenberg uncertainty relation remains valid. Let us note that the problem is not only the violation of the uncertainty principle, but the fact that $H(x, p, t)$ (13.1) is not the energy of the particle and it does not give us the correct equation for the time evolution of the momentum operator.

In addition to the damped harmonic oscillator, analytic solutions can be found for the motion of a particle moving in a viscous medium, but in the absence of a conservative force, or when an external force which is just a function of time t, i.e. $-mg(t)$ acts on the particle. For the first case the wave function in momentum space is given by [5]

$$\phi(p) = \left(\frac{a_0}{\pi}\right)^{\frac{1}{4}}$$

$$\times \quad \exp\left[-\frac{a_0 m\lambda + i\hbar\left(1 - e^{-\lambda t}\right)}{2\hbar^2 m\lambda}\left(p - m\dot{\xi}\right)^2 - \frac{i}{\hbar}\left\{p\xi - \int^t L dt\right\}\right],$$

$$\tag{13.48}$$

where L is the Lagrangian for a particle in viscous medium, a_0 is a constant and $\xi(t)$ is the solution of

$$m\ddot{\xi} + m\lambda\dot{\xi} = 0. \tag{13.49}$$

In this case for $t < \frac{1}{\lambda}$ the width of the wave packet grows as $|a_0 + \frac{i\hbar t}{m}|^{-1}$ which is like the increase in the width when there is no dissipation. However for larger values of t the width of the wave packet becomes constant.

The same wave function $\phi(p)$ Eq. (13.48) can be used when the particle is subject to a constant force $-mg(t)$ with the provision that both L and $\xi(t)$ should be replaced by the corresponding functions for this motion.

13.2   Coherent States of a Damped Oscillator

There are a number of ways of obtaining a wave packet solution for a linear Schrödinger equation describing a damped harmonic oscillator. Among these the coherent state formulation which is closely related to the classical motion is of special interest [8]-[14].

Let us examine the coherent state solution for a one-dimensional wave equation $i\frac{\partial \psi}{\partial t} = \hat{H}\psi$, where

$$\hat{H} = \frac{1}{2}\left[-\frac{\partial^2}{\partial x^2}e^{-\lambda t} + \omega_0^2 x^2 e^{\lambda t}\right] - fe^{\lambda t}x, \qquad (13.50)$$

and where for the sake of simplicity we have used the units where $\hbar = m = 1$. In this expression for $\hat{H}$, f is a constant external force which acts on the particle.

As we have seen earlier the Hamiltonian (13.50) has the following symmetry property expressed by (13.12) [8]. There are also two independent first integrals of motion linear in x and $-i\frac{\partial}{\partial x}$ for this system similar to $I_{\pm}$ that was introduced earlier

$$\hat{I}(t) = a(t)x - i\xi(t)\frac{\partial}{\partial x} + \delta(t), \qquad (13.51)$$

where $\xi(t)$ is a solution of (5.20), and $a(t)$ and $\delta(t)$ will be defined later. Thus $\hat{I}(t)$ satisfies the equation

$$\left[\left(i\frac{\partial}{\partial t} - \hat{H}\right), \hat{I}(t)\right] = 0. \qquad (13.52)$$

From (13.51) and (13.52) it follows that

$$\dot{a}(t) = \omega_0^2 e^{\lambda t}\xi(t), \qquad (13.53)$$

$$\dot{\xi}(t) = e^{-\lambda t}a(t), \qquad (13.54)$$

and

$$\dot{\delta} = -fe^{\lambda t}\xi(t). \qquad (13.55)$$

Thus $\hat{I}(t)$ can be written as

$$\hat{I}(t) = -i\xi(t)\frac{\partial}{\partial x} - \dot{\xi}(t)e^{\lambda t}x - f\int^t e^{\lambda t'}\xi(t')\,dt'. \qquad (13.56)$$

Since Eq. (5.20) admits two linearly independent solutions $\xi_{\pm}(t)$, therefore there are two independent first integrals of motion $\hat{I}_{\pm}(t)$.

We choose the two independent solutions of (5.20) to be

$$\xi_{\pm} = \frac{1}{\sqrt{\omega}}\exp\left[\left(\pm i\omega - \frac{\lambda}{2}\right)t\right], \quad \omega = \sqrt{\omega_0^2 - \frac{\lambda^2}{4}}, \qquad (13.57)$$

so that

$$e^{\lambda t}\left(\dot{\xi}_+\xi_- - \dot{\xi}_-\xi_+\right) = 2i. \tag{13.58}$$

Now we can define two mutually Hermitian conjugate first integrals of motion $\hat{A}(t)$ and $\hat{A}^\dagger(t)$ satisfying the commutation relation

$$\left[\hat{A}(t), \hat{A}^\dagger(t)\right] = 1, \tag{13.59}$$

where $\hat{A}(t)$ and $\hat{A}^\dagger(t)$ are given by

$$\hat{A}(t) = \frac{i}{\sqrt{2}}\left[-i\xi_+(t)\frac{\partial}{\partial x} - \dot{\xi}_+(t)e^{\lambda t}x\right] + \frac{\delta(t)}{\sqrt{2}}, \tag{13.60}$$

and

$$\hat{A}^\dagger(t) = -\frac{i}{\sqrt{2}}\left[i\xi_-(t)\frac{\partial}{\partial x} - \dot{\xi}_-(t)e^{\lambda t}x\right] + \frac{\delta^*(t)}{\sqrt{2}}, \tag{13.61}$$

with

$$\delta(t) = -\frac{if}{\sqrt{\omega}}\frac{\exp\left[\left(i\omega + \frac{\lambda}{2}\right)t\right]}{\left(i\omega + \frac{\lambda}{2}\right)}. \tag{13.62}$$

By writing $\hat{A}(t)$ in the expanded form

$$\begin{aligned}
\hat{A}(t) &= \frac{1}{\sqrt{2\omega}}\left[\exp\left\{\left(i\omega - \frac{\lambda}{2}\right)t\right\}\frac{\partial}{\partial x} - i\left(i\omega - \frac{\lambda}{2}\right)\exp\left\{\left(i\omega + \frac{\lambda}{2}\right)t\right\}x\right] \\
&\quad - \frac{if}{\sqrt{2\omega}}\frac{\exp\left[\left(i\omega + \frac{\lambda}{2}\right)t\right]}{\left(i\omega + \frac{\lambda}{2}\right)},
\end{aligned} \tag{13.63}$$

we observe that $\hat{A}(t)$ and $\hat{A}^\dagger(t)$ satisfy the symmetry properties

$$\hat{A}\left(t + \frac{2i\pi}{\lambda}\right) = -e^{-\frac{2\pi\omega}{\lambda}}\hat{A}(t), \tag{13.64}$$

and

$$\hat{A}^\dagger\left(t + \frac{2i\pi}{\lambda}\right) = -e^{\frac{2\pi\omega}{\lambda}}\hat{A}^\dagger(t). \tag{13.65}$$

We also observe that the eigenfunctions of the operator $\hat{A}^\dagger(t)\hat{A}(t)$ are [15]

$$\begin{aligned}
\psi_n(x,t) &= \left(\frac{1}{2^n n!\sqrt{\pi\hbar\xi_-\xi_+}}\right)^{\frac{1}{2}}\left(\frac{\xi_+}{\sqrt{\xi_-\xi_+}}\right)^{n+\frac{1}{2}} \\
&\quad \times H_n\left(\frac{x}{\sqrt{\hbar\xi_-\xi_+}}\right)\exp\left[\frac{ie^{\lambda t}\dot{\xi}_-x^2}{2\hbar\xi_-}\right].
\end{aligned} \tag{13.66}$$

The coherent state wave functions which we denote by $\psi(\alpha, x, t)$ are the eigenstates of $\hat{A}(t)$, i.e.

$$\hat{A}(t)\psi(\alpha, x, t) = \alpha\psi(\alpha, x, t), \tag{13.67}$$

and at the same time satisfy the equation [8]

$$\hat{A}^\dagger(t)\psi(\alpha,x,t) = \frac{\partial}{\partial\alpha}\left[\exp\left(\frac{|\alpha|^2}{2}\right)\psi(\alpha,x,t)\right]\exp\left(-\frac{|\alpha|^2}{2}\right). \tag{13.68}$$

The wave function $\psi(\alpha,x,t)$ which is the solution of (13.67)-(13.68) is given by

$$
\begin{aligned}
\psi(\alpha,x,t) \;=\;& \left(\frac{\omega}{\pi}\right)^{\frac{1}{4}}\exp\left[-\frac{i}{2}\left(\omega+\frac{i\lambda}{2}\right)t - \frac{1}{2}\left(\omega+\frac{i\lambda}{2}\right)e^{\lambda t}x^2\right.\\
&+ \;\sqrt{2\omega}\,\alpha x e^{\left(\frac{\lambda}{2}-i\omega\right)t} - \frac{1}{2}\alpha^2 e^{-2i\omega t} - \frac{1}{2}|\alpha|^2 - \sqrt{\omega}\,x e^{\left(\frac{\lambda}{2}-i\omega\right)t}\delta(t)\\
&- \;\alpha\sqrt{2\omega}\,e^{\left(\frac{\lambda}{2}-i\omega\right)t}z(t) - \frac{\omega}{2}z^2(t)e^{\lambda t} + i\phi(t)\Big],
\end{aligned}
\tag{13.69}
$$

where $\phi(t)$ and $z(t)$ are defined by

$$\phi(t) = \frac{\omega}{2}\,\mathrm{Re}\int^t e^{-2i\omega t}\delta^2(t)dt, \tag{13.70}$$

and

$$z(t) = -\mathrm{Re}\left[\frac{\delta(t)}{\sqrt{\omega}}e^{-\left(\frac{\lambda}{2}+i\omega\right)t}\right]. \tag{13.71}$$

Returning to the symmetry property of the Hamiltonian expressed by (13.13) we observe that the wave function $u(\alpha,x,t)$ defined by

$$u(\alpha,x,t) = \psi(\alpha,x,t)\exp\left(\frac{1}{2}|\alpha|^2\right) \tag{13.72}$$

has a symmetry property similar to $\hat{H}$

$$u\left(\alpha,x,t+\frac{2\pi i}{\lambda}\right) = u\left(-\alpha e^{\frac{2\pi\omega}{\lambda}},x,t\right)\exp\left(\frac{i\pi}{2}+\frac{\pi\omega}{\lambda}\right). \tag{13.73}$$

An important feature of the wave packet (13.69) is that the product of the widths $\langle x^2\rangle$ and $\langle p^2\rangle$ is independent of time,

$$\langle x^2\rangle = \int_\infty^\infty \psi^* x^2\psi dx = \frac{e^{-\lambda t}}{2\omega}, \tag{13.74}$$

and

$$\langle p^2\rangle = \int_\infty^\infty \psi^*\left(-\frac{\partial^2}{\partial x^2}\right)\psi dx = \frac{\omega_0^2 e^{\lambda t}}{2\omega}, \tag{13.75}$$

therefore

$$\langle x^2\rangle\langle p^2\rangle = \frac{\omega_0^2}{4\omega^2} \geq \frac{1}{4}. \tag{13.76}$$

Thus this solution of the damped harmonic oscillator does not violate the uncertainty principle.

13.3 Squeezed State of a Damped Harmonic Oscillator

In modern optics, in addition to the coherent state which is the minimum uncertainty state, the concept of squeezed state plays an important role [16]. The squeezed state may be regarded as a generalization of the idea of the coherent state. We know that there are fluctuations in conjugate variables (e.g. x and p) of an oscillator or a field and that these fluctuations impose a limit on the accuracy of the measurements of the conjugate variables. Now by reducing the fluctuations of one of the variables and increasing the fluctuations of the other, we can obtain fluctuations less than those in vacuum. Thus for the undamped harmonic oscillator the uncertainty area in the generalized coordinate phase space (x, p) for a coherent state is a circle whereas for a squeezed state is an ellipse. In this section we will study the squeezed state of a damped oscillator. For a review of the squeezed state the reader is referred to the standard books on quantum optics [16].

Consider the creation and annihilation operators $\hat{A}^\dagger(t)$ and $\hat{A}(t)$ defined by Eqs. (13.60)-(13.61) and let us define a new set of operators $\hat{A}^\dagger_{r\theta}(t)$ and $\hat{A}_{r\theta}(t)$ related to $\hat{A}^\dagger(t)$ and $\hat{A}(t)$ by the Bogolyubov transformation [17] [18]

$$\hat{A}_{r\theta}(t) = \mu^* \hat{A}(t) - \nu^* \hat{A}^\dagger(t), \tag{13.77}$$

and

$$\hat{A}^\dagger_{r\theta}(t) = \mu \hat{A}^\dagger(t) - \nu \hat{A}^\dagger(t). \tag{13.78}$$

Here μ and ν are complex numbers, and thus the transformation depends on four parameters. By requiring that this transformation be unitary, i.e.

$$\left[\hat{A}^\dagger_{r\theta}(t), \ \hat{A}_{r\theta}(t) \right] = 1, \tag{13.79}$$

and using the fact that $\left[\hat{A}^\dagger(t), \ \hat{A}(t) \right] = 1$, and the other commutators are zero, we find the following constraint on the parameters μ and ν of the transformation

$$|\mu|^2 - |\nu|^2 = 1. \tag{13.80}$$

Of the four parameters in the transformation, one is an overall phase which can be absorbed in the wave function and the other is determined by the constraint given by (13.80). Thus we are left with a two parameter family of transformation which we write as

$$\mu = \cosh r, \quad \text{and} \quad \nu = e^{i\theta} \sinh r. \tag{13.81}$$

The generator of this unitary transformation is $U(\zeta)$;

$$U(\zeta) = \exp \left[\frac{1}{2} \left(\zeta \hat{A}^{\dagger 2}(t) - \zeta^* \hat{A}^2(t) \right) \right], \tag{13.82}$$

where

$$\zeta = -e^{i\theta} r. \tag{13.83}$$

The action of the operator $U(\zeta)$ is to change $\hat{A}^\dagger(t)$ and $\hat{A}(t)$ to $\hat{A}^\dagger_{r\theta}(t)$ and $\hat{A}_{r\theta}(t)$;

$$\hat{A}_{r\theta}(t) = U(\zeta)\hat{A}(t)U^\dagger(\zeta), \tag{13.84}$$

and

$$\hat{A}^\dagger_{r\theta}(t) = U(\zeta)\hat{A}^\dagger(t)U^\dagger(\zeta). \tag{13.85}$$

Noting that

$$U^\dagger(\zeta) = U^{-1}(\zeta) = U(-\zeta), \tag{13.86}$$

and using Baker-Hausdorff formula (12.103) we have [16]

$$
\begin{aligned}
\hat{A}_{r\theta}(t) &= U(\zeta)\hat{A}(t)U^\dagger(\zeta) \\
&= \hat{A}(t) - \left[\frac{1}{2}\zeta^* \hat{A}^2(t) - \frac{1}{2}\zeta \hat{A}^{\dagger 2}(t) \,,\, \hat{A}(t)\right] + \cdots = \hat{A}(t)\left(1 + \frac{r^2}{2!} + \cdots\right) \\
&\quad + e^{i\theta}\hat{A}^\dagger(t)\left(r + \frac{r^3}{3!} + \cdots\right) = \hat{A}(t)\cosh r + \hat{A}^\dagger(t)e^{i\theta}\sinh r. \tag{13.87}
\end{aligned}
$$

Thus we conclude that $U(\zeta)$ is the unitary operator which transforms $\hat{A}(t)$ to $\hat{A}_{r,\theta}(t)$, and $\hat{A}^\dagger(t)$ to $\hat{A}^\dagger_{r,\theta}(t)$ according to the Bogolyubov transformation Eqs. (13.77)-(13.78).

We can determine the wave function for the squeezed state (r, θ) in a representation where the generalized number operator [19] [20]

$$N_{r,\theta} = \hat{A}^\dagger_{r\theta}(t)\hat{A}_{r\theta}(t), \tag{13.88}$$

is diagonal and has the eigenvalue n [15]

$$N_{r,\theta}\psi(t, r, \theta) = n\psi(t, r, \theta). \tag{13.89}$$

We note that the integer n in this case does not represent the number of quanta, since in the Bogolyubov transformation the eigenvalues of $\hat{A}^\dagger(t)\hat{A}(t)$ are not the same as those of $\hat{A}^\dagger_{r\theta}(t)\hat{A}_{r\theta}(t)$, and the vacuum of the coherent state is different from the vacuum defined by $\hat{A}_{r\theta}(t)|0\rangle = 0|0\rangle$ [18].

It should be pointed out that in this approach the canonical momentum p which satisfies the equation $\ddot{p} - \lambda\dot{p} + \omega_0^2 p = 0$ has been used and that is why (Δp^2) grows in time as $e^{\lambda t}$. But as we have seen in models of Dekker and Ullersma the momentum operator satisfies the time-reversed equation of the previous equation, i.e. $\ddot{p} + \lambda\dot{p} + \omega_0^2 p = 0$. Later we will observe that an approximate solution of the time-dependent squeezed states based on the density matrix formulation gives us exponential decays for x as well as p. Solving this eigenvalue equation in coordinate space with $\hat{A}_{r,\theta}(t)$ and $\hat{A}^\dagger_{r,\theta}(t)$ given by (13.77)-(13.78) and $\hat{A}(t)$ and $\hat{A}^\dagger(t)$ defined by (13.60)-(13.61) and by setting $\delta = 0$, we find the wave function to be

$$\psi(x, t; r, \theta) = \frac{1}{\sqrt{2^n n!}} \left(\frac{\alpha_{r,\theta}}{\sqrt{\pi}} \right)^{\frac{1}{2}} \exp\left[-i\Theta_{r,\theta}(t) \left(n + \frac{1}{2} \right) \right]$$
$$\times \ H_n\left(\alpha_{r,\theta} x \right) \exp\left(-B_{r,\theta} x^2 \right), \tag{13.90}$$

where

$$\alpha_{r,\theta} = \sqrt{\frac{\omega e^{\lambda t}}{\hbar}} \left[\cosh 2r + \sinh 2r \cos(2\omega t + \theta) \right]^{-\frac{1}{2}}, \tag{13.91}$$

$$B_{r,\theta} = \frac{\omega e^{\lambda t}}{2\hbar} \left[\frac{e^{i\omega t} \cosh r - e^{-i(\omega t + \theta)} \sinh r}{e^{i\omega t} \cosh r + e^{-i(\omega t + \theta)} \sinh r} + i\frac{\lambda}{2\omega} \right], \tag{13.92}$$

and

$$\tan \Theta_{r,\theta} = \frac{\sin \omega t - \tanh r \sin(\omega t + \theta)}{\cos \omega t + \tanh r \cos(\omega t + \theta)}. \tag{13.93}$$

In the limit of $r \to 0$, and $\theta \to 0$, this wave function reduces to the one given by Eq. (13.10).

The uncertainty relation for the squeezed states can be obtained from the wave function $\psi(x, t; r, \theta)$, Eq. (13.90);

$$\langle \Delta x \rangle^2 \langle \Delta p \rangle^2 = \int_{-\infty}^{\infty} x^2 |\psi_n(x, t; r, \theta)|^2 \, dx$$
$$\times \int_{\infty}^{\infty} \psi_n(x, t; r, \theta) \left(-\frac{\hbar^2}{2} \frac{\partial^2}{\partial x^2} \right) \psi_n^*(x, t; r, \theta) dx$$
$$= \frac{\hbar^2}{4 \cos^2\left(\frac{\delta_\lambda}{2} \right)} \left(n + \frac{1}{2} \right)^2 \left[\cosh 2r + \sinh 2r \cos(2\omega t + \theta) \right]^2$$
$$\times \ \left[\cosh 2r - \sinh 2r \cos(2\omega t + \theta + \delta_\lambda) \right], \tag{13.94}$$

where

$$\cos \delta_\lambda = \frac{1 - \frac{\lambda^2}{4\omega^2}}{1 + \frac{\lambda^2}{4\omega^2}}. \tag{13.95}$$

13.4 Quantization of a System with Variable Mass

In §4.7 we studied the classical Hamiltonian formulation for the motion of a particle with variable mass, and now we want to consider the possibility of quantizing such a motion. Since there we assumed mc^2 (with $c = 1$) to be the variable mass of the particle (in units of energy) with its conjugate θ, therefore in quantum theory m should be regarded as a dynamical variable i.e. a "mass operator" [21] [22]. If we work in a representation where m is diagonal then θ becomes a differential operator. However there are two points that we must keep in mind:

(1) According to a theorem by Bargman, different mass states in non-relativistic quantum theory cannot be superimposed (super-selection rule), but the theorem is not applicable to this problem [23].

(2) Since we want to have positive values of m, therefore its conjugate, $\theta = i\hbar \frac{\partial}{\partial m}$ will not be a self-adjoint operator in the domain $0 \leq m < \infty$ exactly as in the case of the time and the energy operators [24]-[26].

We write the quantized version of the classical Hamiltonian, Eq. (4.79) as

$$H = -\frac{\hbar^2}{2m}\frac{\partial^2}{\partial x^2} + \frac{1}{2}m\omega_0^2 x^2 + \frac{\lambda\hbar}{i}\left(m\frac{\partial}{\partial m} + \frac{1}{2}\right),\qquad (13.96)$$

where we have used the symmetrized form of the operator $m\theta$;

$$m\theta \rightarrow \frac{1}{2}\left(m\theta + \theta m\right) = i\hbar\left(m\frac{\partial}{\partial m} + \frac{1}{2}\right).\qquad (13.97)$$

The Schrödinger equation for the Hamiltonian (13.96) is given by

$$H(x,m,t)\psi(x,m,t) = i\hbar\frac{\partial\psi(x,m,t)}{\partial t}.\qquad (13.98)$$

In order to separate the variables we introduce a length

$$L(m) = \left(\frac{\hbar}{m\omega_0}\right)^{\frac{1}{2}},\qquad (13.99)$$

and also introduce the dimensionless variable $y = x/L(m)$. Using these we write the wave equation as

$$H\psi = \left\{\frac{\hbar\omega_0}{2}\left(-\frac{\partial^2}{\partial y^2} + y^2 + \frac{\lambda}{i\omega_0}y\frac{\partial}{\partial y}\right) + \frac{\lambda\hbar}{i}\left(m\frac{\partial}{\partial m} + \frac{1}{2}\right)\right\}\psi = i\hbar\frac{\partial\psi}{\partial t}.\qquad (13.100)$$

This differential equation is separable and has a solution of the form

$$\psi(y,m,t) = Y(y)M(m)\exp\left(-\frac{i\mathcal{E}t}{\hbar}\right),\qquad (13.101)$$

where $Y(y)$ and $M(m)$ satisfy the two differential equations

$$\left(-\frac{d^2}{dy^2} + y^2 + \frac{\lambda}{i\omega_0}y\frac{d}{dy}\right)Y(y) = 2LY(y),\qquad (13.102)$$

and

$$\left(m\frac{d}{dm} + \frac{1}{2}\right)M(m) = \frac{i}{2}\beta M(m).\qquad (13.103)$$

In (13.101) $\mathcal{E}$ is related to the separation constant β by

$$\mathcal{E} = \hbar \left(\omega_0 L + \frac{\lambda}{2} \beta \right). \tag{13.104}$$

Equation (13.103) can be integrated to yield the result

$$M = \frac{1}{\sqrt{4\pi m}} \left(\frac{m}{m_0} \right)^{\frac{i\beta}{2}}, \tag{13.105}$$

with m_0 being a constant. We can eliminate the first order derivative in (13.102) by writing

$$Y(y) = \exp \left(-\frac{i\lambda}{2\omega} y \right) U(y), \tag{13.106}$$

and substituting for $Y(y)$ in (13.102) to find the equation for $U(y)$;

$$\left[-\frac{d^2}{dy^2} + \left(\frac{\omega^2}{\omega_0^2} \right) y^2 \right] U(y) = \left(2L + \frac{\lambda}{2\omega_0} \right) U(y), \tag{13.107}$$

where $\omega^2 = \omega_0^2 - \frac{\lambda^2}{4}$. This equation can be solved in terms of the Hermite polynomials:

$$U(u) = N_n H_n(u) \exp \left(-\frac{u^2}{2} \right). \tag{13.108}$$

The new variable u is given by

$$u = \left(\frac{\omega_0}{\omega} \right)^{\frac{-1}{2}} y = \left(\frac{m\omega}{\hbar} \right)^{\frac{1}{2}} x, \tag{13.109}$$

and N_n is the normalization constant. For the eigenvalues of Eq. (13.107) we have

$$2L + \frac{\lambda}{2i\omega_0} = 2 \left(\frac{\omega}{\omega_0} \right) \left(n + \frac{1}{2} \right), \tag{13.110}$$

and thus the eigenvalues are

$$\mathcal{E}(n, \beta) = \left(n + \frac{1}{2} \right) \hbar\omega + \frac{\beta\hbar\lambda}{2} + \frac{i\hbar\lambda}{4}. \tag{13.111}$$

Since β is the separation constant and is arbitrary, the complex solution is found by superimposing different $\psi(x, m, t, \beta)$ s;

$$\begin{aligned}
\psi(x, m, t) &= \sum_n \int \frac{A_n(\beta)}{\sqrt{4\pi m}} \left(\frac{m}{m_0} \right)^{\frac{i\beta}{2}} \exp \left[-\frac{m}{2\hbar} \left(\omega + \frac{i\lambda}{2} \right) x^2 \right] \\
&\quad \times N_n H_n \left[\left(\frac{m\omega}{\hbar} \right)^{\frac{1}{2}} x \right] \exp \left[-\frac{i\mathcal{E}(n, \beta)t}{\hbar} \right] d\beta.
\end{aligned} \tag{13.112}$$

Let $\tilde{A}_n(\xi)$ be the Fourier transform of $A_n(\beta)$, i.e.

$$\tilde{A}_n(\xi) = \frac{1}{\sqrt{2\pi}} \int A_n(\beta) e^{i\beta\xi} d\beta, \tag{13.113}$$

then for a given n we have

$$\frac{1}{\sqrt{m}} \int A_n(\beta) \left(\frac{m}{m_0}\right)^{\frac{i\beta}{2}} \exp\left(-i\frac{\beta\lambda t}{2}\right) d\beta = \frac{\tilde{A}_n}{\sqrt{2\pi m}} \left[\frac{1}{2}\ln\left(\frac{m}{m_0}\right) - \frac{\lambda t}{2}\right]. \tag{13.114}$$

This result shows that the center of each wave packet for fixed n moves about the classical value according to the relation $m = m_0 e^{\lambda t}$.

For the case when $A_n(\beta)$ is constant and is equal to $\frac{1}{\sqrt{m_0}}$ we find

$$\int \frac{L(m)}{\sqrt{mm_0}} dy\,dm \int \exp\left\{i\beta\left[\frac{1}{2}\ln\left(\frac{m}{m_0}\right) - \frac{\lambda t}{2}\right]\right\} d\beta$$

$$= 2\pi \left(\frac{\hbar}{m_0\omega_0}\right)^{\frac{1}{2}} \int \frac{1}{m}\delta\left[\frac{1}{2}\ln\left(\frac{m}{m_0}\right) - \frac{\lambda}{2}\right] dy\,dm$$

$$= 4\pi L(m_0) \int \delta\left(m - m_0 e^{\lambda t}\right) dy\,dm. \tag{13.115}$$

Thus by replacing m by $m_0 e^{\lambda t}$ in (13.112) we find the wave function for the damped oscillator we obtained earlier from the solution of the time-dependent Schrödinger equation, (13.10).

13.5 The Schrödinger-Langevin Equation for Linear Damping

In this section we want to derive a wave equation which describes the motion of a charged particle in the presence of damping while it is moving in an external electromagnetic field.

We start with the general form of the Schrödinger equation for such a motion [27]

$$i\hbar\frac{\partial\psi}{\partial t} = \left[\frac{1}{2m}\sum_{k=1}^{3}\left(-i\hbar\frac{\partial}{\partial x_k} - \frac{e}{c}A_k\right)^2 + V + e\phi + V_L\right]\psi, \tag{13.116}$$

where V_L represents the damping term to be specified later. We know that the classical motion for this particle is given by the equation

$$m\ddot{x}_k = -\frac{\partial V}{\partial x_k} - e\frac{\partial\phi}{\partial x_k} - \frac{e}{c}\frac{\partial A_k}{\partial t} - \lambda\dot{x}_k + \frac{e}{c}\sum_{j=1}^{3}\left(\frac{\partial A_j}{\partial x_k} - \frac{\partial A_k}{\partial x_j}\right)\dot{x}_j, \quad k = 1, 2, 3. \tag{13.117}$$

In the absence of damping, Pauli has shown that (13.116) and (13.117) are related to each other [28]. Here we follow Pauli's method to find the damping potential which we assume to be real, a functional of ψ and ψ^* and furthermore makes (13.116) the Schrödinger equation the quantum analogue of the classical equation (13.117). First we define the current $\mathbf{J}$ with components J_1, J_2 and J_3, where

$$J_k = \frac{1}{2m} \left[\psi^* \left(-i\hbar \frac{\partial}{\partial x_k} - \frac{e}{c} A_k \right) \psi - \psi \left(-i\hbar \frac{\partial}{\partial x_k} + \frac{e}{c} A_k \right) \psi^* \right], \quad (13.118)$$

and then from (13.116) and its complex conjugate we obtain the equation of continuity

$$\frac{\partial}{\partial t} \left(\psi^* \psi \right) + \nabla \cdot \mathbf{J} = 0. \quad (13.119)$$

Next we differentiate J_k with respect to time;

$$m \frac{\partial J_k}{\partial t} = \frac{1}{2} \left[(H\psi)^* \frac{\partial \psi}{\partial x_k} - \psi^* \frac{\partial}{\partial x_k} (H\psi) + H\psi \frac{\partial \psi^*}{\partial x_k} - \psi \frac{\partial}{\partial x_k} (H\psi^*) \right], \quad (13.120)$$

where H is the Hamiltonian used in the Schrödinger equation (13.116). Simplifying (13.120) we get

$$m \frac{\partial J_k}{\partial t} = -\sum_{j=1}^{3} \frac{\partial T_{kj}}{\partial x_j} + \left(-\frac{\partial V}{\partial x_k} - \frac{\partial V_L}{\partial x_k} - \frac{e}{c} \frac{\partial A_k}{\partial t} \right) \psi^* \psi + \frac{e}{c} \sum_{j=1}^{3} B_{kj} J_j, \quad (13.121)$$

where in Eq. (13.121) T_{kj} is the symmetric tensor given by

$$\begin{aligned}
T_{kj} &= \frac{\hbar^2}{4m} \left[\left(-\psi^* \frac{\partial^2 \psi}{\partial x_j \partial x_k} - \psi \frac{\partial^2 \psi^*}{\partial x_j \partial x_k} + \frac{\partial \psi}{\partial x_j} \frac{\partial \psi^*}{\partial x_k} + \frac{\partial \psi}{\partial x_k} \frac{\partial \psi^*}{\partial x_j} \right) \right. \\
&+ \left. \frac{2ie}{\hbar c} \left\{ A_k \left(\psi^* \frac{\partial \psi}{\partial x_j} - \psi \frac{\partial \psi^*}{\partial x_j} \right) + A_j \left(\psi^* \frac{\partial \psi}{\partial x_k} - \psi \frac{\partial \psi^*}{\partial x_k} \right) \right\} \right] \\
&+ \frac{e^2}{mc^2} A_k A_j \psi^* \psi, \quad (13.122)
\end{aligned}$$

and B_{kj} is the antisymmetric tensor

$$B_{jk} = \frac{\partial A_j}{\partial x_k} - \frac{\partial A_k}{\partial x_j}. \quad (13.123)$$

Now we find the time derivative of the expectation value of x_k

$$\frac{d}{dt} \langle x_k \rangle = \frac{d}{dt} \int \psi^* x_k \psi d^3 x = \int \left\{ \frac{\partial \psi^*}{\partial t} x_k \psi + \frac{\partial \psi}{\partial t} x_k \psi^* \right\} d^3 x, \quad (13.124)$$

where $d^3 x$ is the element of volume. Again by substituting for $\frac{\partial \psi^*}{\partial t}$ and $\frac{\partial \psi}{\partial t}$ from the Schrödinger equation we find

$$\frac{d}{dt} \langle x_k \rangle = \int J_k d^3 x, \quad (13.125)$$

and from this we calculate $\frac{d^2\langle x_k \rangle}{dt^2}$

$$\frac{d^2}{dt^2} \langle x_k \rangle = \int \frac{\partial J_k}{\partial t} d^3 x. \tag{13.126}$$

By combining (13.121) and (13.126) we have

$$\frac{d^2}{dt^2} \langle x_k \rangle = \int \left[- \left(\frac{\partial V}{\partial x_k} + \frac{\partial V_L}{\partial x_k} + \frac{e}{c} \frac{\partial A_k}{\partial t} \right) \psi^* \psi + \frac{e}{c} \sum_{j=1}^{3} B_{kj} J_j \right] d^3 x, \tag{13.127}$$

where we have used the divergence theorem,

$$\int \sum_{j=1}^{3} \frac{\partial T_{jk}}{\partial x_j} d^3 x = 0, \tag{13.128}$$

to get the simplified equation (13.127). By comparing Eqs. (13.127) and (13.117) and noting that $\frac{d^2}{dt^2}\langle x_k \rangle$ is given by (13.126), we find that the two are similar in structure if the components of the dissipative force ∇V_L have the following expectation values:

$$\int \psi^* \frac{\partial V_L}{\partial x_k} \psi d^3 x = \lambda \int \left[\frac{-i\hbar}{2} \left(\psi^* \frac{\partial \psi}{\partial x_k} - \psi \frac{\partial \psi^*}{\partial x_k} \right) - \frac{e}{c} A_k \psi^* \psi \right] d^3 x. \tag{13.129}$$

This equation is satisfied for an arbitrary wave function provided that V_L which is a real function is given by

$$V_L = \frac{\lambda \hbar}{2i} \ln \left(\frac{\psi}{\psi^*} \right) - \frac{\lambda e}{c} \sum_{k=1}^{3} \int A_k dx_k. \tag{13.130}$$

The nonlinear differential equation (13.116) with V_L given by (13.130) apart for a time-dependent term which comes from normalization of the wave function is called the Scrödinger-Langevin equation [29]-[31]. This equation was first found by Kostin [29] who used the Heisenberg equation of motion

$$\dot{P} + \lambda P + \omega_0^2 Q = 0, \quad \dot{Q} = P, \tag{13.131}$$

for the position and momentum of a damped harmonic oscillator, Eqs. (10.48)-(10.49) to derive this equation. As we will see the equations of motion for Q and P, Eqs. (13.131), have the same form in classical and in quantum mechanics, the quantal equations were derived for a harmonic oscillator interacting with a bath by Ford *et al.* [32]. Later we will see that we can derive the Schrödinger-Langevin equation directly from the modified form of the Hamilton-Jacobi equation.

The presence of the term

$$\left(-\frac{\lambda e}{c} \sum_k \int A_k dx_k \right), \tag{13.132}$$

in Eq. (13.130) shows that the minimal coupling scheme for the Schrödinger-Langevin equation does not work, i.e. there is an additional term arising from the coupling between the electromagnetic potential and the damping force. If we write this equation as

$$i\hbar\frac{\partial\psi}{\partial t} = \frac{1}{2}\left[-i\hbar\nabla - \frac{e}{c}\mathbf{A}(\mathbf{r},t)\right]^2 \psi + (V(\mathbf{r}) + e\phi(\mathbf{r},t))\,\psi$$
$$+ \frac{\lambda\hbar}{2i}\left[\ln\frac{\psi}{\psi^*}\right]\psi - \left(\frac{\lambda e}{c}\int^{\mathbf{r}}\mathbf{A}(\mathbf{r}',t)\cdot d\mathbf{r}'\right)\psi \qquad (13.133)$$

then we can easily see that (13.133) remains invariant under the usual gauge transformation

$$\mathbf{A} \to \mathbf{A}' = \mathbf{A} + \nabla\chi, \qquad (13.134)$$

$$\phi \to \phi' = \phi - \frac{1}{c}\frac{\partial\chi}{\partial t}, \qquad (13.135)$$

and

$$\psi \to \psi' = \exp\left(\frac{ie\chi}{\hbar c}\right)\psi, \qquad (13.136)$$

where $\chi(\mathbf{r},t)$ is an arbitrary function of $\mathbf{r}$ and t.

In the presence of a magnetic field with the vector potential $\mathbf{A}(\mathbf{r},t)$, the solution of (13.133) is related to the solution in the absence of the field by the relation

$$\psi(\mathbf{r},t) = \psi(\mathbf{r},t,\mathbf{A}=0)\exp\left[\frac{ie}{\hbar c}\int^{\mathbf{r}}\mathbf{A}(\mathbf{r}',t)\cdot d\mathbf{r}'\right]. \qquad (13.137)$$

13.6 An Extension of the Madelung Formulation

Consider the one-dimensional time-dependent Schrödinger equation and let us write the wave function in terms of a variable amplitude $A(x,t)$ and a variable phase function $S(x,t)$, i.e.

$$\psi(x,t) = A(x,t)\exp\left[\frac{iS(x,t)}{\hbar}\right]. \qquad (13.138)$$

Substituting this wave function in the Schrödinger equation and equating the real and imaginary parts we find

$$\frac{\partial S}{\partial t} + \frac{1}{2m}\left(\frac{\partial S}{\partial x}\right)^2 + V = \left(\frac{\hbar^2}{2mA}\right)\frac{\partial^2 A}{\partial x^2}, \qquad (13.139)$$

and

$$m\frac{\partial A}{\partial t} + \left(\frac{\partial A}{\partial x}\right)\left(\frac{\partial S}{\partial x}\right) + \frac{1}{2}A\left(\frac{\partial^2 S}{\partial x^2}\right) = 0. \qquad (13.140)$$

By taking the derivative of (13.139) with respect to x we find the equation of motion of the probability fluid [33]-[35], whereas Eq. (13.140) expresses the equation of continuity (Madelung's hydrodynamical formulation of wave mechanics). Now let us study the following generalization of Eqs. (13.139) and (13.140);

$$\frac{\partial S}{\partial t} \rightarrow \frac{\partial S}{\partial t} + f(x, t), \tag{13.141}$$

and

$$\frac{\partial S}{\partial x} \rightarrow \frac{\partial S}{\partial x} + g(x, t), \tag{13.142}$$

where f and g are functions of x and t to be determined later. With these transformations Eqs. (13.139) and (13.140) are replaced by

$$\frac{\partial S}{\partial t} + f + \frac{1}{2m} \left(\frac{\partial S}{\partial x} + g \right)^2 + V = \left(\frac{\hbar^2}{2mA} \right) \frac{\partial^2 A}{\partial x^2}, \tag{13.143}$$

and

$$m \frac{\partial A}{\partial t} + \left(\frac{\partial A}{\partial x} \right) \left(\frac{\partial S}{\partial x} + g \right) + \frac{1}{2} A \left(\frac{\partial^2 S}{\partial x^2} + \frac{\partial g}{\partial x} \right) = 0. \tag{13.144}$$

If we take the current density to be

$$J = \frac{A^2}{m} \left(\frac{\partial S}{\partial x} + g \right), \tag{13.145}$$

then the continuity equation becomes

$$\frac{\partial A^2}{\partial t} + \frac{\partial J}{\partial x} = 0, \tag{13.146}$$

and this follows from (13.144). We note that these equations reduce to the original forms (13.139) and (13.140) when $g = 0$. The mechanical momentum mv in this case is given by

$$mv = \frac{\partial S}{\partial x} + g. \tag{13.147}$$

Now let us take the classical limit of this generalization of the hydrodynamical model. As $\hbar \rightarrow 0$ we have

$$m \frac{dv}{dt} = -\frac{\partial V}{\partial x} + F_f, \tag{13.148}$$

where

$$F_f = \left(\frac{\partial g}{\partial t} - \frac{\partial f}{\partial x} \right). \tag{13.149}$$

Thus F_f in (13.149) represents an additional force which acts on the particle. From Ehrenfest theorem it follows that the quantum mechanical expectation values satisfy an equation similar to (13.148);

$$m \frac{d}{dt} \langle v \rangle = -\left\langle \frac{\partial V}{\partial x} \right\rangle + \langle F_f \rangle, \tag{13.150}$$

where the symbol $\langle\ \rangle$ denotes the expectation value.

The following three cases are of special interest:

(a) First we choose f and g in such a way that

$$\langle F_f \rangle = -\lambda m \frac{d}{dt}\langle x \rangle. \tag{13.151}$$

Next we derive the classical form of the Hamilton-Jacobi (H-J) equation which is the counterpart of the wave equation with the additional term $\langle F_f \rangle$. In the limit of $\hbar \to 0$, Eqs. (13.143) and (13.144) decouple and (13.143) becomes the H-J equation. For simplicity let us assume that $g = 0$, then we have

$$\frac{\partial S}{\partial t} + f + \frac{1}{2m}\left(\frac{\partial S}{\partial x}\right)^2 + V = 0. \tag{13.152}$$

We note that the only additional term in this equation is $f(x, t)$. Since we have set $g = 0$, therefore from (13.147), (13.149) and (13.151) it follows that

$$\frac{\partial f}{\partial x} = \lambda \frac{\partial S(x, t)}{\partial x}, \tag{13.153}$$

or

$$f(x, t) = \lambda S(x, t) + W(t), \tag{13.154}$$

where $W(t)$ is an arbitrary function of time. Thus the resulting equation is the same as the modified H-J equation found earlier, Eq. (5.5).

(b) For our second case we choose f and g to be:

$$f(x) = \lambda(1 - a)\langle p \rangle (x - \langle x \rangle) - \frac{1}{2}\lambda^2 m a^2 (x - \langle x \rangle)^2, \tag{13.155}$$

and

$$g(x) = \lambda m a(x - \langle x \rangle), \tag{13.156}$$

where p is the momentum operator and a is an arbitrary parameter. Since $\langle g \rangle = 0$, we have $m\frac{d}{dt}\langle x \rangle = \langle p \rangle$, and thus the frictional force in this case is

$$F_f = -\lambda m \frac{d}{dt}\langle x \rangle + \lambda^2 m a^2 (x - \langle x \rangle). \tag{13.157}$$

Let us note that the expectation value of the last term in (13.157) is zero, and therefore (13.157) leads to (13.151).

(c) For the third possibility consider the case where $g = 0$ and $f(x, t) = \lambda(S - \langle S \rangle)$, and thus a force of friction

$$F_f = -\lambda \frac{\partial S}{\partial x} = -\lambda m v, \tag{13.158}$$

acts on the particle. This choice gives us Kostin's nonlinear wave equation (or Schrödinger-Langevin equation) [29] and at the same time $\langle F_f \rangle$ satisfies

(13.151). One can also generate nonlinear damping force proportional to v^n by choosing $g = 0$ and

$$f(x,t) = \gamma_n \int^x \left|\frac{\partial S}{\partial x}\right|^{n-1} \frac{\partial S}{\partial x} dx,$$ (13.159)

for which the damping force is

$$F_f = -\gamma_n |v|^{n-1} v.$$ (13.160)

The extension of the Madelung formulation which leads to the Schrödinger-Langevin equation can also be used to study the stability of the stationary solutions of this equation. To this end we choose $\tilde{\rho}$ as the mass density $\tilde{\rho} = mA^2 = m\psi^*\psi$ and $\mathbf{v} = \frac{1}{m}\nabla S$, and we write the real and imaginary parts of the Schrödinger-Langevin equation as two equations of conservation and momentum balance in hydrodynamics [36] [37]

$$\frac{\partial \tilde{\rho}}{\partial t} + \nabla \cdot (\tilde{\rho}\mathbf{v}) = 0,$$ (13.161)

and

$$\frac{\partial (\tilde{\rho}\mathbf{v})}{\partial t} + \nabla \cdot (\tilde{\rho}\mathbf{v}\mathbf{v} + P_q) = \tilde{\rho}(-\nabla V + \mathbf{f}_L).$$ (13.162)

In these equations $\mathbf{v}\mathbf{v}$ denotes the tensor product of the two vectors [37], and P_q which is the quantum analogue of the pressure tensor is determined from the solution of the equation

$$\nabla \cdot P_q = \frac{\tilde{\rho}}{m}\nabla V_q,$$ (13.163)

where V_q which is called "the quantum potential" [38] is given by

$$V_q = -\frac{\hbar^2}{4m\tilde{\rho}}\left(\nabla^2\tilde{\rho} - \frac{(\nabla\tilde{\rho})^2}{2\tilde{\rho}}\right).$$ (13.164)

The damping force $\mathbf{f}_L = -\nabla V_L$ corresponds to the potential

$$\left[-\frac{\lambda\hbar i}{2}\ln\left(\frac{\psi}{\psi^*}\right)\right],$$ (13.165)

in the Schrödinger-Langevin equation and is equal to $\mathbf{f}_L = -\lambda\mathbf{v}$.

For the stationary solutions of the Schrödinger equation in the absence of damping the pair of variables $(\tilde{\rho}, \mathbf{v})$ will satisfy the condition

$$\mathbf{v}_s = 0,$$ (13.166)

and

$$V_q(\tilde{\rho}_s) + V = E_s = \text{constant}.$$ (13.167)

Now let us study the stability of the stationary solutions of Eqs. (13.161)-(13.162) when these solutions are given by (13.166) and (13.167). We note that

$-\lambda \int \tilde{\rho} \mathbf{v}^2 d^3r \leq 0$ and we assume that both $\tilde{\rho}(\mathbf{r})$ and $\mathbf{v}$ are smooth functions of the coordinates and that the density $\tilde{\rho}(\mathbf{r})$ satisfies the asymptotic condition

$$\lim_{|\mathbf{r}| \to \infty} \tilde{\rho}(\mathbf{r}) e^{|\mathbf{r}|} \to 0. \tag{13.168}$$

For simplicity we choose the units $\hbar = m = 1$, and in this set of units we examine the Liapunov functional [36] [39]

$$Li(\tilde{\rho}, \mathbf{v}) = \int \tilde{\rho} \left[\frac{v^2}{2} + \frac{1}{2} \left(\frac{\nabla \tilde{\rho}}{2\tilde{\rho}} \right)^2 + V(r) - E_s \right] d^3r. \tag{13.169}$$

The variation of $Li(\tilde{\rho}, \mathbf{v})$ can be written as

$$
\begin{aligned}
\delta Li(\tilde{\rho}, \mathbf{v}) &= \int \left[\tilde{\rho} \mathbf{v} \cdot \delta \mathbf{v} + \left(\frac{v^2}{2} + V(r) + V_q - E_s \right) \delta \tilde{\rho} \right] d^3r \\
&+ \oint \frac{1}{4} (\nabla \ln \tilde{\rho}) \delta \tilde{\rho} \cdot d\mathbf{S},
\end{aligned}
\tag{13.170}
$$

where we have used the divergence theorem and have written the last term in (13.170) as a surface integral over a sphere of radius R. As the asymptotic condition (13.168) shows this last term tends to zero as $R \to \infty$. Thus for the stationary solutions according to Eqs. (13.166) and (13.167) $\delta Li(\tilde{\rho}, \mathbf{v})$ must be zero.

If we calculate the time derivative of Li using the equations for $\tilde{\rho}$ and $\mathbf{v}$ we find

$$
\begin{aligned}
\frac{dLi(\tilde{\rho}, \mathbf{v})}{dt} &= -\oint \left[\tilde{\rho} \mathbf{v} \left(\frac{v^2}{2} + V(r) + V_q - E_s \right) + \frac{1}{4} \nabla(\ln \tilde{\rho})(\nabla \cdot \tilde{\rho} \mathbf{v}) \right] \cdot d\mathbf{S} \\
&+ \int \tilde{\rho} \mathbf{v} \cdot \mathbf{f}_d d^3r \leq 0.
\end{aligned}
\tag{13.171}
$$

The inequality in (13.171) follows from the fact that the surface integral in (13.171) vanishes as $R \to \infty$. Thus dLi/dt is a negative semi-definite quantity.

Next we calculate the second variation of $Li(\tilde{\rho}, \mathbf{v})$. By calculating this second variation between $(\delta \tilde{\rho}, \delta \mathbf{v})$ and $(\delta' \tilde{\rho}, \delta' \mathbf{v})$ we find

$$
\int [\delta \mathbf{v}, \delta \tilde{\rho}, \nabla \delta \tilde{\rho}]
\begin{bmatrix}
\tilde{\rho} & \mathbf{v} & 0 \\
\mathbf{v} & \frac{(\nabla \tilde{\rho})^2}{4 \tilde{\rho}^3} & -\frac{\nabla \tilde{\rho}}{4 \tilde{\rho}^2} \\
0 & -\frac{\nabla \tilde{\rho}}{4 \tilde{\rho}^2} & \frac{1}{4 \tilde{\rho}}
\end{bmatrix}
\begin{bmatrix}
\delta' \mathbf{v} \\
\delta' \tilde{\rho} \\
\nabla \delta' \tilde{\rho}
\end{bmatrix}
d^3r.
\tag{13.172}
$$

The minors of the 3×3 matrix in (13.172) are all non-negative for the stationary solution; therefore the linearized system is marginally stable [36].

Let us emphasize that the continuity of $(\tilde{\rho}, \mathbf{v})$ and the asymptotic condition (13.168) are satisfied by the stationary solutions of the harmonic oscillator and Coulomb potentials. In particular the stability condition for the harmonic oscillator has been discussed in detail by Ván and Fülöp [36]. For a damped system the Liapunov functional expresses the difference between the energy of a given state and the energy of the stationary state, and can be expressed in terms of $(\tilde{\rho}, \mathbf{v})$ as we had in (13.169) or in terms of the wave function [36].

13.7   Quantization of a Modified Hamilton-Jacobi Equation for Damped Systems

A different way of arriving at the Schrödinger-Lagevin equation equation is to quantize a modified form of the Hamilton-Jacobi equation [40]-[42]. Here we start with the time-dependent Hamiltonian of Kanai-Cardirola, Eq. (4.24), and we use the minimal coupling rule to account for the interaction of the charged particle with the electromagnetic field. The Hamiltonian in this case is

$$H = \frac{1}{2m} \left[\mathbf{p} - \frac{e}{c}\mathbf{A}(\mathbf{r}, t) \right]^2 e^{-\lambda t} + [V(\mathbf{r}) + e\phi(\mathbf{r}, t)] e^{\lambda t}. \qquad (13.173)$$

The Hamilton-Jacobi equation for this Hamiltonian is given by

$$I_1(\zeta) = \frac{1}{2m} \left[\nabla\zeta - \frac{e}{c}\mathbf{A}(\mathbf{r}, t) \right]^2 + [V(\mathbf{r}) + e\phi(\mathbf{r}, t)] e^{2\lambda t} + e^{\lambda t}\frac{\partial\zeta}{\partial t} = 0. \quad (13.174)$$

The passage from this Hamilton-Jacobi to the wave equation can be achieved by using a method due to Schrödinger [43]. In this method we write the real function $\zeta(\mathbf{r}, t)$ in terms of a complex function $u(\mathbf{r}, t)$ where

$$\zeta(\mathbf{r}, t) = \frac{\hbar}{2i} \ln\left(\frac{u(\mathbf{r}, t)}{u^*(\mathbf{r}, t)} \right), \qquad (13.175)$$

and then find the extremum of the functional

$$uu^* I_1 \left(u, u^* \right), \qquad (13.176)$$

using the variational principle. The functional $I_1(\zeta)$ is given by (13.174) with ζ being replaced by u and u^*, Eq. (13.175). For the present problem we get the wave equation

$$i\hbar\frac{\partial u(\mathbf{r}, t)}{\partial t} = \frac{1}{2m} \left[-i\hbar\nabla - \frac{e}{c}\mathbf{A}(\mathbf{r}, t) \right]^2 e^{-\lambda t} u(\mathbf{r}, t) + [V(\mathbf{r}) + \phi(\mathbf{r}, t)] e^{\lambda t} u(\mathbf{r}, t).$$
$$(13.177)$$

This result can also be obtained by using the standard quantization rule, i.e. by replacing $\mathbf{p}$ by $-i\hbar\nabla$ in the Hamiltonian (13.173). But as we have seen in §(13.1) the rule of association $\mathbf{p} \rightarrow -i\hbar\nabla$ gives us unacceptable result for the velocity-position uncertainty. Therefore let us make the following transformation and replace ζ by S where

$$S(\mathbf{r}, t) = \zeta(\mathbf{r}, t)e^{-\lambda t}, \qquad (13.178)$$

and derive the following equation for the Jacobi function $S(\mathbf{r}, t)$ [41] [42]

$$I_2(S) = \frac{1}{2m} \left[\nabla S - \frac{e}{c}\mathbf{A}(\mathbf{r}, t)e^{-\lambda t} \right]^2 + (V(\mathbf{r}) + e\phi(\mathbf{r}, t)) + \frac{\partial S}{\partial t} + \lambda S = 0. \quad (13.179)$$

This Hamilton-Jacobi equation for a time independent electromagnetic potentials $\phi(\mathbf{r})$ and $\mathbf{A}(\mathbf{r})$ will remain invariant under time translation $t \rightarrow t + t_0$, whereas (13.174) will not.

Again we want to use the Schrödinger method of quantization and for this we express S in terms of ψ and its complex conjugate

$$S = -\frac{i\hbar}{2} \ln\left(\frac{\psi}{\psi^*}\right), \tag{13.180}$$

and try to find the extremum of the functional

$$\psi\psi^* I_2\left(\psi, \psi^*\right) = \frac{\hbar^2}{2m}\left(\nabla + \frac{e\mathbf{A}}{i\hbar c}\right)\psi \cdot \left(\nabla + \frac{e\mathbf{A}}{i\hbar c}\right)\psi^*$$
$$+ \left[V + e\phi + \frac{\lambda\hbar}{2i}\ln\left(\frac{\psi}{\psi^*}\right)\right]\psi\psi^* + \frac{\hbar}{2i}\left[\psi^*\frac{\partial\psi}{\partial t} - \psi\frac{\partial\psi^*}{\partial t}\right], \tag{13.181}$$

subject to the normalization condition

$$\int \psi^*(\mathbf{r}, t)\psi(\mathbf{r}, t)d^3r = 1. \tag{13.182}$$

The normalization of the wave function in the case of a dissipative system should be imposed as a constraint on the variational problem. Thus we try to find the extremum of the functional

$$J\left(\psi, \psi^*\right) = \int \psi^*\psi I_2\left(\psi, \psi^*\right)d^3r\,dt + \eta \int \psi^*\psi\,d^3r\,dt, \tag{13.183}$$

where η is a complex Lagrange's multiplier. The Euler-Lagrange equation for the functional in (13.183) gives us two equations for ψ and ψ^*. For the wave function ψ we have the nonlinear equation

$$i\hbar\frac{\partial\psi(\mathbf{r}, t)}{\partial t} = \frac{1}{2m}\left[-i\hbar\nabla - \frac{e}{c}\mathbf{A}(\mathbf{r}, t)e^{-\lambda t}\right]^2\psi(\mathbf{r}, t)$$
$$+ \frac{\lambda\hbar}{2i}\left[\ln\left(\frac{\psi}{\psi^*}\right)\right]\psi + \left[V(\mathbf{r}) + \phi(\mathbf{r}, t)\right]\psi(\mathbf{r}, t) + \left(\eta - \frac{\lambda\hbar}{2i}\right)\psi, \tag{13.184}$$

and we have the complex conjugate of this equation for ψ^*. If we impose the normalization condition (13.182) and assume the boundary condition

$$\psi(\mathbf{r}, t) \to 0, \quad \text{as} \quad r \to \infty, \tag{13.185}$$

then from (13.184) and its complex conjugate it follows that

$$\eta - \eta^* - \frac{\lambda\hbar}{i} = 0, \quad \text{or} \quad \text{Im } \eta = -\frac{1}{2}\lambda\hbar. \tag{13.186}$$

In addition to the conservative force $-\nabla V(\mathbf{r})$ and the electromagnetic force there can also be a random force $\mathbf{F}(t)$ acting on the particle. This force arises from the interaction with the heat bath and is only a function of time. To include this random force we add a term $-\mathbf{r} \cdot \mathbf{F}(t)$ to the equation (13.184) which now takes the form

$$
\begin{aligned}
i\hbar \frac{\partial \psi(\mathbf{r}, t)}{\partial t} &= \frac{1}{2m} \left[-i\hbar\nabla - \frac{e}{c}\mathbf{A}(\mathbf{r}, t)e^{-\lambda t} \right]^2 \psi(\mathbf{r}, t) \\
&+ \frac{\lambda\hbar}{2i} \left[\ln \left(\frac{\psi}{\psi^*} \right) \right] \psi + [V(\mathbf{r}) + \phi(\mathbf{r}, t) - \mathbf{r} \cdot \mathbf{F}(t)] \psi(\mathbf{r}, t).
\end{aligned}
\tag{13.187}
$$

For this wave equation the law of conservation of probability can be derived the same way as for the ordinary Schrödinger and the result is

$$
\frac{\partial}{\partial} \left(\psi^*\psi \right) + \nabla \cdot \mathbf{J} = 0,
\tag{13.188}
$$

where

$$
\mathbf{J} = \frac{i\hbar}{2m} \left(\psi\nabla\psi^* - \psi^*\nabla\psi \right) - \frac{e}{mc}\psi^* \mathbf{A} e^{-\lambda t}\psi.
\tag{13.189}
$$

From Eq. (13.184) we can deduce that the gauge transformation is

$$
\mathbf{A}' = \mathbf{A} + \nabla\chi(\mathbf{r}, t)e^{-\lambda t}, \quad \phi' = \phi - \frac{1}{c}e^{-\lambda t}\frac{\partial\chi}{\partial t},
\tag{13.190}
$$

and this reduces to the standard form as λ goes to zero. The wave function in this new gauge is related to the wave function in the old gauge by the relation

$$
\psi' = \psi \exp\left[\frac{ie}{\hbar c}\chi(\mathbf{r}, t)e^{-\lambda t} \right].
\tag{13.191}
$$

This derivation of the Schrödinger-Langevin equation from the modified form of the Hamilton-Jacobi equation poses some difficulties. In a very careful analysis of the Schrödinger quantization method applied to the linear dissipative systems Wagner [44] has shown that the choice of a complex Lagrange's multiplier as given in Eq. (13.186) will result in having an equation for ψ^* which is not the same as that of taking the complex conjugate of the differential equation for ψ. This is a serious problem in the above derivation. However Wagner [44] has demonstrated that the same method of quantization can be used for the functional $I(S)$ rather than $I_2(S)$ of Eq. (13.179) where

$$
I(S) = e^{\lambda t}\left[\frac{1}{2m}\left(\nabla S\right)^2 + V(\mathbf{r}) + \frac{\partial S}{\partial t} + \lambda S \right] = 0.
\tag{13.192}
$$

Here for simplicity we have set $\mathbf{A}$ and ϕ in $I(S)$ equal to zero. Now the functional $\psi^*\psi I\left(\psi, \psi^*\right)$ becomes

$$
\begin{aligned}
\psi\psi^* I\left(\psi, \psi^*\right) &= e^{\lambda t}\left[\frac{\hbar^2}{2m}\left(\nabla\psi \cdot \nabla\psi^*\right) + \left\{ V + \frac{\lambda\hbar}{2i}\ln\left(\frac{\psi}{\psi^*}\right) \right\}\psi\psi^* \right] \\
&+ \frac{\hbar}{2i}\left(\psi^*\frac{\partial\psi}{\partial t} - \psi\frac{\partial\psi^*}{\partial t} \right)e^{\lambda t}.
\end{aligned}
\tag{13.193}
$$

The Euler-Lagrange equation for ψ is given by

$$\frac{i\hbar}{2}\left[e^{\lambda t}\frac{\partial\psi}{\partial t} + \frac{\partial}{\partial t}\left(e^{\lambda t}\psi\right)\right] = \left(-\frac{\hbar^2}{2m}\nabla^2\psi + V\psi\right)e^{\lambda t}$$
$$+ \frac{i\hbar\lambda}{2}\psi\left[1 - \ln\left(\frac{\psi}{\psi^*}\right)\right]e^{\lambda t}, \quad (13.194)$$

with its complex conjugate for ψ^*. By multiplying (13.194) by $e^{-\lambda t}$ we find the Schrödinger-Langevin equation (13.187) without the electromagnetic potentials. Thus by introducing the multiplying factor $e^{\lambda t}$ in $I(S)$ we can resolve the difficulty associated with the complex Lagrange's multiplier.

The second problem with the Schrödinger-Langevin equation is the way that the minimal coupling rule modifies this equation. A comparison between (13.133) and (13.187) shows that the magnetic potential appears with a factor $e^{-\lambda t}$ in the latter equation but not in the former, and in neither of these equations the minimal coupling prescription is equivalent to the addition of the Lorentz force to the equation of motion.

Let us examine the conditions under which Eq. (13.133) was derived. We started with the Heisenberg equations of motion for the operator Q and P, Eqs. (13.131), however those equations were obtained for a central harmonic oscillator interacting with a heat bath. Therefore Eqs. (13.131) are valid when the particle is subject to a harmonic force or a random force given by the potential $(-QF(t))$ (in three-dimensional problem by $(-\mathbf{r}\cdot\mathbf{F}(t))$). When we try to include arbitrary conservative and or electromagnetic forces in the Schrödinger-Langevin equation then there is no guarantee that the resulting equation yields acceptable results.

In principle we can add a purely time-dependent potential $U(t)$ in the Schrödinger-Langevin equation and this addition will only affect the phase of the wave function. Let us write (13.187) without the electromagnetic potentials, but with the potential $U(t)$ as

$$i\hbar\frac{\partial\psi}{\partial t} = -\frac{\hbar^2}{2m}\nabla^2\psi + [V(\mathbf{r}) - \mathbf{r}\cdot\mathbf{F} + U(t)]\psi + \frac{\lambda\hbar}{2i}\left[\ln\left(\frac{\psi}{\psi^*}\right)\right]\psi. \quad (13.195)$$

Now we determine $U(t)$ by requiring that the expectation value of the total energy $\langle E(t)\rangle$ be equal to the sum of the expectation values of kinetic and potential energies, i.e.

$$\langle E(t)\rangle = \left\langle\frac{1}{2}m\dot{\mathbf{r}}^2\right\rangle + \langle V(\mathbf{r})\rangle - \langle\mathbf{r}\cdot\mathbf{F}(t)\rangle. \quad (13.196)$$

Next we calculate $\langle E(t)\rangle$ and $\langle\frac{1}{2}m\dot{\mathbf{r}}^2\rangle$,

$$\langle E(t)\rangle = i\hbar\int\psi^*\frac{\partial\psi}{\partial t}d^3r, \quad (13.197)$$

$$\left\langle\frac{1}{2}m\dot{\mathbf{r}}^2\right\rangle = \frac{\hbar^2}{2m}\int(\nabla\psi^*\cdot\nabla\psi)\,d^3r. \quad (13.198)$$

Substituting (13.197) and (13.198) in (13.196) and making use of the wave equation (13.195) we find $U(t)$;

$$U(t) = -\frac{\hbar\lambda}{2i} \int |\psi|^2 \ln\left(\frac{\psi}{\psi^*}\right) d^3r. \qquad (13.199)$$

This completes our derivation of the Schrödinger-Langevin equation from the modified Hamilton-Jacobi equation. The fact that the superposition principle does not hold is a drawback of this nonlinear equation. However the type of logarithmic nonlinearity $-\frac{\lambda\hbar}{2i}\left[\ln\left(\frac{\psi}{\psi^*}\right)\right]\psi$ occurring in the equation has the following important property:

In Chapter 5 we discussed the separability of the Jacobi function $S(\mathbf{r}_1, \mathbf{r}_2, t)$ which can be written as the sum of two terms $S_1(\mathbf{r}_1, t)$ and $S_2(\mathbf{r}_2, t)$, Eq. (5.37). For any linear wave equation, derived from Kanai-Caldirola Hamiltonian, this separability is well-known. However in the case of a nonlinear wave equation such as Schrödinger-Langevin equation this property must be verified. In the case of a nonlinear wave equation the lack of any correlation between the motion of the two particles implies that the total wave function must factorize;

$$\psi_{12}(\mathbf{r}_1, \mathbf{r}_2, t) = \psi_1(\mathbf{r}_1, t)\psi_2(\mathbf{r}_2, t). \qquad (13.200)$$

To show this we write the Schrödinger-Langevin equation for two noninteracting particles moving in a dissipative environment

$$-\frac{\hbar^2}{2}\left(\frac{1}{m_1}\nabla_1^2 + \frac{1}{m_2}\nabla_2^2\right)\psi_{12} + [V_1(\mathbf{r}_1) + V_2(\mathbf{r}_2)]\psi_{12}$$
$$+\frac{\lambda\hbar}{2i}\left[\ln\left(\frac{\psi_{12}}{\psi_{12}^*}\right)\right]\psi_{12} = i\hbar\frac{\partial\psi_{12}}{\partial t}, \qquad (13.201)$$

then we observe that the solution of this equation with logarithmic nonlinearity can be factorized as a product $\psi_{12} = \psi_1(\mathbf{r}_1, t)\psi_{12}(\mathbf{r}_2, t)$ where each factor satisfies the nonlinear equation

$$-\frac{\hbar^2}{2m_i}\nabla_i^2\psi_i + V_i(\mathbf{r}_i)\psi_i + \frac{\lambda\hbar}{2i}\left[\ln\left(\frac{\psi_i}{\psi_i^*}\right)\right]\psi_i = i\hbar\frac{\partial}{\partial t}\psi_i, \quad i = 1, 2. \qquad (13.202)$$

13.8   Exactly Solvable Cases of the Schrödinger -Langevin Equation

Having discussed some of the deficiencies of the nonlinear Schrödinger-Langevin equation we now want to study few solvable examples. But before considering specific cases, we observe that because of the nonlinearity of this equation, and the inapplicability of the superposition principle, we cannot find the general solution and also we cannot ascertain the uniqueness of the solution that we have found. For the linear partial differential equations the method of separation of variables is an important technique. In the present case we note that the time-dependent part can be factored out if we assume that the wave function is a product of the form $\psi(\mathbf{r}, t) = \phi(\mathbf{r})T(t)$. Thus if we write

$$\psi(\mathbf{r}, t) = \phi(\mathbf{r}) \exp\left[\frac{i\mathcal{E}}{\lambda\hbar}\left(e^{-\lambda t} - 1\right) - \frac{i}{\hbar}e^{-\lambda t}\int_0^t e^{\lambda t'} U\left(t'\right) dt'\right], \quad (13.203)$$

and substitute $\psi(\mathbf{r}, t)$ in the Schrödinger-Langevin equation

$$i\hbar\frac{\partial \psi}{\partial t} = -\frac{\hbar^2}{2m}\nabla^2\psi + [V(\mathbf{r}) + U(t)]\,\psi + \frac{\lambda\hbar}{2i}\left[\ln\left(\frac{\psi}{\psi^*}\right)\right]\psi, \quad (13.204)$$

we find that $\phi(\mathbf{r})$ satisfies the nonlinear partial differential equation

$$-\frac{\hbar^2}{2m}\nabla^2\phi + V(\mathbf{r})\phi + \frac{\lambda\hbar}{2i}\left[\ln\left(\frac{\phi}{\phi^*}\right)\right]\phi = \mathcal{E}\phi. \quad (13.205)$$

The factors in (13.203) are chosen in such a way that in the limit $\lambda \to 0$, the solution goes over to that of the Schrödinger equation. However a separable solution of the type given by (13.203) may not be an interesting one.

Now let us consider the following two cases where the wave function can be obtained analytically:

(1) The motion of a particle in a viscous medium subject to the action of the random force $\mathbf{F}(t)$, but in the absence of the conservative force $(-\nabla V)$. We can solve the resulting equation by writing

$$\psi(\mathbf{r}, t) = \exp\left[\frac{iS(\mathbf{r}, t)}{\hbar}\right]\exp\left[-\frac{i}{\hbar}e^{-\lambda t}\int_0^t e^{\lambda t'} U\left(t'\right) dt'\right], \quad (13.206)$$

and substituting it in (13.195) to find the following equation for S

$$\frac{\partial S}{\partial t} = \frac{1}{2m}\left[i\hbar\nabla^2 S - (\nabla S)^2\right] + \mathbf{r} \cdot \mathbf{F}(t) - \frac{\lambda}{2}\left(S + S^*\right). \quad (13.207)$$

In the classical limit of $\hbar \to 0$ and with the assumption that S is real, this equation becomes the Hamilton-Jacobi equation for the time-dependent force $\mathbf{F}(t)$

$$\frac{\partial S}{\partial t} = -\frac{1}{2m}(\nabla S)^2 + \mathbf{r} \cdot \mathbf{F}(t) - \lambda S, \quad (13.208)$$

which can be solved to yield

$$S = \mathbf{r} \cdot \mathbf{p}(t) + \theta(t), \quad (13.209)$$

where $\mathbf{p}(t)$ and $\theta(t)$ are given by

$$\mathbf{p}(t) = \mathbf{p}_0 e^{-\lambda t} + e^{-\lambda t}\int_0^t \mathbf{F}\left(t'\right) e^{\lambda t'} dt', \quad (13.210)$$

and

$$\theta(t) = \theta(0)e^{-\lambda t} - \frac{1}{2m} \int_0^t |\mathbf{p}\,(t')|^2 \, e^{\lambda t'} \, dt'. \tag{13.211}$$

Since S is a linear function of $\mathbf{r}$, therefore $\nabla^2 S = 0$ and hence S is an exact solution of (13.207). Equation (13.210) shows how the momentum of the particle changes in time. We also note that $\mathbf{p}(t)$ is an eigenvalue of

$$-i\hbar\nabla\psi(\mathbf{r}, t) = \mathbf{p}(t)\psi(\mathbf{r}, t), \tag{13.212}$$

where $\psi(\mathbf{r}, t)$ is given by (13.206).

We can also find a solution for the Schrödinger-Langevin equation in the viscous medium when $V = 0$ which is the analogue of the spreading of a Gaussian wave packet [45].

Let us consider the one-dimensional motion of the wave packet

$$\Psi(x, t) = N e^{ik(t)x} \exp\left[-\frac{(x - \xi(t))^2}{2a}\right], \tag{13.213}$$

where the time-dependent wave number $k(t)$ is given by

$$k(t) = k_0 e^{-\lambda t}, \tag{13.214}$$

and $\xi(t)$ is a solution of $\ddot{\xi}(t) + \lambda\dot{\xi}(t) = 0$, i.e.

$$\xi(t) = \xi_0 + \frac{\hbar k_0}{m\lambda}\left(1 - e^{-\lambda t}\right). \tag{13.215}$$

When $\lambda \to 0$, the complex number $a = a_R + ia_I$ is given by

$$a = 2\left(\Delta x_0\right)^2 + \frac{i\hbar t}{m}, \tag{13.216}$$

with Δx_0 being the initial width of the wave packet. Now if we substitute (13.213) in the Schrödinger-Langevin equation we obtain two coupled first order differential equations for a_R and a_I;

$$\frac{da_R}{dt} = \frac{2\lambda a_R a_I^2}{|a|^2}, \tag{13.217}$$

and

$$\frac{da_I}{dt} = \frac{\hbar}{m} - \lambda a_I \frac{\left(a_R^2 - a_I^2\right)}{|a|^2}. \tag{13.218}$$

The asymptotic solutions of these equations for large t are:

$$a_I \to \frac{\hbar}{m\lambda}, \quad |a| \approx a_R \to \frac{2\hbar}{m}\sqrt{\frac{t}{\lambda}}. \tag{13.219}$$

Thus in a viscous medium the width of the wave packet grows as $t^{\frac{1}{2}}$, whereas for $\lambda = 0$, according to (13.216) the width grows linearly as a function of time. This result can be interpreted in the following way: Because the velocity component at the leading edge of the packet is greater than the packet's group velocity, the wave packet spreads. Now in the viscous medium the leading edge experiences a larger force of friction than the center of the packet. A larger viscous force on the leading edge and a smaller one at the trailing edge causes a retardation in the spreading of the wave packet and hence a slower rate for increase in $|a(t)|$. (2) The Schrödinger-Langevin equation for the problem of damped harmonic oscillator can be written as

$$i\hbar \frac{\partial \psi(x,t)}{\partial t} = -\frac{\hbar^2}{2m} \frac{\partial^2 \psi(x,t)}{\partial x^2} + \frac{1}{2} m\omega_0^2 x^2 \psi(x,t) + \frac{\lambda \hbar}{2i} \left(\ln \frac{\psi}{\psi^*} \right) \psi(x,t). \quad (13.220)$$

As we have noted before, the classical action S_J, Eq. (5.19), is given by a complex quadratic function of x which depends on $(x - \xi(t))$ as well as $mx\dot{\xi}(t)$. Let us write the wave function as

$$\psi_n(x,t) = \phi_n[x - \xi(t)] \exp \left\{ -\frac{iE_n}{\lambda \hbar} + i\hbar \left[mx\dot{\xi}(t) + C(t) \right] \right\}. \quad (13.221)$$

In this relation $\phi_n(x)$ is the harmonic oscillator wave function

$$\phi_n(x) = \left(\frac{m\omega_0}{\hbar \pi} \right)^{\frac{1}{4}} \frac{1}{\sqrt{2^n n!}} H_n \left[\left(\frac{m\omega_0}{\hbar} \right)^{\frac{1}{2}} x \right] \exp \left(-\frac{m\omega_0}{2\hbar} x^2 \right), \quad (13.222)$$

and E_n the eigenvalue of $\phi_n(x)$,

$$E_n = \hbar \omega_0 \left(n + \frac{1}{2} \right). \quad (13.223)$$

The time-dependent terms $\xi(t)$ and $C(t)$ are given by (5.20) and (5.21) respectively [42] [46] [47].

13.9   Harmonically Bound Radiating Electron and the Schrödinger-Langevin Equation

We have seen (§2.3) that the equation of motion for the non-relativistic radiating electron is given by Eq. (2.17). Loinger [48] has shown that the quantal position operator of the electron, X, satisfies an equation which is formally analogous to the classical Dirac-Lorentz equation, i.e.

$$\frac{d^2 X}{dt^2} + \nu^2 X - \tau \left(\frac{d^3 X}{dt^3} \right) = \frac{1}{m} F(t), \quad (13.224)$$

where $F(t)$ is the external force, m is the mass of the electron and τ is the constant defined by equation (2.18). As in the classical problem we transform (13.224) by the Bopp [2] transformation (Chapter 2), i.e.

$$x = \frac{(a\tau)^{\frac{1}{2}}}{\nu} \left(\dot{X} - \frac{1}{a}\ddot{X} \right), \tag{13.225}$$

to find a quadratic equation for the operator x:

$$\ddot{x} + \lambda\dot{x} + \omega_0^2 x = \frac{1}{\sqrt{a\tau}m\nu} \frac{dF}{dt}, \tag{13.226}$$

where λ is the positive root of (2.30) and ω_0^2 is given by

$$\omega_0^2 = \frac{\nu^2}{a\nu}. \tag{13.227}$$

Let us emphasize the fact that x in (13.225) is not simply the position operator, but a generalized coordinate which is a linear combination of velocity and acceleration of the particle. If we assume that $m[x, \dot{x}] = i\hbar$ then as we have seen earlier a Heisenberg-Langevin equation like (13.226) is related to the Schrödinger-Langevin equation

$$i\hbar \frac{\partial\psi}{\partial t} = -\frac{\hbar^2}{2m} \frac{\partial^2\psi}{\partial x^2} + \frac{1}{2}m\omega_0^2 x^2\psi + \frac{\lambda\hbar}{2i}\left(\ln \frac{\psi}{\psi^*} \right)\psi - \frac{x\dot{F}(t)}{m\nu\sqrt{a\tau}}. \tag{13.228}$$

A solution of (13.228) which is of the form of a wave packet is given by [50] [51]

$$\psi_n(x,t) = \phi_n(x - \xi(t)) \exp\left[-\frac{iE_n}{\hbar\lambda} + \frac{i}{\hbar}\left(mx\dot{\xi}(t) + C(t) \right) \right], \tag{13.229}$$

where in this case $C(t)$ is given by (5.21) and $\xi(t)$ is the solution of the inhomogeneous differential equation

$$\ddot{\xi}(t) + \lambda\xi(t) + \omega_0^2\xi(t) = \frac{\dot{F}(t)}{m\nu\sqrt{a\tau}}. \tag{13.230}$$

A direct way of quantizing the classical problem of radiating electron can be formulated in the following way:
We start by constructing a Hamiltonian of Ostrogradsky type. Since the equation of motion is of the third order, therefore the corresponding phase space is six-dimensional. By introducing a boson, an anti-boson and an additional "ghost" operator according to the Feshbach and Tikochinsky method of §12.4 we can determine the eigenvalues of the harmonically bound electron [4].

We can also use the fourth order equation, (4.54), for a radiating electron and write the Schrödinger equation for the Hermitian Hamiltonian operator obtained from (4.57) which we write as

$$i\hbar \frac{\partial}{\partial t}\left(e^{iS}\psi \right) = \hat{H}\left(e^{iS}\psi \right), \tag{13.231}$$

or

$$
i\hbar \frac{\partial \psi}{\partial t} = \exp(-iS) \sum_i \left(-i\hbar \frac{\partial}{\partial q_i} Q_i \right) \psi - \frac{8}{m\tau^2} \exp(-iS) \exp\left(\frac{4t}{\tau} \right)
$$
$$
\times \sum_i \left(-\hbar^2 \frac{\partial^2 \psi}{\partial Q_i^2} \right) + \exp(-iS) \exp\left(-\frac{4t}{\tau} \right) V(q,t)\psi + \hbar \frac{\partial S}{\partial t} \psi,
$$

$$(13.232)$$

where S is the Hermitian operator

$$
S = -\frac{it}{\tau} \sum_i \left(q_i \frac{\partial}{\partial q_i} + \frac{\partial}{\partial q_i} q_i + Q_i \frac{\partial}{\partial Q_i} + \frac{\partial}{\partial Q_i} Q_i \right). \tag{13.233}
$$

We can find the continuity equation for the wave equation (13.232) by defining

$$
\rho(q_i, t) = \int |\psi(q_i, Q_i, t)|^2 \, dQ_1 dQ_2 dQ_3, \tag{13.234}
$$

and

$$
j_k(q_i, t) = \int |\psi(q_i, Q_i, t)|^2 \, Q_k dQ_1 dQ_2 dQ_3, \tag{13.235}
$$

as the probability and current density respectively. Then from (13.232) and its complex conjugate we obtain

$$
\frac{\partial \rho}{\partial t} + \nabla_{\mathbf{q}} \cdot \mathbf{j} = -\frac{8i\hbar}{m\tau^2} \exp\left(\frac{4t}{\tau} \right) \int_{S(Q_i)} (\psi^* \nabla_{\mathbf{Q}} \psi - \psi \nabla_{\mathbf{Q}} \psi^*) \cdot d\mathbf{S}(Q_i), \tag{13.236}
$$

where S is a closed surface in Q space. By requiring that ψ and $\nabla \psi$ vanish at the boundary of Q space, Eq. (13.236) reduces to the standard form of the continuity equation.

The conditions that ψ and $\nabla \psi$ tend to zero as $Q \to \infty$ means that the probability of finding a particle with infinite velocity is zero [49]

13.10   Other Phenomenological Nonlinear Potentials for Dissipative Systems

In addition to the Schrödinger-Langevin equation that we obtained from the quantization of a classical damped systems, we can find other equations in a similar way [5].

Let us start with the classical equations of motion with a damping term proportional to the momentum of the particle,

$$
\frac{dP}{dt} + \lambda P + \frac{dV(x)}{dx} = 0, \quad P = m\frac{dX}{dt}, \tag{13.237}
$$

where P is the mechanical momentum of the particle. The total energy of the particle, E, which is given by $E = \frac{P^2}{2m} + V(x)$ decreases in time, and from (13.237) we find the rate of energy loss to be

$$\frac{dE}{dt} = -\frac{\lambda}{m}P^2. \tag{13.238}$$

The quantal equations corresponding to (13.237) according to Ehrenfest's theorem are

$$\left\langle \frac{dp}{dt} \right\rangle + \lambda \langle p \rangle + \left\langle \frac{dV(x)}{dx} \right\rangle = 0, \quad \langle p \rangle = m \left\langle \frac{dx}{dt} \right\rangle, \tag{13.239}$$

where x and p are the position and momentum operators and where the expectation values of these operators are defined by the relation

$$\langle O \rangle = \int_{-\infty}^{\infty} \psi^*(x,t) O(x,p) \psi(x,t) dx. \tag{13.240}$$

In Eq. (13.240) $\psi(x,t)$ is a wave packet solution of the time-dependent Schrödinger equation

$$i\hbar \frac{\partial \psi(x,t)}{\partial t} = H\psi(x,t). \tag{13.241}$$

We assume that the Hamiltonian can be written as the sum of kinetic energy and the potential energy plus an additional term $\lambda W(x,t)$ which accounts for the dissipative force, i.e.

$$H = \frac{p^2}{2m} + V(x) + \lambda W(x,t). \tag{13.242}$$

Now we want to determine $W(x,t)$ in such a way that using the Heisenberg equation of motion $\frac{dp}{dt} = -\frac{i}{\hbar}[p,H]$ and the definition of the expectation value (13.240), we obtain (13.239).
Following these steps we find that

$$\left\langle \frac{\partial W(x,t)}{\partial x} \right\rangle = \langle p \rangle. \tag{13.243}$$

But this relation does not yield a unique solution for $W(x)$, and as we will see we can construct a family of $W(x,t)$ s with the property (13.243).
    The simplest solution of (13.243) is

$$\int \psi^* \frac{\partial W(x,t)}{\partial x} \psi dx = -\frac{i\hbar}{2} \int \left[\psi^* \frac{\partial \psi}{\partial x} - \frac{\partial \psi^*}{\partial x} \psi \right] dx, \tag{13.244}$$

from which we obtain

$$W(x,t) = -\frac{i\hbar}{2} \left[\ln\left(\frac{\psi}{\psi^*}\right) - \left\langle \ln\left(\frac{\psi}{\psi^*}\right) \right\rangle \right]. \tag{13.245}$$

The second term which is a function of time is added so as to make the expectation value of W equal to zero. This choice of $W(x,t)$ gives us the nonlinear Schrödinger-Langevin equation which we have discussed in §13.5. Next let us examine some other possibilities for $W(x,t)$. We want W to be Hermitian and from (13.243) we conclude that W must be a general quadratic function of x, p, $\langle x \rangle$ and $\langle p \rangle$. Consider the Süssmann-Hasse dissipative potential

$$W_a(x,t) = a \left[\frac{1}{2}(xp + px) - \langle x \rangle p \right] + (1 - a)\langle p \rangle \left(x - \langle x \rangle \right), \qquad (13.246)$$

which depends on the real parameter a, and satisfies (13.243). This parameter, a, can be a constant or a function of time. The mechanical momentum of the system

$$m\dot{x} = m\frac{\partial H}{\partial p} = p + m\lambda a(x - \langle x \rangle), \qquad (13.247)$$

also depends on a. However for this damping potential the expectation value of W_a is given by

$$\langle W_a(x,t) \rangle = a \left[\frac{1}{2}\langle xp + px \rangle - \langle x \rangle\langle p \rangle \right] \approx 0, \qquad (13.248)$$

whereas for $W(x,t)$ this expectation value is exactly zero.

As a specific example of the damping potential W_a let us consider the damped harmonic oscillator, $V(x) = \frac{1}{2}m\omega^2 x^2$. Here the wave function also depends on a;

$$\begin{aligned}
\psi_n(x,t) &= N_n \exp\left[-i\omega_a \left(n + \frac{1}{2} \right) t - \left(\frac{m}{2\hbar} \right) (\omega_a + ia\lambda) x^2 \right] \\
&\quad \times H_n \left(\sqrt{\frac{m\omega_a}{\hbar}} x \right),
\end{aligned} \qquad (13.249)$$

where N_n is the normalization constant

$$N_n = \left(\frac{m\omega_a}{\pi\hbar} \right)^{\frac{1}{4}} \frac{1}{\sqrt{2^n n!}}, \qquad (13.250)$$

and ω_a is given by [5]

$$\omega_a = \begin{cases} \omega_0 & \text{for } a = 0 \\ \left(\omega_0^2 - \frac{\lambda^2}{4} \right)^{\frac{1}{2}} & \text{for } a = \pm\frac{1}{2} \end{cases}. \qquad (13.251)$$

In addition to the damped harmonic oscillator the problem of motion of a particle when $V(x) = 0$ is also exactly solvable [5].

13.11 Scattering in the Presence of Frictional Forces

In the theory of heavy-ion collisions, e.g. collision of ^{84}Kr with ^{209}Bi, certain features of the scattering can be explained if we assume the presence of a viscous medium around the heavy nuclei (See Chapter 3). Dissipation of the relative kinetic energy and relative angular momentum and the trapping of the nuclei are among the phenomena which can be attributed to a simple damping force [52]. This force must be nonzero for a finite separation between the target and the projectile i.e. $\lambda \to \lambda(\mathbf{r})$ and $\lambda(\mathbf{r}) \to 0$ as $|\mathbf{r}| \to \infty$, or it must act for a finite time, i.e. $\lambda \to \lambda(t), \lambda(t) \to 0$ as $t \to \pm\infty$. For the latter case we can study the problem within the framework of the time-dependent scattering theory.

Let us consider a Hamiltonian of the form

$$H = \frac{1}{2m}\mathbf{p}^2 \exp\left[-\Lambda(t)\right] + V(r) \exp\left[\Lambda(t)\right], \qquad (13.252)$$

where

$$\Lambda(t) = \int_{-\infty}^{t} \lambda\left(t'\right) dt'. \qquad (13.253)$$

Here it is assumed that $\lambda(t)$ is non-negative and that it goes to zero as $t \to \pm\infty$ fast enough so that $\Lambda(t)$ is well-defined for all values of t. The equation of motion found from (13.252) is explicitly time-dependent;

$$\ddot{\mathbf{r}} + \lambda(t)\dot{\mathbf{r}} + \frac{1}{m}\nabla V(r) = 0. \qquad (13.254)$$

Using the Hamiltonian (13.252) we find the modified Hamilton-Jacobi equation

$$\frac{1}{2m}\left(\nabla S\right)^2 + \frac{\partial S}{\partial t} + V(r) + \lambda(t)S - U(t) = 0, \qquad (13.255)$$

where $U(t)$ is the time-dependent potential introduced in Eq. (13.195). This classical equation can be quantized exactly as before with the result that

$$-\frac{\hbar^2}{2m}\nabla^2\psi + \frac{\hbar}{i}\frac{\partial\psi}{\partial t} + [V(r) - U(t)]\psi + \frac{\hbar\lambda(t)}{2i}\ln\left(\frac{\psi}{\psi^*}\right)\psi = 0, \qquad (13.256)$$

and with

$$U(t) = -\frac{\hbar\lambda(t)}{2i}\int \psi^* \ln\left(\frac{\psi}{\psi^*}\right)\psi d^3r. \qquad (13.257)$$

As is well-known in the theory of time-dependent scattering, we use the "adiabatic switching", i.e. we replace $V(r)$ by $V(r)e^{-\varepsilon|t|}$, where ε is a small positive number, to get the correct asymptotic forms for the incoming and outgoing waves as $t \to -\infty$ and $t \to +\infty$ respectively [53]. The solution of (13.256) for large $|t|$ is

$$\psi(\mathbf{r}, t) \to \exp\left\{\frac{i}{\hbar}\left[\mathbf{r} \cdot \mathbf{q}e^{-\Lambda(t)} + e^{-\Lambda(t)}\Phi(t)\right]\right\}, \qquad (13.258)$$

where

$$\Phi(t) = B + \int_{-\infty}^{t} \left(U(t') e^{\Lambda(t')} - \frac{\mathbf{q}^2}{2m} e^{-\Lambda(t')} \right) dt', \tag{13.259}$$

and $\mathbf{q}$ and B are constants.

When $t \to -\infty$, (13.258) reduces to

$$\psi(\mathbf{r}, t) \to \exp \left[\frac{i}{\hbar} (\mathbf{q} \cdot \mathbf{r}) \right], \tag{13.260}$$

and this shows that $\mathbf{q}$ is the initial momentum of the particle. Using the asymptotic forms of the wave function (13.258) as $t \to \pm\infty$ we can calculate the kinetic energy per unit volume as $t \to -\infty$ and as $t \to +\infty$:

$$\lim_{t \to -\infty} \left(\frac{\hbar^2}{2m} \nabla \psi^* \cdot \nabla \psi \right) = \frac{q^2}{2m}, \tag{13.261}$$

and

$$\lim_{t \to +\infty} \left(\frac{\hbar^2}{2m} \nabla \psi^* \cdot \nabla \psi \right) = \frac{q^2}{2m} \exp[-2\Lambda(\infty)]. \tag{13.262}$$

Thus the energy per unit volume dissipated in collision is

$$\frac{q^2}{2m} \left(1 - e^{-2\Lambda(\infty)} \right). \tag{13.263}$$

Since Eq. (13.256) is nonlinear, the complete solution of the scattering problem can only be found numerically. Among the numerical technique which can be applied to this problem are the method of quasi-linearization of Bellman and Kalaba [41] [54] or the method of Immele *et al.* [45] for the one-dimensional motion of a wave packet.

13.12 Application of the Noether Theorem: Linear and Nonlinear Wave Equations for Dissipative Systems

In this section we compare the conserved quantities for the two types of the wave equation which we have derived from a single Lagrangian for the damped motion linear in velocity [55] [44].

Let us recall that from the Lagrangian (4.11) we obtained the linear equation, Eq. (13.4), and from the same Lagrangian and the corresponding Hamiltonian, via the modified Hamilton-Jacobi equation we found the Schrödinger-Langevin equation (13.187). The Lagrangian densities for the linear and for the Schrödinger-Langevin wave equations are given by

$$\mathcal{L}_L = i\hbar\psi^*\dot{\psi} - \left(\frac{\hbar^2}{2m} \right) \nabla\psi^* \cdot \nabla\psi e^{-\lambda t} - V(\mathbf{r})e^{\lambda t}\psi^*\psi, \tag{13.264}$$

and

$$\mathcal{L}_S = e^{\lambda t}\left[i\hbar\psi^*\dot\psi - \left(\frac{\hbar^2}{2m}\right)\nabla\psi^* \cdot \nabla\psi - V(\mathbf{r})\psi^*\psi\right]$$
$$+ e^{\lambda t}\left[\frac{i\hbar\lambda}{2}\psi^*\psi\left\{\ln\left(\frac{\psi}{\psi^*}\right)+1\right\}\right], \tag{13.265}$$

where the subscripts L and S refer to the linear and the Schrödinger-Langevin equations respectively. The momentum density for the two cases are different, that is

$$\pi_L(\mathbf{r},t) = \frac{\partial \mathcal{L}_L}{\partial \dot\psi} = i\hbar\psi^*(\mathbf{r},t), \tag{13.266}$$

and

$$\pi_S(\mathbf{r},t) = \frac{\partial \mathcal{L}_S}{\partial \dot\psi} = i\hbar e^{\lambda t}\psi^*(\mathbf{r},t). \tag{13.267}$$

The corresponding Hamiltonian densities for the two wave fields obtained from the corresponding Lagrangians are

$$\mathcal{H}_L = \pi_L\dot\psi - \mathcal{L}_L = \left(\frac{\hbar^2}{2m}\right)\nabla\psi^* \cdot \nabla\psi e^{-\lambda t} + V(\mathbf{r})e^{\lambda t}\psi^*\psi, \tag{13.268}$$

and

$$\mathcal{H}_S = e^{\lambda t}\left[\left(\frac{\hbar^2}{2m}\right)\nabla\psi^* \cdot \nabla\psi + V(\mathbf{r})\psi^*\psi - \frac{i\hbar\lambda}{2}\psi^*\psi\left\{\ln\left(\frac{\psi}{\psi^*}\right)+1\right\}\right]. \tag{13.269}$$

Both of the Lagrangian densities $\mathcal{L}_L$ and $\mathcal{L}_S$ reduce to the Lagrangian $\mathcal{L}$ of the Schrödinger field when λ tends to zero. The same is true of the Hamiltonian densities $\mathcal{H}_L$ and $\mathcal{H}_S$, i.e. they both tend to $\mathcal{H}$ as $\lambda \to 0$.

For simplicity we only consider the motion of a free particle in a viscous medium. In this case we set $V(\mathbf{r}) = 0$ in our equations for $\mathcal{H}_L$ and $\mathcal{H}_S$.

First let us study the conservation laws for the ψ_L field i.e. the linear wave equation [44]. These are given by the following relations:
(1) Conservation of the energy which for $V(\mathbf{r}) = 0$ is just the conservation of the kinetic energy

$$\frac{d}{dt}\int\left[\left(\frac{\hbar^2}{2m}\right)\nabla\psi^* \cdot \nabla\psi\right]d^3r = \frac{d}{dt}\left(Ee^{2\lambda t}\right) = 0. \tag{13.270}$$

(2) Conservation of linear momentum

$$\frac{d}{dt}\int\left(\frac{\hbar}{2i}\right)\{\psi^*\nabla\psi - \psi\nabla\psi^*\}d^3r = \frac{d}{dt}\left(\mathbf{P}e^{\lambda t}\right) = 0. \tag{13.271}$$

(3) Conservation of angular momentum

$$\frac{d}{dt}\int\left(\frac{\hbar}{2i}\right)\mathbf{r}\wedge\{\psi^*\nabla\psi - \psi\nabla\psi^*\}d^3r = \frac{d}{dt}\left(\mathbf{L}e^{\lambda t}\right) = 0. \tag{13.272}$$

(4) Motion of the center of the wave packet (or the center of mass for a system of particles)

$$\frac{d}{dt} \int \left[\left(\frac{\hbar}{2i}\right) [\psi^* \nabla \psi - \psi \nabla \psi^*] \left(\frac{1-e^{-\lambda t}}{\lambda}\right) - mr\psi^* \psi \right] d^3r$$

$$= \frac{d}{dt} \left(\mathbf{P} \frac{e^{\lambda t} - 1}{\lambda} - m\mathbf{R} \right) = 0. \tag{13.273}$$

For the Schrödinger-Langevin equation we have the following conservation laws:

(1) Conservation of the (kinetic) energy

$$\frac{d}{dt} \int \left[\left(\frac{\hbar^2}{2m}\right) \nabla \psi^* \cdot \nabla \psi - \frac{i\hbar\lambda}{2} \psi^* \psi \ln\left(\frac{\psi}{\psi^*}\right) \right] e^{\lambda t} d^3r$$

$$= \frac{d}{dt} \left[\left\{ E - \frac{i\hbar\lambda}{2} \int \psi^* \psi \ln\left(\frac{\psi}{\psi^*}\right) \right\} e^{\lambda t} d^3r \right] = 0. \tag{13.274}$$

(2) Conservation of linear momentum

$$\frac{d}{dt} \int \left(\frac{\hbar}{2i}\right) \{\psi^* \nabla \psi - \psi \nabla \psi^*\} e^{\lambda t} d^3r = \frac{d}{dt} \left(\mathbf{P} e^{\lambda t} \right) = 0. \tag{13.275}$$

(3) Conservation of angular momentum

$$\frac{d}{dt} \int \left(\frac{\hbar}{2i}\right) \mathbf{r} \wedge [\psi^* \nabla \psi - \psi \nabla \psi^*] e^{\lambda t} d^3r = \frac{d}{dt} \left(\mathbf{L} e^{\lambda t} \right) = 0. \tag{13.276}$$

(4) Conservation of the motion of the center of the wave packet (or the center of mass for a system of particles)

$$\frac{d}{dt} \int \left[\left(\frac{\hbar}{2i}\right) \{\psi^* \nabla \psi - \psi \nabla \psi^*\} \left(\frac{e^{\lambda t} - 1}{\lambda}\right) - mr\psi^* \psi \right] d^3r$$

$$= \frac{d}{dt} \left(\mathbf{P} \frac{e^{\lambda t} - 1}{\lambda} - m\mathbf{R} \right) = 0. \tag{13.277}$$

In addition to these we have the invariance of the Lagrangian densities $\mathcal{L}_L$ and $\mathcal{L}_S$ with respect to a global change of phase of the wave function,

$$\psi(\mathbf{r}, t) \rightarrow e^{i\alpha} \psi(\mathbf{r}, t) \quad \text{for the } L \text{ field,} \tag{13.278}$$

and

$$\psi(\mathbf{r}, t) \rightarrow \exp\left(i\alpha e^{-\lambda t}\right) \psi(\mathbf{r}, t) \quad \text{for the } S \text{ field.} \tag{13.279}$$

For both fields, this invariance implies the conservation of probability [44]

$$\frac{d}{dt} \int \rho(\mathbf{r}, t) d^3r = 0. \tag{13.280}$$

13.13   Wave Equation for Impulsive Forces Acting at Certain Intervals

In Chapter 4 we studied a Hamiltonian formulation of a classical system expressed by first order linear difference equation (4.187). The difference equation results from simplifying assumptions regarding the nature of impulsive forces, i.e. using them in the form of constraints at times $t = t_n$, on the dynamics of otherwise free motion of a particle (or a system).

To find the quantum equivalent of Eq. (4.187) we note that the Hamiltonian (4.197) is of the Kanai-Cardirola type, and that the commutation relation

$$[x,p] = [x,P]\exp\left\{-\int_0^t f\left(t'\right)dt'\right\} = i\hbar\exp\left\{-\int_0^t f\left(t'\right)dt'\right\} \qquad (13.281)$$

violates the position-velocity uncertainty unless $\rho = 1$. Therefore as we have seen in this chapter the direct quantization of (4.197) leads to unacceptable results. However we can use Kostin's method and start with the Schrödinger equation

$$i\hbar\frac{\partial\psi}{\partial t} = -\frac{\hbar^2}{2m}\left(\frac{\partial^2\psi}{\partial x^2}\right) + (V_L(x,t) - xg(t))\,\psi, \qquad (13.282)$$

and follow the same steps to find that

$$\langle p\rangle = \frac{\hbar}{2i}\int_{-\infty}^{\infty}\left(\psi^*\frac{\partial\psi}{\partial x} - \psi\frac{\partial\psi^*}{\partial x}\right)dx, \qquad (13.283)$$

and

$$\left\langle\frac{\partial V_L(x,t)}{\partial x}\right\rangle = f(t)\langle p\rangle. \qquad (13.284)$$

From Eqs. (13.283) and (13.284) we obtain $V_L(x,t)$ to be

$$V_L(x,t) = \frac{\hbar f(t)}{2i}\ln\left(\frac{\psi(x,t)}{\psi^*(x,t)}\right) + W(t). \qquad (13.285)$$

By imposing the condition that

$$\langle E(t)\rangle = \frac{1}{2m}\left\langle p^2\right\rangle - g(t)\langle x\rangle, \qquad (13.286)$$

we find $W(t)$;

$$W(t) = -\frac{\hbar}{2i}f(t)\frac{\int_{-\infty}^{\infty}|\psi(x,t)|^2\ln\left(\frac{\psi(x,t)}{\psi^*(x,t)}\right)dx}{\int_{-\infty}^{\infty}|\psi(x,t)|^2\,dx}. \qquad (13.287)$$

Thus Eq. (13.282) is the Schrödinger-Langevin equation for the classical Hamiltonian (4.197) [56].

13.14 Classical Limit for the Time-Dependent Problems

In the classical limit when $\hbar \to 0$, the Schrödinger equation as well as the wave equations for linear and nonlinear damping reduces to the Hamilton-Jacobi equation. Now we want to study a different formulation of the same problem based on the concept of distribution function of the positions and momenta of an ensemble of identical particles [57]. Consider an ensemble of identical particles whose initial positions ξ_0 are distributed with a distribution function $\mathcal{F}(\xi_0)$ with all of the initial momenta being equal to p_0. Starting from the classical Hamilton-Jacobi equation one can show that the probability density $P(x, t)$ for this ensemble is given by

$$P(x,t) = \frac{\partial^2 S}{\partial p_0 \partial x} \left| \mathcal{F} \left(\frac{\partial S}{\partial p_0} \right) \right|^2, \qquad (13.288)$$

where $S(x, p_0, t)$ is the Hamilton principal function. The fact that the above ensemble is defined for a fixed p_0 is problematic. Therefore we choose a Gaussian form for the distribution function $\mathcal{F}(\xi_0)$ and calculate the wave packet for the free particle $\psi_F(x,t)$ and its Fourier transform $\phi_F(x,t)$. From these we obtain one of the possible quantum mechanical distribution functions for this problem which is in the form of the product [58]

$$D(x,p,t=0) = |\psi_F(x,0)|^2 \, |\phi_F(p,0)|^2. \qquad (13.289)$$

This distribution function does not have a fixed momentum, p_0, but a Gaussian distribution about p_0 [59].

Now we start with the following definition of the classical limit:
Let us consider an ensemble with a given initial distribution of positions and momenta and then using Liouville theorem we determine $D(x,p,t)$ by replacing x and p from the solution of the classical equations of motion, i.e.

$$D(x,p,t) = D[x(t=0), p(t=0)], \qquad (13.290)$$

where $x(t=0)$ and $p(t=0)$ are the initial coordinates of the particle written in terms of $x(t)$ and $p(t)$. This gives us the distribution function at the time t.

Consider two ensembles, one classical and the other quantum-mechanical, and let us assume that the initial distribution of position and momentum are the same for both ensembles. Then the classical limit of the quantum-mechanical ensembles at a given time t is defined as that of the classical ensemble at t. This definition can be used with any distribution function such as the one given by (13.289) or by others like the Wigner distribution function [60].

Next we apply this idea to the problem of the damped harmonic oscillator where the wave function for the ground state is given by Eq. (13.221)

$$\psi_0(x,t) = \left(\frac{m\omega_0}{\hbar\pi} \right)^{\frac{1}{4}} \exp\left[-\frac{m\omega_0}{2\hbar} (x - \xi(t))^2 \right] \times$$

$$\times \quad \exp\left[-\frac{i\omega_0}{2\lambda} + i\hbar\left(mx\dot{\xi} + C(t)\right)\right]. \tag{13.291}$$

Thus the initial quantum mechanical distribution function at $t = 0$ is given by

$$P(x, t = 0) = |\psi_0(x, t = 0)|^2 = \left(\frac{m\omega_0}{\hbar\pi}\right)^{\frac{1}{2}} \exp\left[-\frac{m\omega_0}{\hbar}(x - \xi_0)^2\right], \tag{13.292}$$

and at a later time t, $P(x, t)$ becomes

$$P(x, t) = \left(\frac{m\omega_0}{\hbar\pi}\right)^{\frac{1}{2}} \exp\left[-\frac{m\omega_0}{\hbar}(x - \xi(t))^2\right]. \tag{13.293}$$

The corresponding classical phase space density function $D(x, p, t = 0)$ is the product of two Gaussian functions

$$D(x, p, t = 0) = \frac{1}{\pi} \exp\left\{-\left(\frac{m\omega_0}{\hbar}\right)(x - \xi_0)^2\right\} \exp\left\{-\left(\frac{1}{\omega_0 m\hbar}\right)(p - p_0)^2\right\}. \tag{13.294}$$

The classical distribution function at time t can be found from the solution of the classical equation of motion. Thus by solving the differential equation for $x = \xi(t)$, Eq. (5.20), with the initial conditions $x(0)$ and $p(0) = m\dot{x}(0)$ and then inverting the result we find $x(0)$ and $p(0)$,

$$x(0) = e^{-\frac{\lambda}{2}t}\left\{x\left(\cos\omega t - \frac{\lambda}{2\omega}\sin\omega t\right) - \frac{p}{m\omega}\sin\omega t\right\} \tag{13.295}$$

and

$$p(0) = e^{-\frac{\lambda}{2}t}\left\{p\left(\cos\omega t + \frac{\lambda}{2\omega}\sin\omega t\right) + m\omega_0 x\sin\omega t\right\}. \tag{13.296}$$

Using these classical results we determine $D(x, p, t)$;

$$D(x, p, t) = D(x(0), p(0)) = \frac{1}{\pi}\exp\left\{-\left(\frac{m\omega_0}{\hbar}\right)(x(0) - \xi_0)^2\right\}$$

$$\times \quad \exp\left\{-\left(\frac{1}{\omega_0 m\hbar}\right)(p(0) - p_0)^2\right\}, \tag{13.297}$$

where $x(0)$ and $p(0)$ are given by (13.295) and (13.296) respectively. Now if we calculate the integral

$$P_c(x, t) = \int D(x, p, t)dp, \tag{13.298}$$

and if this $P_c(x, t)$ is identical to the quantum mechanical result (13.293), we say that $D(x, p, t)$ is the classical limit of the quantum distribution function. When $\lambda = 0$, i.e. no damping, then $\xi(t) = \xi_0 \cos\omega_0 t$, and (13.298) and (13.293) are the same. But with $\lambda \neq 0$ the classical limit in this sense does not exist.

Bibliography

[1] E.H. Kerner, Can. J. Phys. 36, 371 (1958).

[2] F. Bopp, Zeit. Angw. Phys. 14, 699 (1962).

[3] W.K.H. Stevens, Pro. Phys. Soc. Lond. 72, 1072 (1958).

[4] B-G. Englert, Ann. Phys. (NY) 129, 1 (1980).

[5] R.W. Hasse, J. Math. Phys. 16, 2005 (1975).

[6] I.K. Edwards, Am. J. Phys. 47, 153 (1979).

[7] V.V. Dodonov and V.I. Man'ko, Nuovo Cimento, 44B, 265 (1978).

[8] V.V. Dodonov and V.I. Man'ko, Phys. Rev. A20, 550 (1979).

[9] P. Cardirola, Lett. Nuovo Cimento, 20, 589 (1977).

[10] V.V. Dodonov and V.I. Man'ko in *Invariants and the Evolution of the Nonstationary Quantum Systems*, Edited by M.A. Markov (Nova Scientific, Commack, 1989) p. 154.

[11] T. Toyoda and K. Wildermuth, Phys. Rev. A22, 2391 (1980).

[12] M. Razavy, Nuovo Cimento, 64B, 396 (1981).

[13] R. Remaud and E.H. Hernandez, Physica 103A, 35 (1980).

[14] G. Ghosh and R.W. Hasse, Phys. Rev. A24, 1621 (1981).

[15] S.P. Kim, J. Phys. A36, 12089 (2003).

[16] J-S Peng and G-X Li, *Introduction to Modern Optics*, (World Scientific, Singapore, 1998) p. 160.

[17] N.N. Bogolyubov, J. Phys. USSR, 9, 23 (1947).

[18] See for instance A.S. Davydov, *Quantum Mechanics*, translated by D. ter Harr, (Pergamon Press, 1965) p. 610.

[19] H.P. Yuen, Phys. Rev. A13, 2226 (1976).

[20] R.W. Henry and S.C. Glotzer, Am. J. Phys. 56, 318 (1988).

[21] D.M. Greenberger J. Math. Phys. 20, 762 (1979).

[22] D.M. Greenberger J. Math. Phys. 20, 771 (1979).

[23] See for instance, F.A. Kaempfer, *Concepts in Quantum Mechanics*, (Acadmic Press, New York, N.Y. 1965).

[24] M. Razavy, Nuovo Cimento, 63B, 271 (1969).

[25] F. Englemann and E. Fick, Nuovo Cimento, 12, 63 (1959).

[26] E. Fick and F. Englemann, Z. Phys. 175, 271 (1964).

[27] A. Pimpale and M. Razavy, Phys. Rev. A 36, 2739 (1987).

[28] W. Pauli, *General Principles of Quantum Mechanics*, translated by P. Achuthan and K. Vankatesan (Springer-Verlag, Berlin, 1980) pp. 28-30.

[29] M.D. Kostin, J. Chem. Phys. 57, 3589 (1972).

[30] M.D. Kostin, J. Stat. Phys. 12, 145 (1975).

[31] H. Dekker, Phys. Rep. 80, 1 (1981).

[32] G.W. Ford, M. Kac and P. Mazur, J. Math. Phys. 6, 504 (1965).

[33] W. Stoker and K. Albrecht, Ann. Phys. (NY) 117, 436 (1979).

[34] E. Madelung, Z. Phys. 40, 322 (1926).

[35] J. Mehta and H. Rechenberg, *The Historical Development of Quantum Theory*, vol. 5, Part 2 (Springer-Verlag, New York, N.Y. 1987) p. 856.

[36] P. Ván and T. Fülöp, Phys. Lett. A323, 374 (2004).

[37] See for example Z.U.A. Warsi, *Fluid Dynamics*, (CRC Press, Boca Raton, 1992) p. 38.

[38] P.R. Holland, *The Quantum Theory of Motion*, (Cambridge University Press, Cambridge, 1993).

[39] L. Meirovitch, *Methods of Analytical Dynamics*, (McGraw-Hill, New York, 1970) §6.7.

[40] M. Razavy, Z. Physik B26, 201 (1977).

[41] M. Razavy, Can. J. Phys. 56, 311 (1978).

[42] M. Razavy, Can. J. Phys. 56, 1372 (1978).

[43] E. Scrödinger, Ann. Phys. 79, 361 (1926).

[44] H-J Wagner, Z. Physik, B95, 261 (1994).

[45] J.D. Immele, K-K. Kan and J.J. Griffin, Nucl. Phys. A241, 47 (1975).

[46] J. Messer, Acta Physica Aust. 50, 75 (1997).

[47] B. K. Skagerstam, J. Math. Phys. 18, 308 (1977).

[48] A. Loinger, Nuovo Cimento, 2, 511 (1955).

[49] G. Valentini, Nuovo Cimento, 19, 1280 (1961).

[50] I.R. Senitzky, Phys. Rev. 95, 1115 (1954).

[51] M. Razavy, Lett. Nuovo Cimento, 24, 293 (1979).

[52] W. Nörenberg and H.A. Weidenmüller, *Introduction to the Theory of Heavy-Ion Collisions*, (Springer-Verlag, New York, 1980) §2.6.

[53] See, for example, C.J. Joachain, *Quantum Collision Theory*, (North-Holland, Amsterdam, 1975) p. 304.

[54] R.E. Bellman and R.E. Kalaba, *Quasilinearization and Nonlinear Boundary Value Problems*, (Rand Corp., Santa Monica, 1965).

[55] K. Yasue, Ann. Phys. (NY) 114, 479 (1978).

[56] M. Razavy, Hadronic J. 17, 515 (1994).

[57] L.S. Brown, Am. J. Phys. 40, 371 (1972).

[58] L. Cohen, J. Math. Phys. 7, 781 (1966).

[59] D. Home and S. Sengupta, Am. J. Phys. 51, 265 (1983).

[60] E.P. Wigner, Phys. Rev. 40, 749 (1932).

Chapter 14

Density Matrix and the Wigner Distribution Function

In Chapters 3, 4 and 5 we discussed the analytical dynamics of simple damped systems in terms of Lagrangian, Hamiltonian and Hamilton-Jacobi formulations. Since the quantization of dissipative motion based on variational method are, as we have seen, problematic, there have been attempts to circumvent some of the difficulties by the direct use of the classical (Newtonian) equations of motion. The Yang-Feldman method discussed in §12.2 is one way of quantizing the second law of motion directly. In this chapter we want to study another powerful method of studying dissipative system in quantum theory which is based on the idea of the density matrix (or the density operator) and its connection to the classical Liouville theorem [1]-[3].

14.1 Classical Distribution Function for Nonconservative Motions

According to the Liouville theorem, for a conservative system the distribution function or the density in phase space which here we denote by $w_c(x, p, t)$ satisfies the equation [4] [5]

$$\frac{dw_c}{dt} = \frac{\partial w_c}{\partial t} + \{w_c, H\} = 0, \qquad (14.1)$$

where $\{w_c, H\}$ is the Poisson bracket,

$$\{w_c, H\} = \sum_j \left(\frac{\partial w_c}{\partial x_j} \frac{\partial H}{\partial p_j} - \frac{\partial w_c}{\partial p_j} \frac{\partial H}{\partial x_j} \right). \qquad (14.2)$$

An extension of (14.1) to the non-conservative system has been suggested by Gerlich [6], by Steep [7] and by Bolivar [8].

We will study this generalization of the Liouville theorem for the simple case of a one-dimensional system where the equations of motion are given by

$$\dot{x} = X_1(x, p, t), \qquad (14.3)$$

and

$$\dot{p} = X_2(x, p, t). \qquad (14.4)$$

Since (14.1) is an equation expressing the conservation of $w_c(x, p, t)$ as a function of time, we generalize the Liouville theorem so that the new form also expresses a conservation law, i.e.

$$\frac{\partial w_c}{\partial t} + \text{div}\,(w_c \mathbf{X}) = 0, \qquad (14.5)$$

where

$$\text{div}(\mathbf{X}) = \frac{\partial X_1}{\partial x} + \frac{\partial X_2}{\partial p}. \qquad (14.6)$$

Equation (14.5) has the correct limit when damping becomes negligible. Thus for the motion of a single particle in the absence of dissipation $X_1 = p/m$, and $X_2 = X_2(x, t)$, therefore $\text{div}(\mathbf{X}) = 0$, and

$$\frac{dw_c}{dt} = \frac{\partial w_c}{\partial t} + \frac{p}{m} \frac{\partial w_c}{\partial x} + X_2(x, t) \frac{\partial w_c}{\partial p} = 0, \qquad (14.7)$$

which is the same as Eq. (14.1).

The function $w_c(x, p, t)$ is the analogue of the Wigner distribution function $W(x, p, t)$ which will be introduced later [9]. To establish the connection between $w_c(x, p, t)$ and the quantum mechanical wave function, we introduce a new function $\chi \left(x + \frac{1}{2}\delta\xi, x - \frac{1}{2}\delta\xi \right)$ by [8] [10] [11]

$$\chi \left(x + \frac{1}{2}\delta\xi, x - \frac{1}{2}\delta\xi \right) = \int_{-\infty}^{\infty} w_c(x, p, t) \exp \left(\frac{ip\delta\xi}{\ell} \right) dp, \qquad (14.8)$$

where ℓ is a constant having the dimension of action and $\delta\xi$ is a small but nonzero quantity. First let us obtain the equation that χ satisfies when the system is conservative and the Hamiltonian is given by $H = \frac{1}{2m}p^2 + V(x)$ and therefore $X_2(x) = -dV(x)/dx$. Substituting for X_2 in (14.7), multiplying the resulting equation by $\exp \left(\frac{ip\delta\xi}{\ell} \right)$ and integrating over all p we find

$$-\frac{\partial \chi}{\partial t} + \frac{i\ell}{m} \frac{\partial^2 \chi}{\partial x \partial(\delta\xi)} - \frac{i}{\ell} \delta V(x)\chi = 0, \qquad (14.9)$$

where we have used the condition

$$\left| w_c(x,p,t) \exp\left(\frac{ip\delta\xi}{\ell}\right) \right|_{p=-\infty}^{p=\infty} = 0, \qquad (14.10)$$

which is expected from a probability distribution function. In Eq. (14.9) $\delta V(x)$ stands for the expression

$$\delta V(x) = \frac{\partial V(x)}{\partial x}\delta\xi = V\left(x + \frac{1}{2}\delta\xi\right) - V\left(x - \frac{1}{2}\delta\xi\right). \qquad (14.11)$$

Now if we change the variables to

$$y = x + \frac{1}{2}\delta\xi, \quad \text{and} \quad y' = x - \frac{1}{2}\delta\xi, \qquad (14.12)$$

then we can write (14.9) as

$$\left[\frac{\ell^2}{2m}\left(\frac{\partial^2}{\partial y^2} - \frac{\partial^2}{\partial y'^2}\right) - \{V(y) - V(y')\}\right]\chi(y,y',t) = -i\ell\frac{\partial}{\partial t}\chi(y,y',t). \qquad (14.13)$$

Thus for a conservative system, if we assume that $\chi(y,y',t)$ can be written as

$$\chi(y,y',t) = \psi(y,t)\,\psi^*(y',t), \qquad (14.14)$$

then $\psi^*(y',t)$ and $\psi(y,t)$ each satisfy the Schrödinger equation if we set $\ell = \hbar$.

This method of quantization which starts with the classical distribution function $w_c(x,p,t)$ and for conservative systems leads to the Schrödinger equation has been advanced as a way of circumventing problems associated with using generalized (non-Cartesian) coordinates [10]. The same method can be applied to quantize non-conservative systems. For instance if we choose

$$X_1 = \frac{p}{m}, \quad \text{and} \quad X_2 = f_d(x,p,t) - \frac{\partial V(x,t)}{\partial x}, \qquad (14.15)$$

then the equation for w_c takes the form

$$\frac{\partial w_c}{\partial t} + \frac{p}{m}\frac{\partial w_c}{\partial x} - \frac{\partial V(x,t)}{\partial x}\frac{\partial w_c}{\partial p} = \frac{\partial}{\partial p}[f_d(x,p,t)w_c]. \qquad (14.16)$$

Following the same steps outlined earlier we find the following equation for $\chi(y,y',t)$

$$i\hbar\frac{\partial}{\partial t}\chi(y,y',t) = -\frac{\hbar^2}{2m}\left(\frac{\partial^2}{\partial y^2} - \frac{\partial^2}{\partial y'^2}\right)\chi(y,y',t) + \{V(y) - V(y')\}$$
$$\times \quad \chi(y,y',t) - i\hbar I(y,y',t)\chi(y,y',t), \qquad (14.17)$$

where

$$\hbar I(y,y',t)\chi(y,y',t) = \int \frac{\partial}{\partial p}[f_d(x,p,,t)w_c]\exp\left(\frac{ip\delta\xi}{\hbar}\right)dp, \qquad (14.18)$$

and y and y' are given by (14.12).

In general the operator I cannot be obtained in closed form, and even if $I\chi$ can be found analytically, the function χ will not be separable, i.e, it cannot be written as $\psi^*(y',t)\psi(y,t)$. However for some simple cases such as linear damping, $f_d = \lambda p$, one can find $I(y',y,t)$ as a differential operator. For this type of damping the last term in (14.17) takes the form [12]

$$i\hbar I(y,y',t)\chi(y,y',t) = \frac{i\hbar\lambda}{2}(y-y')\left(\frac{\partial}{\partial y} - \frac{\partial}{\partial y'}\right)\chi(y,y',t). \qquad (14.19)$$

Thus for this case χ satisfies a partial differential equation of second order in y and y', and first order in time.

We note that in this approach we can bypass the construction of the classical Lagrangian and the Hamiltonian for the damped system, but like other cases the result is not unique and one can find different wave equations even for the simple case of linear damping [8].

The initial condition needed for the solution of the differential equation for χ can be obtained from the initial condition imposed on $w_c(x,p,t)$. Since $w_c(x,p,t)$ is the classical distribution function, for the motion of a single particle at $t=0$, it is given by

$$w_c(x,p,0) = \delta(x-x_0)\delta(p-p_0). \qquad (14.20)$$

But this initial form of w_c is not suitable for finding the quantum mechanical distribution function $\chi(y,y',t)$, since at $t=0$ the exact values of x and p are given in violation of the uncertainty principle. Thus instead of (14.20) we choose a Gaussian form for w_c;

$$w_c(x,p,0) = \frac{1}{\pi\ell}\exp\left[-\frac{\varepsilon}{\ell}(x-x_0)^2\right]\exp\left[-\frac{1}{\varepsilon\ell}(p-p_0)^2\right], \qquad (14.21)$$

where ε is an arbitrary positive constant having the dimension of (mass/time). By substituting $w_c(x,p,0)$ in (14.8) and then setting $\ell = \hbar$, we find the initial value of χ to be

$$\chi(y,y',0) = \sqrt{\frac{\varepsilon}{\pi\hbar}}\exp\left[-\frac{\varepsilon}{2\hbar}(y-y_0)^2\right]\exp\left[-\frac{\varepsilon}{2\hbar}(y'-y_0')^2\right]. \qquad (14.22)$$

Here we have assumed that the initial form of $\chi(y,y',0)$ is separable, but this does not imply that $\chi(y,y',t)$ will remain as a product of two factors, $\psi^*(y',t)\psi(y,t)$ at some later time.

14.2 The Density Matrix

If a state of the system is given by the wave function $\psi(t)$, then we define the density matrix by the outer product

$$\rho(t) = |\psi(t)\rangle \langle \psi(t)| . \tag{14.23}$$

This is the case when we are considering a pure state [1]. However if the system is not known precisely, but is known to be in a state $|\psi_n(t)\rangle$ with the probability p_n, then we define $\rho(t)$ in the Schrödinger picture by

$$\rho(t) = \sum_n p_n |\psi_n(t)\rangle \langle \psi_n(t)| , \tag{14.24}$$

with

$$\sum_n p_n = 1. \tag{14.25}$$

In either case we have the important result that

$$\text{Trace } \rho(t) = 1. \tag{14.26}$$

This follows from the fact that trace of $\rho(t)$ means summing over the diagonal elements of $\rho(t)$, i.e.

$$\text{Trace } \rho(t) = \sum_j \sum_n p_n \langle \psi_j|\psi_n\rangle \langle \psi_n|\psi_j\rangle = \sum_n p_n = 1. \tag{14.27}$$

In addition to its trace being equal to one, the density matrix has the following properties:

(1) The mean value of an operator $\mathcal{O}$ can be written as

$$\langle \psi_n|\mathcal{O}|\psi_n\rangle = \text{Trace}(\mathcal{O}\rho(t)). \tag{14.28}$$

(2) The operator $\rho(t)$ is positive semi-definite, i.e. for any state A

$$\langle A|\rho(t)|A\rangle = \sum_n p_n|\langle A|\psi_n(t)\rangle|^2 \geq 0. \tag{14.29}$$

(3) For a pure state

$$\rho^2(t) = |\psi(t)\rangle\langle\psi(t)|\psi(t)\rangle\langle\psi| = \rho(t), \tag{14.30}$$

but for mixed states we have

$$\text{Trace } \rho^2(t) = \sum_{j,n} p_j p_n |\langle \psi_j|\psi_n\rangle|^2. \tag{14.31}$$

Since $|\langle\psi_j|\psi_n\rangle|^2 \leq 1$, and $\sum_j p_j = 1$, therefore $\sum_n p_n |\langle\psi_j|\psi_n\rangle| \leq 1$ for any j, and we have the inequality

$$\text{Trace } \rho^2(t) \leq \sum_n p_n = 1. \tag{14.32}$$

The time evolution of the density matrix for an open system can be obtained in the Heisenberg picture by noting that $\rho(t)$ is defined by Eq. (14.23), and $|\psi(t)\rangle$ is the solution of the Schrödinger equation. If at $t = t_0$ the state of the system is given by $|\psi(t_0)\rangle$, then

$$|\psi(t)\rangle = U(t, t_0)|\psi(t_0)\rangle, \tag{14.33}$$

where $U(t, t_0)$ is the unitary time-evolution operator which satisfies the differential equation

$$i\hbar\frac{\partial}{\partial t}U(t, t_0) = H(t)U(t, t_0), \tag{14.34}$$

and is subject to the initial condition

$$U(t_0, t_0) = 1, \tag{14.35}$$

1 being the identity operator. For a time-dependent Hamiltonian of a non-conservative system the formal solution of (14.34) with the boundary condition (14.35) is expressible in terms of an integral equation

$$U(t, t_0) = 1 - \frac{i}{\hbar}\int_{t_0}^{t} H(t')\, U(t', t_0)\, dt'. \tag{14.36}$$

This integral equation can be solved by iteration

$$U(t, t_0) = 1 + \sum_{n=1}^{\infty}\left(\frac{-i}{\hbar}\right)^n \int_{t_0}^{t} H(t_n)dt_n \int_{t_0}^{t_n} H(t_{n-1})dt_{n-1}\cdots$$

$$\times \int_{t_0}^{t_2} H(t_1)dt_1. \tag{14.37}$$

In general the operators $H(t_1), H(t_2)\cdots$ do not commute for different values of time, so that the indicated order must be maintained. Thus the operators in (14.37) are arranged in order of increasing time as we go from right to left. Equation (14.37) can also be written as [13] [14]

$$U(t, t_0) = 1 + \sum_{n=1}^{\infty}\left(\frac{-i}{\hbar}\right)^n \frac{1}{n!}\int_{t_0}^{t} dt_n \int_{t_0}^{t} dt_{n-1}$$

$$\times \cdots\int_{t_0}^{t} T\left[H(t_n)\cdots H(t_1)\right] dt_1, \tag{14.38}$$

where T denotes the time-ordering symbol, i.e. we arrange the n factors $H(t_i)$ in the order of increasing time from right to left. We can also write (14.38) formally as an exponential operator

$$U(t, t_0) = T \left\{ \exp \left[-\frac{i}{\hbar} \int_{t_0}^t H(t_1) dt_1 \right] \right\}. \qquad (14.39)$$

From the definition of $\rho(t)$ in terms of the wave function Eq. (14.24), we have

$$\rho(t) = \sum_n U(t, t_0) |\psi_n(t_0)\rangle \langle \psi_n(t_0)| U^\dagger(t, t_0), \qquad (14.40)$$

or

$$\rho(t) = U(t, t_0) \rho(t_0) U^\dagger(t, t_0). \qquad (14.41)$$

By differentiating this equation using (14.34) we obtain the following equation for $\partial\rho/\partial t$

$$i\hbar \frac{\partial \rho(t)}{\partial t} = [\rho(t), H]. \qquad (14.42)$$

14.3   Phase Space Quantization of Dekker's Hamiltonian

As an example of the density matrix formulation of the motion of a damped oscillator we want to study the complex Hamiltonian of Dekker (§4.5). Our starting point will be the classical density function $\rho(q, \pi, t)$ (or $w_c(x, p, t)$ of §14.1) which satisfies the continuity equation

$$\frac{\partial \rho}{\partial t} + \frac{\partial}{\partial q}(\dot{q}\rho) + \frac{\partial}{\partial \pi}(\rho\dot{\pi}) = 0. \qquad (14.43)$$

For $\dot{q}$ and $\dot{\pi}$ we can substitute from the canonical relations

$$\dot{q} = \frac{\partial H}{\partial \pi} = \{q, H\}, \quad \dot{\pi} = -\frac{\partial H}{\partial q} = \{\pi, H\}, \qquad (14.44)$$

where $\{q, H\}$ and $\{\pi, H\}$ are the Poisson brackets. This gives us the equation of continuity in terms of the Hamiltonian of the system. Since for any function F of the dynamical variables q and π we have [4]

$$\{q, F\} = \frac{\partial}{\partial \pi} F, \quad \text{and} \quad \{\pi, F\} = -\frac{\partial}{\partial q} F, \qquad (14.45)$$

therefore we can write the equation for $\partial\rho/\partial t$ in terms of the double Poisson brackets;

$$\frac{\partial \rho}{\partial t} = \{\pi, \{q, H\} \rho\} - \{\rho \{H, \pi\}, q\}. \qquad (14.46)$$

The quantal analogue of (14.46) can be found by using the Dirac rule of association between the classical Poisson bracket and the quantum mechanical commutator [15];

$$\{u, v\} \rightarrow \frac{1}{i\hbar} [u, v].$$ (14.47)

In §12.6 we discussed the ambiguities resulting from this rule of association, but this ambiguity is not the only source of difficulty in making transition from classical to quatum mechanics, since here we have the additional problem of the non-Hermiticity of the Hamiltonian. Now assuming the validity of this rule of association we obtain the master equation [16]

$$\frac{\partial \hat{\rho}}{\partial t} = -\frac{1}{\hbar^2} [\pi, [q, \mathsf{H}] \hat{\rho}] + \frac{1}{\hbar^2} [\hat{\rho} [\mathsf{H}, \pi], q],$$ (14.48)

where now $\hat{\rho}, \pi, q$ and H are all operators. The Hamiltonian H is the sum of two terms $\mathsf{H} = H_1 + i\Gamma$, where H_1 and Γ are Hermitian operators, $H_1^\dagger = H_1$. Substituting for H in (14.48) and simplifying the result we find the operator equation for $(\partial \hat{\rho}/\partial t)$

$$\frac{\partial \hat{\rho}}{\partial t} = -\frac{i}{\hbar} [H_1, \hat{\rho}] - \frac{i}{\hbar^2} [\pi, [q, \Gamma] \hat{\rho}] - \frac{i}{\hbar^2} [\hat{\rho} [\Gamma, \pi], q],$$ (14.49)

where H_1 and Γ are given by

$$\mathsf{H} = H_1 + i\Gamma = -i\omega\pi q - \frac{\lambda}{2}\pi q.$$ (14.50)

Using (14.50), Eq. (14.49) simplifies to

$$\frac{\partial \hat{\rho}}{\partial t} = -\frac{\omega}{\hbar} [\pi q, \hat{\rho}] + \frac{i\lambda}{2\hbar} ([\pi, q\hat{\rho}] + [\hat{\rho}\pi, q]).$$ (14.51)

This equation can also be written in terms of creation and annihilation operators,

$$a^\dagger = -i\sqrt{\frac{2}{\hbar}}\pi, \quad a = \frac{1}{\sqrt{2\hbar}}q.$$ (14.52)

Noting that the Hamiltonian is

$$\mathsf{H} = \hbar \left(\omega - \frac{i\lambda}{2} \right) a^\dagger a,$$ (14.53)

then the master equation becomes

$$\frac{\partial \hat{\rho}}{\partial t} = -i\omega [a^\dagger a, \hat{\rho}] - \frac{\lambda}{2} ([a^\dagger, a\hat{\rho}] + [\hat{\rho}a^\dagger, a]).$$ (14.54)

We can now write (14.54) in terms of the original operators defined by the quantum analogues of (4.66) and (4.72):

$$q = \frac{1}{\sqrt{\omega}} \left[p + \left(\frac{\lambda}{2} - i\omega \right) x \right],$$ (14.55)

and

$$\pi = \frac{i}{2\sqrt{\omega}} \left[p + \left(\frac{\lambda}{2} + i\omega \right) x \right], \tag{14.56}$$

Substituting for q and π in (14.51) the master equation takes the form

$$\frac{\partial \hat{\rho}}{\partial t} = -\frac{i}{\hbar} [H_1, \hat{\rho}] - \frac{\lambda}{4\hbar\omega} [p, [p, \hat{\rho}]] - \frac{\lambda^2}{8\hbar\omega} [x, [p, \hat{\rho}]]$$

$$- \frac{\lambda^2}{8\hbar\omega} [p, [x, \hat{\rho}]] - \frac{\lambda\omega_0^2}{4\hbar\omega} [x, [x, \hat{\rho}]] - \frac{i\lambda}{2\hbar} ([x, \hat{\rho}p] - [p, \hat{\rho}x]), \tag{14.57}$$

where

$$H_1 = \frac{1}{2} p^2 + \frac{\lambda}{4} (px + xp) + \frac{1}{2} \omega_0^2 x^2. \tag{14.58}$$

We have observed that if the state of the system can be described by a single wave function, then the density matrix satisfies Eq. (14.42). But for the damped harmonic oscillator with the Hamiltonian (14.58), $(\partial\hat{\rho}/\partial t)$ in Eq. (14.57) has additional terms proportional to λ and λ^2, therefore one can conclude that for such a damped system the master equation cannot be related to a single wave function [16].

14.4 Density Operator and the Fokker-Planck Equation

We have seen that the density matrix, $\hat{\rho}(t)$, in the Schrödinger picture satisfies Eq. (14.54) which we write as

$$i\hbar \frac{\partial \hat{\rho}(t)}{\partial t} = \hbar\omega \left[a^\dagger a, \ \hat{\rho}(t) \right] + [V(t), \ \hat{\rho}(t)]. \tag{14.59}$$

In this equation $V(t)$ is defined by

$$V(t) = -\frac{i\hbar\lambda}{2} \left(a^\dagger a \hat{\rho}(t) + \hat{\rho}(t) a^\dagger a - 2a\hat{\rho}(t) a^\dagger \right). \tag{14.60}$$

We can transform $\hat{\rho}(t)$ and write it in the interaction picture as the operator $\hat{\rho}_I(t)$ [2];

$$\hat{\rho}(t) = U_0^\dagger(t) \hat{\rho}_I(t) U_0(t), \tag{14.61}$$

where $U_0(t)$ is given by

$$U_0(t) = \exp \left[-i\omega t (a^\dagger a) \right]. \tag{14.62}$$

By substituting (14.61) and (14.62) in (14.59) and simplifying we obtain the following result for the operator $\hat{\rho}_I(t)$;

$$i\hbar \frac{\partial \hat{\rho}_I(t)}{\partial t} = [V_I(t), \ \hat{\rho}_I(t)], \tag{14.63}$$

with the commutator $[V_I(t), \hat{\rho}_I]$ given by

$$[V_I(t), \hat{\rho}_I(t)] = U_0^\dagger V(t) U_0(t) = \frac{i\hbar\lambda}{2} \left(a^\dagger a \hat{\rho}_I(t) + \hat{\rho}_I(t) a^\dagger a - 2a\hat{\rho}_I(t)a^\dagger \right). \quad (14.64)$$

We can transform (14.63) to a Fokker-Planck equation using the coherent state representation for $\hat{\rho}_I(t)$, i.e. we write it as [17]

$$\hat{\rho}_I(t) = \int P(\alpha, t) |\alpha\rangle\langle\alpha| \, d^2\alpha, \quad (14.65)$$

where $|\alpha\rangle$s are the coherent states,

$$a|\alpha\rangle = \alpha|\alpha\rangle. \quad (14.66)$$

Using the following correspondences:

$$a\hat{\rho}_I(t) \longleftrightarrow \alpha P(\alpha, t), \quad (14.67)$$

$$a^\dagger \hat{\rho}_I(t) \longleftrightarrow \left(\alpha^* - \frac{\partial}{\partial\alpha} \right) P(\alpha, t), \quad (14.68)$$

$$\hat{\rho}_I(t)a^\dagger \longleftrightarrow P(\alpha, t)\alpha^*, \quad (14.69)$$

and

$$\hat{\rho}_I(t)a \longleftrightarrow \left(\alpha - \frac{\partial}{\partial\alpha^*} \right) P(\alpha, t), \quad (14.70)$$

we obtain the Fokker-Planck equation

$$\frac{\partial P(\alpha, t)}{\partial t} = \frac{\lambda}{2} \left(\frac{\partial}{\partial\alpha}\alpha + \frac{\partial}{\partial\alpha^*}\alpha^* \right) P(\alpha, t). \quad (14.71)$$

In addition to Eq. (14.71) there are other representations which do not have classical analogues. Here it is easier to work with $Q(\alpha, \alpha^*)$ defined in terms of the density matrix by [1]

$$Q(\alpha, \alpha^*) = \langle \alpha | \hat{\rho}_I | \alpha \rangle. \quad (14.72)$$

From the normalization of the density operator, i.e.

$$\text{Trace}\{\hat{\rho}_I\} = \text{Trace} \int |\alpha\rangle\langle\alpha|\hat{\rho}_I \, d^2\alpha = \int \langle\alpha|\hat{\rho}_I|\alpha\rangle d^2\alpha = 1, \quad (14.73)$$

we find the normalization of $Q(\alpha, \alpha^*)$

$$\int Q(\alpha, \alpha^*) \, d^2\alpha = 1. \quad (14.74)$$

We can express the averages of anti-normally ordered products of creation and annihilation operators as an integral over $Q(\alpha, \alpha^*)$. Thus

$$\langle a^r a^{\dagger s} \rangle = \text{Trace}\left[a^r a^{\dagger s} \hat{\rho}_I \right] = \int \alpha^r \alpha^{*s} Q(\alpha, \alpha^*) \, d^2\alpha. \quad (14.75)$$

Similarly, relations like Eqs. (14.67)-(14.70) for $P(\alpha, \alpha^*)$, can be obtained from the definition of $Q(\alpha, \alpha^*)$;

$$a\hat{\rho}_I \longleftrightarrow \left(\alpha + \frac{\partial}{\partial \alpha^*}\right) Q(\alpha, \alpha^*), \qquad (14.76)$$

$$a^\dagger \hat{\rho}_I \longleftrightarrow \alpha^* Q(\alpha, \alpha^*), \qquad (14.77)$$

$$\hat{\rho}_I a \longleftrightarrow \alpha Q(\alpha, \alpha^*), \qquad (14.78)$$

and

$$\hat{\rho}_I a^\dagger \longleftrightarrow \left(\alpha^* + \frac{\partial}{\partial \alpha}\right) Q(\alpha, \alpha^*). \qquad (14.79)$$

From these relations and Eq. (14.63) we find the partial differential equation satisfied by $Q(\alpha, \alpha^*, t)$.

$$\frac{\partial Q(\alpha, \alpha^*, t)}{\partial t} = \left[\frac{\lambda}{2}\left(\frac{\partial}{\partial \alpha}\alpha + \frac{\partial}{\partial \alpha^*}\alpha^*\right) + \lambda\frac{\partial^2}{\partial \alpha \partial \alpha^*}\right] Q(\alpha, \alpha^*, t). \qquad (14.80)$$

As an interesting application of (14.80) we find the time evolution of the squeezed state $|\chi, \zeta\rangle$ for the damped harmonic oscillator where unlike the squeezed states of Caldirola-Kanai Hamiltonian both (Δx^2) and (Δp^2) decay in time.

Consider the combined action of the squeezing operator $U(\zeta)$, and the displacement operator $D(\chi)$ defined by

$$D(\chi) = \exp\left[-\frac{1}{2}|\chi|^2\right] e^{\chi a^\dagger(t)} e^{\chi^* a(t)} \qquad (14.81)$$

on the vacuum $|0\rangle$. If we choose $\theta = \pi$, U becomes a function of r, and we denote the resulting state by $|\chi, r\rangle$,

$$|\chi, r\rangle = D(\chi)U(r)|0\rangle. \qquad (14.82)$$

From the Baker-Hausdorff theorem we find the effect of the operator $U(r)$ on the annihilation (creation) operator $a\ (a^\dagger)$

$$U^\dagger(r)a(t)U(r) = a(t)\cosh r + a^\dagger(t)\sinh r. \qquad (14.83)$$

Similarly, we have for the $D(\chi)$ operator

$$D^\dagger(\chi)a(t)D(\chi) = a(t) + \chi. \qquad (14.84)$$

Noting that $a(t)$ which is now dependent on r and will be denoted by $a_r(t)$ is a complex operator satisfying the commutation relation $\left[a_r^\dagger(t),\ a_r(t)\right] = 1$, we can write it in terms of the real and imaginary operators X_1 and X_2:

$$a_r(t) = X_1 + iX_2, \qquad (14.85)$$

where these satisfy the commutation relation

$$[X_1, X_2] = \frac{i}{2}. \tag{14.86}$$

The variances in X_i s are defined by

$$(\Delta X_i^2) = \langle X_i^2 \rangle - \langle X_i \rangle^2, \quad i = 1, 2. \tag{14.87}$$

Thus from (14.86) it follows that the squeezed states are states of minimum uncertainty, i.e

$$\Delta X_1 \Delta X_2 = \frac{1}{4}, \tag{14.88}$$

but the uncertainties in X_1 and X_2 are given by

$$\Delta X_1 = \frac{1}{2} e^{-r}, \quad \text{and} \quad \Delta X_2 = \frac{1}{2} e^r. \tag{14.89}$$

Compared with the coherent state where $\Delta X_1 = \Delta X_2 = \frac{1}{2}$, here we have reduced fluctuations in one and increased fluctuations in the other, so that the product $\Delta X_1 \Delta X_2$ remains unchanged. Using these properties of the squeezed state $|\chi, r\rangle$ the representation of $Q(\alpha, \alpha^*, 0)$ can be found and it is of the form [17]-[19]

$$Q(\alpha, \alpha^*, 0) = \frac{1}{\pi \cosh r} \exp \left[-\frac{1}{2} (z - \langle z \rangle)^T \sigma^{-1} (z - \langle z \rangle) \right], \tag{14.90}$$

with

$$z^T = [\alpha, \alpha^*], \quad \langle z \rangle^T = [\langle a \rangle, \langle a^\dagger \rangle] = [\chi, \chi^*]. \tag{14.91}$$

Here T denotes the transpose and σ is defined by the following 2×2 matrix:

$$\sigma = \frac{1}{2} \begin{bmatrix} -\sinh 2r & \cosh 2r + 1 \\ \cosh 2r + 1 & -\sinh 2r \end{bmatrix}. \tag{14.92}$$

When damping is present then we solve Eq. (14.80) with the initial condition (14.90) [17]. The variances of the distribution in this case are:

$$(\Delta X_1^2)(t) = \frac{1}{4} \left[\left(e^{-2r} - 1 \right) e^{-\lambda t} + 1 \right], \tag{14.93}$$

and

$$(\Delta X_2^2)(t) = \frac{1}{4} \left[\left(e^{2r} - 1 \right) e^{-\lambda t} + 1 \right]. \tag{14.94}$$

These relations show that if $r > 0$, then $(\Delta X_1^2)(t) < 1/4$, and $(\Delta X_2^2)(t) > 1/4$, that is the fluctuations for $X_1(t)$ are decreased and for $X_2(t)$ are amplified.

14.5 The Density Matrix Formulation of a Solvable Model

In Chapter 8 we observed that for a harmonic oscillator coupled linearly to a bath of oscillators the classical equations of motion can be solved exactly. Now we want to consider the damped motion of a more general nonlinear system [20]. Again we start with the total Hamiltonian

$$H_T = H_S + H_B + H_I, \tag{14.95}$$

where H_S, H_B and H_I are the Hamiltonians for the system, the bath and the interaction respectively. Here we take the dissipative system S to be a particle and this particle can be in any number of the sites ν, where ν is an integer $-\infty < \nu < +\infty$. We denote the states of the particle by $|\nu\rangle$ s and assume that these states form an orthonormal set;

$$\langle \nu | \nu' \rangle = \delta_{\nu \nu'}. \tag{14.96}$$

We also assume that the Hamiltonian H_S is given by

$$H_S = \Omega S = \frac{1}{2}\Omega \left(s + s^{-1} \right), \tag{14.97}$$

where we have set $\hbar = 1$ and where Ω is a constant and s and s^{-1} are the raising and lowering operators

$$s|\nu\rangle = |\nu + 1\rangle, \quad s^{-1}|\nu\rangle = |\nu - 1\rangle. \tag{14.98}$$

Clearly S is a Hermitian operator, and its eigenfunctions are $|\eta\rangle$ s where

$$|\eta\rangle = \frac{1}{\sqrt{2\pi}} \sum_{\nu=-\infty}^{\infty} e^{i\eta\nu} |\nu\rangle, \quad 0 \leq \eta < 2\pi \tag{14.99}$$

From Eqs. (14.97) and (14.99) we can calculate the matrix elements of H in η-representation and this is given by

$$\langle \eta | H_S | \eta' \rangle = \Omega \cos \eta \, \delta \left(\eta - \eta' \right). \tag{14.100}$$

The bath consists of a collection of harmonic oscillators for which the Hamiltonian is

$$H_B = \sum_n \omega_n a_n^\dagger a_n, \quad \omega_n > 0. \tag{14.101}$$

We also assume that the interaction Hamiltonian is of the form

$$H_I = \frac{1}{2}\Omega \left(s + s^{-1} \right) \sum_n \epsilon_n \left(a_n^\dagger + a_n \right), \tag{14.102}$$

with the coupling constants ϵ_n.

The total Hamiltonian H_T commutes with H_S, and therefore it is diagonal in η-representation

$$\langle \eta | H_T | \eta' \rangle = \delta (\eta - \eta') \{\Omega \cos \eta + H_\eta\}, \qquad (14.103)$$

where in (14.103) H_η is defined as

$$H_\eta = \sum_n H_{\eta,n} = \sum_n \left[\omega_n a_n^\dagger a_n + \cos \eta \, \epsilon_n \left(a_n^\dagger + a_n\right)\right]. \qquad (14.104)$$

We can write $H_{\eta,n}$ in the form

$$H_{\eta,n} = \omega_n \left(a_n^\dagger + u_n\right) \left(a_n + u_n\right) - \omega_n u_n^2, \qquad (14.105)$$

with

$$u_n = \frac{\epsilon_n}{\omega_n} \cos \eta. \qquad (14.106)$$

Having found the matrix elements of H_T, we want to find the master equation giving us the time development of the density matrix $\hat{\rho}_S(t)$. We note that at $t = 0$, the density matrix of the total system is of the form of a product

$$\hat{\rho}_T(0) = \hat{\rho}_S(0) \otimes \hat{\rho}_B, \qquad (14.107)$$

where $\hat{\rho}_S(0)$ is the initial state of S, and $\hat{\rho}_B(0)$ is the initial state of the bath. This density matrix evolves in time, and at some later time $\hat{\rho}_T(t)$ is given by

$$\hat{\rho}_T(t) = \exp \left(-iH_T t\right) \hat{\rho}_T(0) \exp \left(iH_T t\right). \qquad (14.108)$$

The time evolution of the density matrix of S alone can be obtained by averaging $\hat{\rho}_T(t)$ over the bath;

$$\hat{\rho}_S(t) = \text{Trace} \exp \left(-iH_T t\right) \hat{\rho}_S(0) \otimes \hat{\rho}_B \exp \left(iH_T t\right). \qquad (14.109)$$

Next we calculate the matrix elements of $\hat{\rho}_T(t)$ with the help of (14.103) and (14.104);

$$\langle \eta | \hat{\rho}_T(t) | \eta' \rangle = \langle \eta | \hat{\rho}_T(0) | \eta' \rangle e^{-i\Omega t \left(\cos \eta - \cos \eta'\right)} \otimes e^{-iH_\eta t} \hat{\rho}_B(0) e^{iH_{\eta'} t}. \qquad (14.110)$$

The last factor in (14.110) can be determined by noting that

$$e^{-iH_\eta t} \hat{\rho}_B(0) e^{iH_{\eta'} t} \rightarrow \prod_n \left\{ \sum_{N_n} \langle N_n | e^{-iH_{\eta,n} t} e^{iH_{\eta',n} t} | N_n \rangle \right\}. \qquad (14.111)$$

Now we substitute for $H_{\eta,n}$ and $H_{\eta',n}$ from (14.105) to obtain

$$\begin{aligned}
e^{-iH_{\eta,n} t} e^{iH_{\eta',n} t} &= \exp \left[i\omega_n t \left(u_n^2 - u_n'^2\right)\right] \exp \left[-i\omega_n t \left(a_n^\dagger + u_n\right) \left(a_n + u_n\right)\right] \\
&\times \exp \left[i\omega_n t \left(a_n^\dagger + u_n'\right) \left(a_n + u_n'\right)\right].
\end{aligned} \qquad (14.112)$$

By taking the trace of this operator over the bath states we have

$$
\begin{aligned}
\text{Trace } e^{-iH_{\eta,n}t} e^{iH_{\eta',n}t} \;=\; & \exp\left[i\left(\omega_n t - \sin\omega_n t\right)\left(u_n^2 - u_n'^2\right)\right] \\
& \times \; \text{Trace } \exp\left\{u_n'\left[a_n\left(e^{i\omega_n t}-1\right) - a_n^\dagger\left(e^{-i\omega_n t}-1\right)\right]\right. \\
& \left. + \; u_n\left[a_n\left(1-e^{i\omega_n t}\right) - a_n^\dagger\left(1-e^{-i\omega_n t}\right)\right]\right\}. \quad (14.113)
\end{aligned}
$$

Now by writing the trace as a sum over the eigenstates of $a_n^\dagger a_n$, we have the final result for $\hat{\rho}_S$ which is

$$
\begin{aligned}
\langle\eta\,|\hat{\rho}_S(t)|\,\eta'\rangle \;=\; & \langle\eta\,|\hat{\rho}_S(0)|\,\eta'\rangle\, e^{-i\Omega t\left(\cos\eta - \cos\eta'\right)} \\
& \times\; \exp\left[iF(t)\left(\cos^2\eta - \cos^2\eta'\right) - G(t)\left(\cos\eta - \cos\eta'\right)^2\right],
\end{aligned}
$$

$$(14.114)$$

where

$$
F(t) = \sum_n \left(\frac{\epsilon_n^2}{\omega_n}\right)\left(t - \frac{\sin\omega_n t}{\omega_n}\right), \qquad (14.115)
$$

and

$$
G(t) = \sum_n \left(\frac{\epsilon_n^2}{\omega_n^2}\right)\left(1 - \cos\omega_n t\right). \qquad (14.116)
$$

If we differentiate (14.114) with respect to time we obtain an equation like (14.63)-(14.64)

$$
\begin{aligned}
\frac{\partial\hat{\rho}_S(t)}{\partial t} \;=\; & -i\Omega\,[S,\hat{\rho}_S(t)] + i\dot{F}(t)\,[S^2,\hat{\rho}_S(t)] \\
& -\; \dot{G}(t)\left\{S^2\hat{\rho}_S(t) - 2S\hat{\rho}_S(t)S + \hat{\rho}_S(t)S^2\right\}, \qquad (14.117)
\end{aligned}
$$

but here $\dot{F}(t)$ and $\dot{G}(t)$ are time-dependent functions. Let us suppose that the ω_n s are dense on the segment $0 < \omega_n < \infty$, and let us define as in the case of Ullersma's model a strength function $\gamma(\omega)$ by

$$
\gamma(\omega)d\omega = \sum_{\omega < \omega_n < \omega + \Delta\omega} \epsilon^2, \qquad (14.118)
$$

and this strength function must satisfy the integrability condition, viz,

$$
\int_0^\infty \frac{\gamma(\omega)}{\omega}\,d\omega = \text{finite}. \qquad (14.119)
$$

In terms of $\gamma(\omega)$ we can write $F(t)$ and $G(t)$ as

$$
F(t) = \int_0^\infty \frac{\gamma(\omega)}{\omega}\left(t - \frac{\sin(\omega t)}{\omega}\right)d\omega, \qquad (14.120)
$$

and

$$
G(t) = \int_0^\infty \frac{\gamma(\omega)}{\omega^2}\left(1 - \cos(\omega t)\right)d\omega. \qquad (14.121)
$$

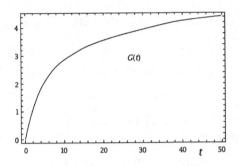

Figure 14.1: The function $G(t)$ calculated from (14.124), is shown as a function of time.

For instance if we choose

$$\gamma(\omega) = \frac{\omega}{\omega^2 + \nu^2}, \tag{14.122}$$

then we can calculate $F(t)$ and $\dot{G}(t)$ analytically:

$$F(t) = \frac{\pi}{2\nu}\left(t - \frac{1 - e^{-\nu t}}{\nu}\right), \tag{14.123}$$

and

$$\dot{G}(t) = \frac{1}{2\nu}\left\{\left[e^{\frac{-\nu t}{2}} Ei\left(\frac{\nu t}{2}\right) - e^{\frac{\nu t}{2}} Ei\left(-\frac{\nu t}{2}\right)\right]\right\}. \tag{14.124}$$

For $\nu t \gg 2$, both $F(t)$ and $G(t)$ are approximately linear functions of t (for the latter see Fig. (14.1)), i.e. the coefficient $\dot{F}(t)$ in Eq. (14.117), becomes time-independent, whereas $\dot{G}(t)$ goes slowly to zero.

14.6   Wigner Distribution Function for the Damped Oscillator

The coherent state representation of $\hat{\rho}$ was just one of a number of ways of converting the operator equation for the density to a c-number equation. There are other ways of transforming Eq. (14.57) to a c-number equation. For instance we can start with Eq. (14.43) and write a master equation for the Wigner distribution function $W(p, x, t)$ in terms of the original coordinate x and momentum p [9]. This distribution function W is related to $\hat{\rho}$ by

$$W(p, x, t) = \frac{1}{2\pi\hbar}\int_{-\infty}^{\infty} e^{\frac{ip y}{\hbar}}\left\langle x - \frac{1}{2}y \,|\hat{\rho}|\, x + \frac{1}{2}y\right\rangle dy. \tag{14.125}$$

We know that in the absence of damping, $\lambda = 0$, the Wigner distribution function for the harmonic oscillator is given by the solution of the partial differential equation

$$\frac{\partial W}{\partial t} = -\frac{\partial}{\partial x}(pW) + \omega_0^2 \frac{\partial}{\partial p}(xW), \quad \lambda = 0, \tag{14.126}$$

and we want to find a similar equation in the presence of damping. Let I_1 denote the contribution of the term

$$\left\{-\left(\frac{\lambda}{4\hbar\omega}\right)[p,[p,\hat{\rho}]]\right\} \tag{14.127}$$

to the distribution function, i.e.

$$I_1 = -\frac{\lambda}{4\hbar\omega}\frac{1}{2\pi\hbar}\int_{-\infty}^{\infty} e^{\frac{ipy}{\hbar}}\left\langle x - \frac{1}{2}y\,|p[p,\hat{\rho}] - [p,\hat{\rho}]p|\,x + \frac{1}{2}y\right\rangle dy. \tag{14.128}$$

We can write I_1, using a complete set of states,

$$\int |r\rangle\langle r|dr = 1, \tag{14.129}$$

as

$$\begin{aligned}
I_1 &= -\frac{\lambda}{8\pi\hbar^2\omega}\int_{-\infty}^{\infty}\exp\left(\frac{ipy}{\hbar}\right)dy\int_{-\infty}^{\infty}ds\int_{-\infty}^{\infty}dr \times \\
&\quad \times \left[\left\langle x - \frac{1}{2}y|p|r\right\rangle\left\{\langle r|p|s\rangle\left\langle s|\hat{\rho}|x + \frac{1}{2}y\right\rangle - \langle r|\hat{\rho}|s\rangle\left\langle s|p|x + \frac{1}{2}y\right\rangle\right\}\right. \\
&\quad - \left\{\left\langle x - \frac{1}{2}y|p|s\right\rangle\langle s|\hat{\rho}|r\rangle - \left\langle x - \frac{1}{2}y\,|\hat{\rho}|\,s\right\rangle\langle s|p|r\rangle\right\} \\
&\quad \left. \times \left\langle r\,|p|\,x + \frac{1}{2}y\right\rangle\right].
\end{aligned} \tag{14.130}$$

We simplify (14.130) by noting that

$$\langle r'\,|p|\,r\rangle = -i\hbar\frac{\partial}{\partial r'}\delta(r' - r). \tag{14.131}$$

Thus substituting for the expectation value of p in (14.130) and using partial integration we obtain

$$\begin{aligned}
I_1 &= \frac{\lambda}{8\pi\omega}\frac{\partial}{\partial x}\int_{-\infty}^{\infty}\exp\left(\frac{ipy}{\hbar}\right)dy\int_{-\infty}^{\infty}\left[\delta\left(s - x + \frac{1}{2}y\right)\frac{\partial}{\partial s}\left\langle s\,|\hat{\rho}|\left(x + \frac{1}{2}y\right)\right\rangle\right. \\
&\quad + \left. \delta\left(s - x - \frac{1}{2}y\right)\frac{\partial}{\partial s}\left\langle\left(x - \frac{1}{2}y\right)|\hat{\rho}|\,s\right\rangle\right]ds.
\end{aligned} \tag{14.132}$$

This expression reduces to

$$I_1 = \frac{\lambda\hbar}{4\omega}\frac{1}{2\pi\hbar}\frac{\partial^2}{\partial x^2}\int_{-\infty}^{\infty}e^{\frac{ipy}{\hbar}}\left\langle x - \frac{1}{2}y\,|\hat{\rho}|\,x + \frac{1}{2}y\right\rangle dy = \frac{\hbar\lambda}{4\omega}\frac{\partial^2 W}{\partial x^2}. \tag{14.133}$$

The contribution of other terms in (14.57) to the equation for distribution function can be calculated in a similar way. The final result is given by the partial differential equation for $W(p, x, t)$ [16]:

$$\frac{\partial W}{\partial t} = -\frac{\partial}{\partial x}(pW) + \omega_0^2 \frac{\partial}{\partial p}(xW) + \lambda \frac{\partial}{\partial p}(pW)$$
$$+ \frac{\lambda \hbar}{4\omega} \frac{\partial^2 W}{\partial x^2} - \frac{\hbar \lambda^2}{4\omega} \frac{\partial^2 W}{\partial x \partial p} + \frac{\hbar \lambda \omega_0^2}{4\omega} \frac{\partial^2 W}{\partial p^2}. \tag{14.134}$$

This equation can be solved approximately (or numerically) for $W(x, p, t)$. For the initial condition we assume that at $t = 0$ there is no coupling between the oscillator and the damping force; therefore

$$W(x, p, 0) = \frac{1}{2\pi\hbar} \int_{-\infty}^{\infty} \psi_n^*\left(x + \frac{y}{2}\right) \psi_n\left(x - \frac{y}{2}\right) \exp\left(\frac{ipy}{\hbar}\right) dy, \tag{14.135}$$

where $\psi_n(x)$ is the normalized harmonic oscillator wave function for the n-th excited state. The integral in (14.135) can be obtained analytically [21]:

$$W(x, p, 0) = \frac{(-1)^n}{\pi\hbar} L_n\left(r^2\right) \exp\left(-\frac{1}{2}r^2\right); \tag{14.136}$$

r^2 in this equation represents the dimensionless quantity

$$r^2 = 2\left(\frac{m\omega_0}{\hbar^2}x^2 + \frac{1}{m\omega_0\hbar}p^2\right), \tag{14.137}$$

and L_n is the Laguerre polynomial. For the ground state of the oscillator (14.136) takes the simple form of

$$W_0(x, p, 0) = \frac{1}{\pi\hbar} \exp\left(-\frac{1}{2}r^2\right), \tag{14.138}$$

which is always non-negative, whereas for the first excited state we find

$$W_1(x, p, 0) = \frac{1}{\pi\hbar}\left(r^2 - 1\right) \exp\left(-\frac{1}{2}r^2\right), \tag{14.139}$$

which is negative around the origin but becomes positive for larger r.

Here we have discussed the density matrix and the Wigner distribution function starting with the Heisenberg equations for the damped harmonic oscillator. We can change our starting point and use the model where a harmonically bound particle is coupled to a field or to a bath of oscillators and for this particle derive both $\hat{\rho}$ and W. Once these are known for the whole system, then one can try to isolate the motion of the central particle. This formulation of the damped motion of the particle will be studied in the next section as well as in §17.1.

14.7 Density Operator for a Particle Coupled to a Heat Bath

When the central particle is linearly coupled to a heat bath of N degrees of freedom and at the same time is subject to an external force, then the Hamiltonian will be quadratic in coordinates and momenta. We can write this Hamiltonian in the compact form of

$$H = \frac{1}{2} z^T B z + C(t) z, \tag{14.140}$$

where z^T and $C(t)$ are row matrices with $2N + 2$ elements

$$z^T = [P, \ p_1, \dots p_N, Q, \ q_1, \dots q_N], \tag{14.141}$$

and

$$C(t) = [C_1(t), C_2(t), \dots C_{2N}(t)]. \tag{14.142}$$

In these equations T denotes the transpose of the matrix and B is a symmetric matrix $B^T = B$.

The equations of motion derived from (14.140) can be written in the matrix form:

$$\frac{dz}{dt} = -ABz - AC(t). \tag{14.143}$$

Here A is a $(2N + 1) \times (2N + 1)$ antisymmetric matrix with constant matrix elements

$$A = \begin{bmatrix} 0 & I_{N+1} \\ -I_{N+1} & 0 \end{bmatrix}, \quad A^2 = -I_{2N+2}, \tag{14.144}$$

where I_{N+1} in an $(N + 1) \times (N + 1)$ unit matrix. Since B is symmetric and A is antisymmetric, therefore Trace $(AB) = 0$.

As an example consider the Hamiltonian given by (10.1) where

$$B = \begin{bmatrix} K_{N+1} & 0 \\ 0 & -I_{N+1} \end{bmatrix}, \tag{14.145}$$

and

$$K_{N+1} = \frac{1}{2} \begin{bmatrix} 2\Omega_0^2 & \epsilon_1, & \epsilon_2 & \cdots & \epsilon_N \\ \epsilon_1 & 2\omega_1^2 & 0 & \cdots & 0 \\ \epsilon_2 & 0 & 2\omega_2^2 & \cdots & 0 \\ & \cdots & \cdots & \cdots & \\ \epsilon_N & 0 & 0 & \cdots & 2\omega_N^2 \end{bmatrix}. \tag{14.146}$$

Now let us consider the first order differential equation (14.143) with B given by (14.145). This represents a central particle which interacts with a heat bath. From what we have seen earlier for a single harmonic oscillator, the Wigner distribution function is of the form given by (14.126) but now it is for $N + 1$ degrees of freedom. Therefore we can write this distribution function as

$$\frac{\partial W}{\partial t} = \sum_{j=1}^{2N+2} \frac{\partial}{\partial z_j} \left[\sum_k \{(AB)_{jk} z_k + A_{jk} C_k\} W \right], \tag{14.147}$$

but the corresponding density matrix will not be positive definite. A way to address this problem is to add the terms

$$\sum_{jk} D_{jk} \frac{\partial^2 W}{\partial z_j \partial z_k},$$ (14.148)

to the right hand side of (14.147) and make it like a Fokker-Planck equation. The coefficients D_{jk} (diffusion coefficients) are independent of the coordinates z_j, but can depend on time. These coefficients are not arbitrary, but are subject to certain conditions, and these conditions ensure that $W(z,t)$ remains a positive distribution function for all times [22].

We can also solve (14.147) for the whole system and then average over the coordinates and momenta of the bath. For this we assume that initially the Wigner distribution function is given as a product

$$W(z,t) = W(P,Q)W(p_1 \ldots p_N; q_1 \ldots q_N)$$ (14.149)

and then solve the partial differential equation (14.147) with this initial condition. Averaging $W(z,t)$ over the bath degrees of freedom gives us $W(P,Q,t)$ for the motion of the central particle [22]-[25].

Bibliography

[1] C.W. Gardiner, *Quantum Noise*, (Springer-Verlag, Berlin 1991) Chapter 2.

[2] W.H. Louisell, *Quantum Statistical Properties of Radiation*, (John Wiley & Sons, New York, 1973).

[3] H.-P. Breuer and F. Petruccione, *The Theory of Open Quantum Systems*, (Oxford University Press, 2002) Chapter 3.

[4] H. Goldstein, *Classical Mechanics*, Second Edition (Addison-Wesley Reading, 1980).

[5] J.B. Marion, *Classical Dynamics of Particles and Systems*, Second Edition, (Academic Press, New York, 1970) §7.14.

[6] G. Gerlich, Physica A, 69, 4586 (1973).

[7] W.-H. Steep, Physica A, 95, 181 (1979).

[8] A.O. Bolivar, Phys. Rev A58, 4330 (1998).

[9] E. Wigner, Phys. Rev. 40, 749 (1932).

[10] L.S.F. Olavo, Physica A, 262, 197 (1999).

[11] A.O. Bolivar, Physica A, 301, 210 (2001).

[12] A.O. Bolivar, Can. J. Phys. 81, 663 (2003).

[13] F.J. Dyson, Phys. Rev. 75, 486 (1949).

[14] J.J. Sakurai, *Modern Quantum Mechanics*, (Addison-Wesley, Menlo Park, California, 1985) p. 325.

[15] P.A.M. Dirac, *The Principles of Quantum Mechanics*, Fourth Edition, (Oxford University Press, 1958).

[16] H. Dekker, Physica, 95A, 311 (1979).

[17] G.J. Milburn and D.F. Walls, Am. J. Phys. 51, 1134 (1983).

[18] H.P. Yuen, Phys. Rev. A13, 2226 (1976).

[19] R.W. Henry and S.C. Glotzer, Am. J. Phys. 56, 318 (1988).

[20] N.G. van Kampen, J. Stat. Phys. 78, 299 (1995).

[21] Y.S. Kim and M.E. Noz, *Phase Space Picture of Quantum Mechanics*, (World Scientific, Singapore, 1991) §3.4.

[22] V.V. Dodonov and O.V. Man'ko, Theor. Math. Phys. vol. 65, 1033 (1986).

[23] V.V. Dodonov and V.I. Man'ko, in *Classical and Quantum Effects in Electrodynamics*, Edited by A.A. Komar (Nova Science, Commack, 1986) p. 53.

[24] V.V. Dodonov, O.V. Man'ko and V.I. Man'ko, J. Russian Laser Res. 16, 1 (1995).

[25] V.V. Dodonov, and V.I. Man'ko, in *Quantum Field Theory, Quantum Mechanics and Quantum Optics*, Preceeding of the Lebedev Physics Institute vol. 187, Edited by V.V. Dodonov, and V.I. Man'ko (Nova Scientific, Commack, 1989) p. 209.

Chapter 15

Path Integral Formulation of a Damped Harmonic Oscillator

An elegant way of quantizing the dissipative systems which is specially suited for the damped harmonic oscillator is by means of the Feynman formulation [1] where one can obtain the propagator and from the propagator determine the time development of the wave function. This same method of path integration can be applied to a central particle under the action of the potential $V(Q)$ but coupled linearly to a bath of harmonic oscillators. We will discuss the latter problem in detail later.

For the construction of the propagator we can use either the Lagrangian or the Hamiltonian, but here like other methods of quantization (Dirac or Schrödinger) the Feynman approach will not lead to a unique result. The non-uniqueness arises from the fact that we can use an infinite set of classically equivalent Lagrangians as the starting point. But even if we decide on a single Lagrangian L, the ordering of $x(t)$ and $\dot{x}(t)$ operators will not be a trivial task in the case of dissipative systems where, as we have noted, there are terms like $x^n(t)\dot{x}^j(t)$ in the Lagrangian. Once it was thought that the path integral method can be used to determine a unique way of ordering the operators $x(t)$ and $\dot{x}(t)$ [2]. But later it was shown that this is not the case and in the path integral formulation like other methods there is no unique way of ordering x and $\dot{x}$ operators [3] [4].

In addition to the ambiguity due to the ordering of x and $\dot{x}$ factors, there is also the problem of choosing the proper classical Lagrangian for the path integral quantization. By "proper Lagrangian" we mean those Lagrangians

223

that are quadratic functions of velocity, since for equivalent Lagrangians like $\frac{1}{6p_0}m^2\dot{x}^3e^{2\lambda t}$, Eq. (3.102), the Feynman method of construction of propagator does not work [1].

Among the many possible ways of constructing the propagator for the damped harmonic oscillator we focus our attention on two simple cases. In the first one we start with the time-dependent Lagrangian for linear damping and find the propagator and using this propagator we calculate the time-dependent wave function. For the second example we determine the propagator for the classical Lagrangian assuming an optical harmonic potential, Eq. (4.64).

15.1   Propagator for the Damped Harmonic Oscillator

There are different ways of constructing the propagator from the classical Lagrangian for one degree of freedom when the damping is linear in velocity, i.e.

$$L\left(x,\dot{x},t\right) = \left[\frac{1}{2}m\dot{x}^2 - V(x)\right]e^{\lambda t}. \tag{15.1}$$

We can start with the definition of the kernel $G\left(x'',x',t'',t'\right)$ in phase space [5]-[11]

$$\begin{aligned}
G\left(x'',x',t'',t'\right) &= \int \exp\left\{\frac{i}{\hbar}\int_{t'}^{t''}[p\dot{x} - H(p,x,t)]\,dt\right\}\mathcal{D}[x]\mathcal{D}[p]\\
&= \lim_{N\to\infty}\int_{-\infty}^{\infty}\cdots\int_{-\infty}^{\infty}\frac{dp'}{\sqrt{2\pi\hbar}}\prod_{k=1}^{N-1}\frac{dp_k dx_k}{\sqrt{2\pi\hbar}}\\
&\quad\times\ \exp\left\{\frac{i}{\hbar}\sum_{k=0}^{N-1}[p_k(x_{k+1} - x_k) - \varepsilon H(p_k,x_k,t)]\right\},
\end{aligned}$$

$$\tag{15.2}$$

and carry out the functional integration in (15.2) to find $G\left(x'',x',t'',t'\right)$. In Eq. (15.2) $H(p,x,t)$ is the Kanai-Caldirola Hamiltonian found from (15.1), $\varepsilon = \frac{t''-t'}{N}$ is the small time increment and $x'' = x\left(t''\right)$ and $x' = x\left(t'\right)$ are the initial and final coordinates respectively. By substituting for $H(p,x,t)$ in (15.2) and integrating over all momenta in the phase space we find [5] [6]

$$G\left(x'',x';t'',t'\right) = \int \exp\left\{\frac{i}{\hbar}\int_{t'}^{t''}L\left(x,\dot{x},t\right)dt\right\}\mathcal{D}[x]$$

$$= \lim_{N\to\infty}\left(\frac{me^{\lambda t'}}{2\pi i\hbar\varepsilon}\right)^{\frac{1}{2}}\prod_{k=1}^{N-1}\left(\frac{me^{\lambda t_k}}{2\pi i\hbar\varepsilon}\right)^{\frac{1}{2}}$$

$$\times \int_{-\infty}^{\infty} \cdots \int_{-\infty}^{\infty} \exp\left\{ \frac{i\varepsilon}{\hbar} \sum_{k=0}^{N-1} \left[\frac{m}{2} \left(\frac{x_{k+1} - x_k}{\varepsilon} \right)^2 - V(x_k) \right] e^{\lambda t_k} \right\} \prod_{k=1}^{N-1} dx_k,$$

(15.3)

where we have used the Gaussian integral [7]-[8]

$$\frac{1}{\sqrt{2\pi\hbar}} \int_{-\infty}^{\infty} \exp\left\{ \frac{i}{\hbar} \left[p_k (x_{k+1} - x_k) - \frac{\varepsilon p_k^2}{2m} e^{-\lambda t_k} \right] \right\} dp_k$$

$$= \left(\frac{m e^{\lambda t_k}}{i\varepsilon} \right) \exp\left[\frac{im\varepsilon e^{\lambda t_k}}{2\hbar} \left(\frac{x_{k+1} - x_k}{\varepsilon} \right)^2 \right],$$

(15.4)

to carry out the integration over momenta.

Now let us consider the damped harmonic oscillator with $V(x) = \frac{1}{2} m\omega_0^2 x^2$ for which (15.3) can be written as

$$G(x'', x'; t'', t') = \lim_{N \to \infty} \left[\prod_{k=1}^{N-1} \left(\frac{m e^{\lambda t_k}}{2\pi i \hbar} \right)^{\frac{1}{2}} \right]$$

$$\times \int_{-\infty}^{\infty} \cdots \int_{-\infty}^{\infty} \exp\left\{ \frac{im\varepsilon}{2\hbar} \left[\left(\frac{1}{\varepsilon^2} \right) \sum_{k=0}^{N-1} e^{\lambda t_k} (x_{k+1} - x_k)^2 - \sum_{k=0}^{N-1} e^{\lambda t_k} \omega_0^2 x_k^2 \right] \right\}$$

$$\times \prod_{k=1}^{N-1} dx_k.$$

(15.5)

It is convenient to change the variable x_k to s_k where

$$s_k = \left(\frac{m}{2\hbar\varepsilon} \right)^{\frac{1}{2}} \exp\left(\frac{\lambda t_k}{2} \right) x_k,$$

(15.6)

and then $G(x'', x', t'', t')$ can be rewritten as

$$G(s'', s'; t'', t') = \lim_{N \to \infty} \frac{1}{(i\pi)^{\frac{N}{2}}} \left(\frac{m e^{\lambda t''}}{2\hbar\varepsilon} \right)^{\frac{1}{2}}$$

$$\times \exp\left\{ i \left[\left(s''^2 + s'^2 \right) - \varepsilon^2 \omega_0^2 s'^2 \right] \right\}$$

$$\times \int_{-\infty}^{\infty} \cdots \int_{\infty}^{\infty} \exp\left[i \left(\sum_{k=1}^{N-1} (1 + e^{\lambda\varepsilon} - \omega_0^2 \varepsilon^2) s_k^2 - 2 \sum_{k=1}^{N-1} e^{\frac{\lambda\varepsilon}{2}} s_k s_{k+1} \right) \right]$$

$$\times \prod_{k=1}^{N-1} ds_k.$$

(15.7)

This multiple integral can be transformed to a Gaussian integral

$$\int_{-\infty}^{\infty} \cdots \int_{\infty}^{\infty} \exp\left[i \left(s^T A s + 2 b^T s \right) \right] \prod_{k=1}^{N-1} ds =$$

$$= (i\pi)^{\frac{N}{2}} \frac{1}{\sqrt{\det A}} \exp\left(-ib^T A^{-1} b\right), \tag{15.8}$$

where the matrix A is given by

$$A = \begin{bmatrix} a_1 & -d & 0 & 0\cdots\cdots 0 & 0 & 0 & 0 \\ -d & a_2 & -d & 0\cdots\cdots 0 & 0 & 0 & 0 \\ 0 & -d & a_3 & -d\cdots\cdots 0 & 0 & 0 & 0 \\ \cdots & \cdots & \cdots & \cdots & \cdots & \cdots & \cdots \\ 0 & 0 & 0 & 0\cdots\cdots -d & a_{N-3} & -d & 0 \\ 0 & 0 & 0 & 0\cdots\cdots 0 & -d & a_{N-2} & -d \\ 0 & 0 & 0 & 0\cdots\cdots 0 & 0 & -d & a_{N-1} \end{bmatrix}. \tag{15.9}$$

The matrix elements a_k and d are functions of ε;

$$a_k = 1 + e^{\lambda\varepsilon} - \omega_0^2 \varepsilon^2, \quad \text{and} \quad d = \exp\left(\frac{\lambda\varepsilon}{2}\right), \tag{15.10}$$

while the column matrix b consists of the following elements

$$b_1 = -s' \exp\left(\frac{\lambda\varepsilon}{2}\right) = -\sqrt{\frac{m}{2\hbar\varepsilon}}\, x' \exp\left(\frac{\lambda(t'+\varepsilon)}{2}\right), \tag{15.11}$$

$$b_k = 0, \quad k = 2, 3 \cdots N - 2, \tag{15.12}$$

and

$$b_{N-1} = -s'' \exp\left(\frac{\lambda\varepsilon}{2}\right) = -\sqrt{\frac{m}{2\hbar\varepsilon}}\, x'' \exp\left[\frac{\lambda}{2}(t''+\varepsilon)\right]. \tag{15.13}$$

Now if we substitute (15.8) and (15.9) in (15.7) and note that as $\varepsilon \to 0$, $\exp\left(-i\varepsilon^2\omega_0^2 s'^2\right) \to 1$ we obtain

$$G\left(s'', s'; t'', t'\right) = \lim_{\varepsilon\to 0} \left(\frac{me^{\lambda t''}}{2\pi i\hbar\varepsilon \det A}\right)^{\frac{1}{2}} \exp\left[iB\left(s'', s'; \varepsilon\right)\right], \tag{15.14}$$

where

$$B\left(s'', s'; \varepsilon\right) = \left(s''^2 + s'^2\right) - b^T(\varepsilon) A^{-1}(\varepsilon) b(\varepsilon). \tag{15.15}$$

Next we find the limit of $\varepsilon \to 0$ of (15.14). First let us find the limit $\varepsilon \det A$ as ε goes to zero. To this end we consider the following determinants:

$$q_{N-1} = \varepsilon a_{N-1},$$

$$q_{N-2} = \varepsilon \begin{vmatrix} a_{N-2} & -d \\ -d & a_{N-1} \end{vmatrix},$$

$$q_{N-3} = \varepsilon \begin{vmatrix} a_{N-3} & -d & 0 \\ -d & a_{N-2} & -d \\ 0 & -d & a_{N-1} \end{vmatrix},$$

$$\cdots\cdots\cdots\cdots$$

$$q_1 = \varepsilon \det A \tag{15.16}$$

and

$$p_1 = \varepsilon a_1,$$

$$p_2 = \varepsilon \begin{vmatrix} a_1 & -d \\ -d & a_2 \end{vmatrix},$$

$$p_3 = \varepsilon \begin{vmatrix} a_1 & -d & 0 \\ -1 & a_2 & -d \\ 0 & -d & a_3 \end{vmatrix},$$

$$\cdots\cdots\cdots\cdots$$

$$p_{N-1} = \varepsilon \det A. \tag{15.17}$$

We can verify that q_k and p_k satisfy the finite difference equations

$$\frac{1}{\varepsilon^2} \left(q_{n+1} - 2q_n + q_{n-1} \right) = -\omega_0^2 q_n - \frac{\lambda}{\varepsilon} \left(q_{n+1} - q_n \right), \tag{15.18}$$

and

$$\frac{1}{\varepsilon^2} \left(p_{n+1} - 2p_n + p_{n-1} \right) = -\omega_0^2 p_n + \frac{\lambda}{\varepsilon} \left(p_n - p_{n-1} \right), \tag{15.19}$$

respectively. In the limit of $\varepsilon \to 0$, Eqs. (15.18) and (15.19) become differential equations

$$\ddot{q} + \lambda \dot{q} + \omega_0^2 q = 0, \tag{15.20}$$

and

$$\ddot{p} - \lambda \dot{p} + \omega_0^2 p = 0. \tag{15.21}$$

These equations are subject to the boundary conditions that at the final time t'',

$$q(t'') = 0, \quad \dot{q}(t'') = -1, \tag{15.22}$$

whereas at the initial time t'

$$p(t') = 0, \quad \dot{p}(t') = 1. \tag{15.23}$$

Thus the solution of (15.20) and (15.21) subject to these conditions are:

$$q(t) = \frac{1}{\omega} \exp\left[-\frac{\lambda}{2}(t - t'') \right] \sin \omega (t'' - t), \tag{15.24}$$

and

$$p(t) = \frac{1}{\omega} \exp\left[-\frac{\lambda}{2}(t' - t) \right] \sin \omega (t - t'), \tag{15.25}$$

where $\omega = \sqrt{\omega_0^2 - \frac{\lambda^2}{4}}$. We note that while $q(t)$ and $p(t)$ are the solutions of canonical equations for coordinates and momentum found from Kanai-Caldirola

Hamiltonian, the initial conditions are different, and that both $q(t)$ and $p(t)$ have the same dimensions.

From the definitions of q_k and p_k we can also find the following relation

$$q_{k+1}p_k - d^2 q_{k+2}p_{k-1} = q_k p_{k-1} - d^2 q_{k+1}p_{k-2}. \tag{15.26}$$

Using this together with the difference equations for q_k and p_k we obtain p_k in terms of q_ks and p_{k-1}, i.e.

$$p_k = \frac{\varepsilon q_1 q_{k+2}}{q_{k+1}q_{k+2}} + d^2 \frac{q_{k+2}}{q_{k+1}} p_{k-1}. \tag{15.27}$$

By continuing this iteration we can write p_k in terms of q_ks only

$$p_k = \varepsilon q_1 q_{k+2} \sum_{j=1}^{k+1} \frac{d^{2(k-j+1)}}{q_j q_{j+1}}. \tag{15.28}$$

Thus $b^T A^{-1} b$ in (15.15) is expressible in terms of q_ks [5]

$$b^T A^{-1} b = \sum_{k=1}^{N-1} \frac{\left(\sum_{j=k}^{N-1} b_j d^j q_{j+1} \right)^2}{q_k q_{k+1} d^{2k}}. \tag{15.29}$$

After a lengthy calculation we obtain the following result;

$$
\begin{aligned}
b^T A^{-1} b &= \left(\frac{m q_2}{2\hbar \varepsilon q_1} \right) e^{\lambda t'} q^2 (t') + \left(\frac{m}{\hbar q_1} \right) e^{\lambda t''} q(t') q(t'') \\
&\quad + \frac{m}{2\hbar} \left(\sum_{k=1}^{N-1} \frac{d^{2(N-k)}}{q_k q_{k+1}} \right) e^{\lambda t''} q^2 (t'').
\end{aligned}
\tag{15.30}
$$

Substituting for $\varepsilon \det A$, $B(s'', s'; \varepsilon)$ and $b^T(\varepsilon) A^{-1}(\varepsilon) b(\varepsilon)$ in the expression for the propagator (15.14), taking the limit as $\varepsilon \to 0$ and simplifying the result we obtain

$$
\begin{aligned}
G(x'', x'; t', t'') &= \left(\frac{m e^{\lambda t''}}{2\pi i \hbar q(t')} \right)^{\frac{1}{2}} \\
&\quad \times \exp \left[\left(\frac{m e^{\lambda t''}}{2i\hbar q(t')} \right) \left[e^{-\lambda(t'' - t')} \dot{q}(t') x'^2 + 2x'x'' + (\lambda q(t') - \dot{p}(t'')) x''^2 \right] \right].
\end{aligned}
\tag{15.31}
$$

We can rewrite (15.31) by substituting for $q(t)$ and $p(t)$ from (15.24) and (15.25)

$$
\begin{aligned}
G(x'', x'; t', t'') &= \left(\frac{m \omega e^{\frac{\lambda}{2}(t'' + t')}}{2\pi i \hbar \sin \omega T} \right)^{\frac{1}{2}} \exp \left[-\frac{\lambda}{2i\omega} \left(X'^2 - X''^2 \right) \right] \\
&\quad \times \exp \left[\left(\frac{im\omega}{2\hbar \sin \omega T} \right) \left\{ \left(X'^2 + X''^2 \right) \cos \omega T - 2X'X'' \right\} \right],
\end{aligned}
\tag{15.32}
$$

where

$$X(t) = \left(\frac{me^{\lambda t}}{2\hbar}\right)^{\frac{1}{2}} x(t) \quad \text{and} \quad T = t'' - t'. \tag{15.33}$$

For $\lambda = 0$, the propagator G reduces to the well-known form of the harmonic oscillator propagator, and only then G will become a function of $T = t'' - t'$.

The time-dependent wave function for this propagator can be obtained from the initial wave function. Suppose that at $t = 0$ we have

$$\psi_n(X, 0) = \psi_n(X, \omega) \exp\left(-\frac{i\lambda X^2}{2\omega}\right), \tag{15.34}$$

where $\psi_n(X, \omega)$ is the normalized n-th eigenfunction of the harmonic oscillator with shifted frequency ω, and $\exp\left(-\frac{i\lambda X^2}{2\omega}\right)$ is a space-dependent phase factor. Using Eq. (15.34) we can calculate the wave function at a later time t from the integral

$$\psi_n(X, t) = \int G(X, X'; t, 0) \, \psi_n(X', 0) \, dX'. \tag{15.35}$$

By substituting for G from (15.32) and for $\psi_n(X', 0)$ from (15.34) and carrying out the integration over X' we find

$$\begin{aligned} \psi_n(x, t) &= N_n \exp\left(\frac{i\mathcal{E}_n t}{\hbar}\right) \exp\left[-\frac{m\omega}{2\hbar}\left(1 + \frac{i\lambda}{2\omega}\right) e^{\lambda t} x^2\right] \\ &\quad \times H_n\left[\sqrt{\frac{m\omega}{\hbar}} e^{\frac{\lambda t}{2}} x\right], \end{aligned} \tag{15.36}$$

where

$$\mathcal{E}_n = \left[\left(n + \frac{1}{2}\right)\omega + \frac{i\lambda}{4}\right]\hbar, \tag{15.37}$$

is the eigenvalue for the n-th excited state, and $N_n = \left(\frac{m\omega}{\pi\hbar}\right)^{\frac{1}{4}} \left(\frac{1}{2^n n!}\right)^{\frac{1}{2}}$. In Eq. (15.36) we have expressed the wave function in terms of the original coordinate x. This result is identical with the wave function obtained from Kanai-Caldirola Hamiltonian, Eq. (13.10) [9]-[11].

The damped motion of a harmonic oscillator in a uniform magnetic field is another problem for which the wave function as well as the kernel G can be obtained exactly [12], provided that the magnetic field increases with time as $H(t) = H_0 e^{\lambda t}$.

15.2 Path Integral Quantization of a Harmonic Oscillator with Complex Spring Constant

In the classical formulation of the damped harmonic oscillator §4.1 we observed that we can use complex time formulation to obtain a very simple Lagrangian as well as the classical action for this motion [13]. As we have seen in §4.1 the action is

$$S(x_2, x_2; \tau_2, \tau_2) = \frac{m\omega}{2\sin\omega(\tau_2 - \tau_2)} \left[(x_2^2 + x_1^2)\cos\omega(\tau_2 - \tau_1) - 2x_1 x_2 \right], \quad (15.38)$$

where $\tau = (1 - i\alpha t)$ and this τ can also be written as $\tau = \left(1 - \frac{i\lambda}{2\omega}\right)t$, with λ being the damping constant. We can change the complex time τ to the real time t by writing

$$\beta = (i + \alpha)\omega, \quad \text{or} \quad \omega\tau = -i\beta t, \quad (15.39)$$

and expressing S in terms of $T = t_2 - t_1$,

$$S(x_2, x_2; T) = -\frac{m\omega}{2i\sinh\beta T} \left[(x_2^2 + x_1^2)\cosh\beta T - 2x_1 x_2 \right]. \quad (15.40)$$

Since the action is quadratic in the coordinates x_2 and x_1, we can write the propagator as [7]

$$G(2,1) = \int_1^2 \exp\left[\frac{i}{\hbar} S(2,1) \right] \mathcal{D}[x], \quad (15.41)$$

where 2 and 1 refer to the space-time coordinates (x_2, t_2) and (x_1, t_1). Since the action (15.38) is identical with the action for a harmonic oscillator, using the method of Gaussian integrals [7] we find $G(2,1)$ to be

$$
\begin{aligned}
G(2,1) \quad &= \quad \frac{\alpha}{[2\pi \sinh\beta T]^{\frac{1}{2}}} \\
&\times \quad \exp\left[\frac{\alpha^2}{2\sinh\beta T} \left\{ (x_2^2 + x_1^2)\cosh\beta(\tau_2 - \tau_1) - 2x_1 x_2 \right\} \right].
\end{aligned}
$$

$$(15.42)$$

From the summation formula [14]

$$
\sum_{k=0}^{\infty} \left(\frac{\lambda}{2}\right)^k \frac{1}{k!} H_k(x) H_k(y) \exp\left[-\frac{1}{2}(x^2 + y^2) \right]
$$

$$
= \frac{1}{\sqrt{1 - \lambda^2}} \exp\left[-\frac{1 + \lambda^2}{2(1 - \lambda)^2}(x^2 + y^2) + \frac{2\lambda}{1 - \lambda^2}xy \right], \quad (15.43)
$$

it follows that $G(2,1)$ can be written as a product of the harmonic oscillator wave functions

$$G(2,1) = \sqrt{\frac{m\omega}{\pi\hbar}} \sum_{n=0}^{\infty} \frac{1}{2^n n!} H_n\left(\sqrt{\frac{m\omega}{\hbar}}x_2\right) H_n\left(\sqrt{\frac{m\omega}{\hbar}}x_1\right)$$

$$\times \exp\left[-\frac{m\omega}{2\hbar}(x_1^2 + x_2^2)\right] \exp\left[-\beta\left(n+\frac{1}{2}\right)T\right]. \qquad (15.44)$$

This expression for the propagator can be used to verify that

$$G(3,1) = \int G(3,2)G(2,1)dx_2, \qquad (15.45)$$

i.e. the amplitudes for the events occurring in succession in time multiply.

Now we want to examine the evolution of a displaced wave packet in time with the help of the propagator $G(2,1)$. We choose the initial wave packet to be

$$\Psi_n(x_1,0) = \left(\frac{m\omega}{\pi\hbar}\right)^{\frac{1}{4}} \frac{1}{\sqrt{2^n n!}} H_n\left[\sqrt{\frac{m\omega}{\hbar}}(x_1+\xi_0)\right] \exp\left[-\frac{m\omega}{2\hbar}(x_1+\xi_0)^2\right],$$
$$(15.46)$$

where ξ_0 is a constant. The wave packet at a later time, t, can be obtained from the propagator G and the initial wave packet Ψ_n;

$$\Phi_n(x,t) = \int_{-\infty}^{\infty} G(x,t;x_1,0)\Psi_n(x_1,0)dx_1, \qquad (15.47)$$

where from the definition of $\Phi_n(x,t)$ it follows that

$$\Phi_n(x,0) = \Psi_n(x,0). \qquad (15.48)$$

By substituting for G and Ψ_n from (15.44) and (15.46) and carrying out the integration over x_1 we find

$$\Phi_n(x,t) = \left(\frac{m\omega}{\hbar\pi}\right)^{\frac{1}{4}} H_n\left[\sqrt{\frac{m\omega}{\hbar}}\left\{x + \xi_0 \cosh\left(i+\frac{\lambda}{2\omega}\right)\omega t\right\}\right]$$

$$\times \exp\left[-\beta\left(n+\frac{1}{2}\right)t - \frac{m\omega}{2\hbar}\left(x + \xi_0 \cos\omega t\, e^{\frac{-\lambda t}{2}}\right)^2 + \frac{m\omega}{4\hbar}\xi_0^2\left(e^{-\lambda t}-1\right)\right]$$

$$\times \frac{1}{\sqrt{2^n n!}} \exp\left[\frac{im\omega}{\hbar}\left\{\frac{1}{4}\xi_0^2 \sin(2\omega t)e^{-\lambda t} + x\xi_0 \sin(\omega t)e^{-\frac{\lambda t}{2}}\right\}\right],$$
$$(15.49)$$

where in this equation we have replaced α by $\frac{\lambda}{2\omega}$. This wave function is not normalized except for $t = 0$. If we want a properly normalized wave packet, we change the propagator by a time-dependent factor. Let $\Gamma(x,t;x_1,0)$ be the new propagator

$$\Gamma_n(x,t;x_1,0) = \frac{1}{D_n(t)}G(x,t;x_1,0), \qquad (15.50)$$

and let us use the same initial wave packet, then the wave packet at a later time is given by

$$\Psi_n(x,t) = \int_{-\infty}^{\infty} \Gamma_n(x,t;x_1,0)\Psi_n(x_1,0)dx_1 = \frac{\Phi_n(x,t)}{D_n(t)}. \qquad (15.51)$$

Since we want $\Psi_n(x,t)$ to be normalized from (15.51) we obtain

$$\langle \Psi_n(x,t)|\Psi_n(x,t)\rangle = \frac{\langle \Phi_n|\Phi_n\rangle}{|D_n(t)|^2} = 1. \qquad (15.52)$$

From Eqs. (15.49) and (15.52) we find $D_n(t)$ to be

$$D_n(t) = \exp\left[-\left(n+\frac{1}{2}\right)\frac{\lambda t}{2} - \frac{m\omega}{4\hbar}\xi_0^2\left(1-e^{-\lambda t}\right)\right]$$
$$\times \left[L_n\left\{-\frac{2m\omega}{\hbar}\xi_0^2\sinh^2\left(\frac{\lambda t}{2}\right)\right\}\right]^{\frac{1}{2}}, \qquad (15.53)$$

where we have used the following integral [15]:

$$\int_{-\infty}^{\infty} e^{-x^2} H_n(x+y)H_n(x+z)dx = 2^n\sqrt{\pi}n!L_n(-2yz). \qquad (15.54)$$

The wave packet at the time t is given by (15.51), and the probability density at this time is

$$|\Psi_n(x,t)|^2 = \sqrt{\frac{m\omega}{\pi\hbar}}\frac{1}{2^n n!}\exp\left[-\frac{m\omega}{\hbar}\left(x+\xi_0\cos(\omega t)e^{\frac{-\lambda t}{2}}\right)^2\right]$$
$$\times \frac{\left|H_n\left[\sqrt{\frac{m\omega}{\hbar}}\left(x+\xi_0\cosh\left[\left(i\omega+\frac{\lambda}{2}\right)t\right]\right)\right]\right|^2}{L_n\left[-\frac{2m\omega}{\hbar}\xi_0^2\sinh^2\left(\frac{\lambda t}{2}\right)\right]}. \qquad (15.55)$$

From this result we can conclude that:

(1) For $n=0$, $H_0=1$ and $L_0=1$, thus we have the damped oscillations of a Gaussian wave packet with the damping constant λ. Asymptotically for large t (15.55) becomes

$$|\Psi_0(x,t)|^2 \rightarrow \sqrt{\frac{m\omega}{\hbar\pi}}\exp\left(-\frac{m\omega x^2}{\hbar}\right). \qquad (15.56)$$

(2) For all values of n, $|\Psi_n(x,t)|^2$ decays to the ground state, $n=0$, and the lifetime of the excited state is proportional to $\frac{2}{\lambda}$.

(3) We can use the Feynman-Kac formula [4] and relate the energy of the ground state to the asymptotic form of the propagator Γ_n. This formula is expressible as

$$E_0 = -\hbar \lim_{t\to\infty}\frac{1}{it}\ln\left\{\int_{-\infty}^{\infty}\Gamma_0(x,t;x,0)dx\right\}. \qquad (15.57)$$

If we substitute for $\Gamma_0(x,t;x,0)$ from (15.50), we find

$$
\begin{aligned}
\int_{-\infty}^{\infty} \Gamma_0(x,t;x,0)dx &= \frac{1}{D_0(t)} \int_{-\infty}^{\infty} G(x,t;x,0)dx \\
&= \frac{\exp\left[-\frac{1}{2}\beta t - \frac{m\omega}{4\hbar}\xi_0^2\left(1-e^{-\lambda t}\right)\right]}{(1-e^{-\beta t})}.
\end{aligned} \tag{15.58}
$$

By substituting (15.58) in (15.57) and taking the limit as $t \to \infty$, we find $E_0 = \frac{1}{2}\hbar\omega$, as expected.

15.3 Modified Classical Action and the Propagator for the Damped Harmonic Oscillator

In addition to the propagator that we found earlier from Kanai-Caldirola Hamiltonian we can also use a form of the classical action that we discussed in Chapter 4. The starting point of this formulation is the Hamiltonian

$$
H = \frac{p^2}{2m} + \frac{m}{2}\omega_0^2 x^2 + \lambda m\dot{\xi}(t)(x - \xi(t)), \tag{15.59}
$$

and the action (5.29) which we now write as

$$
S(x,t) = \frac{m}{2}\omega_0\left[x - \xi(t)\right]^2 \cot\omega_0 t + mx\dot{\xi}(t) + g(t), \tag{15.60}
$$

i.e. we have replaced α by ω_0 where

$$
\omega_0^2 = \left(1+\alpha^2\right)\omega^2 = \left(\omega^2 + \frac{\lambda^2}{4}\right). \tag{15.61}
$$

The time-dependent function $g(t)$ will be determined later. We can write this action for two space-time points (x_2, t_2) and (x_1, t_1);

$$
\begin{aligned}
S(2,1) &= \frac{m\omega_0}{2\sin\omega_0 T}\left[\{(x_1 - \xi_1)^2 + (x_2 - \xi_2)^2\}\cos\omega_0 T\right. \\
&\quad - \left.2(x_1 - \xi_1)(x_2 - \xi_2)\right] + m\left(x_2\dot{\xi}_2 - x_1\dot{\xi}_1\right) + g(t_2) - g(t_1).
\end{aligned} \tag{15.62}
$$

In this relation we have used the following notations: $T = t_2 - t_1$, $\xi_1 = \xi(t_1)$ and $\xi_2 = \xi(t_2)$.

We observe that $S(2,1)$ is a quadratic function of x_1 and x_2, therefore the propagator can be written as [16]

$$
G(2,1) = F(t_2, t_1)\exp\left[\frac{i}{\hbar}S(2,1)\right], \tag{15.63}
$$

where $F(t_2, t_1)$ is defined by the functional integral

$$F(t_2, t_1) = \int_0^0 \exp\left[\frac{i}{\hbar} S(2, 1)\right] \mathcal{D}[x]. \qquad (15.64)$$

For a complete determination of $G(2, 1)$ we can either carry out the functional integration in (15.64) or alternatively try to find $F(t_2, t_1)$ by considering the time evolution of a wave packet. In the latter method for the wave packet of the n-th excited state we choose $\Psi_n(x_1, t)$;

$$\Psi_n(x_1, 0) = \left(\frac{1}{2^n n!} \frac{\alpha_0}{\pi^{\frac{1}{2}}}\right)^{\frac{1}{2}} H_n\left[\alpha_0(x_1 - \xi_1)\right] \exp\left[-\frac{\alpha_0^2}{2}(x - \xi_1)^2\right] \exp\left(\frac{im}{\hbar} x_1 \dot{\xi}_1\right), \qquad (15.65)$$

where we have denoted $\frac{m\omega_0}{\hbar}$ by α_0^2. The last factor in (15.65), $\exp\left(\frac{im}{\hbar} x_1 \dot{\xi}_1\right)$, has been included in $\Psi_n(x_1, 0)$, so that the corresponding factor in the kernel can be cancelled. The wave function at t_2 is given by

$$\Psi_n(x_2, t_2) = \int_{-\infty}^{\infty} G(x_2, t_2; x_1, 0)\Psi_n(x_1, 0)dx_1. \qquad (15.66)$$

Substituting for G and Ψ from (15.63) and (15.65) in (15.66) and carrying out the integration we obtain

$$\begin{aligned}
\Psi_n(x_2, t_2) &= \frac{\alpha_0^{\frac{1}{2}}}{(2^n n!)^{\frac{1}{2}} \pi^{\frac{1}{4}}} H_n\left[\alpha_0(x_2 - \xi_2)\right] \exp\left[-\frac{\alpha_0^2}{2}(x_2 - \xi_2)^2\right] \\
&\quad \times \exp\left\{-i\omega_0\left(n + \frac{1}{2}\right)t_2 + \frac{i}{\hbar} m x_2 \dot{\xi}_2 + \frac{i}{\hbar}[g(t_2) - g(t_0)]\right\},
\end{aligned} \qquad (15.67)$$

provided that

$$F(t_2) = \frac{\alpha_0}{\sqrt{2\pi i \sin(\omega_0 t_2)}}. \qquad (15.68)$$

This is consistent with the form of the initial wave packet (15.65), and $\Psi_n(x_2, t_2)$ is normalized for all values of t_2.

The time-dependent function $g(t)$ which has not been defined so far can be determined from the energy of the particle. For the n-th excited state the energy $E_n(t)$ is

$$E_n(t) = \left\langle \Psi_n \left| i\hbar \frac{\partial}{\partial t} \right| \Psi_n \right\rangle = \left(n + \frac{1}{2}\right)\hbar\omega_0 - m\xi(t)\ddot{\xi}(t) - \dot{g}(t). \qquad (15.69)$$

This energy should be equal to the quantized energy of the oscillator, i.e. $\left(n + \frac{1}{2}\right)\hbar\omega_0$ plus the classical energy of the displaced oscillator. Thus

$$-\left(m\xi(t)\ddot{\xi}(t) + \dot{g}(t)\right) = \frac{1}{2}m\dot{\xi}^2(t) + \frac{1}{2}m\omega_0^2\xi^2(t), \qquad (15.70)$$

where $\xi(t)$ is the solution of Eq. (5.20). From Eq. (15.70) we calculate $g(t)$

$$g(t) = m \int \left(\frac{1}{2}\omega_0^2\xi^2(t) - \frac{1}{2}\dot{\xi}^2(t) + \lambda\xi(t)\dot{\xi}(t) \right) dt. \qquad (15.71)$$

15.4 Path Integral Formulation of a System Coupled to a Heat Bath

Here again we start with the Lagrangian corresponding to the Hamiltonian (10.1), i.e.

$$L = \frac{1}{2}\dot{Q}^2 - V(Q) + \sum_n \left[\frac{1}{2}\left(\dot{q}_n^2 - \omega_n^2 q_n^2\right) - \epsilon_n q_n Q \right]. \qquad (15.72)$$

As in our earlier discussions we want to eliminate the bath degrees of freedom and find an effective Lagrangian for the motion of the central particle [17]. To this end we must choose the initial and final states for the coordinates of the oscillator. Let us consider the paths with a given initial and final coordinates where [18]

$$Q(t=0) = Q_i, \quad \text{and} \quad Q(t=T) = Q_f. \qquad (15.73)$$

Thus the ground state of the n-th oscillator initially is a Gaussian centered about $q_{n_i} = \epsilon_n Q_i/\omega_n^2$ and finally is at $q_{n_f} = \epsilon_n Q_f/\omega_n^2$. Now we calculate the integral

$$\langle Q_f, q_{n_f}|Q_i, q_{n_i}\rangle = \oint_{Q_i}^{Q_f} \mathcal{D}[Q] \prod_n \left[\int \phi_0^* \left(x_{n_f} - q_{n_f}\right) dx_{n_f} \right.$$

$$\times \int \phi_0 \left(x_{n_i} - q_{n_i}\right) dx_{n_i} \oint_{x_{n_i}}^{x_{n_f}} \exp\left\{ \frac{i}{\hbar} \int_0^T L(Q, q_n) dt \right\} \right] \mathcal{D}[q_n]. \qquad (15.74)$$

The integrals over the bath coordinates form a series of Gaussian integrals, and these can be done analytically. We can also write (15.74) as

$$\oint_{Q_i}^{x_{Q_f}} \exp\left\{ \frac{i}{\hbar} \int_0^T L(Q, q_n) dt \right\} \mathcal{D}[Q] \prod_n C_{q_n}, \qquad (15.75)$$

where suppressing the index n in C_{q_n} we have

$$C_q = \int \left(\frac{\omega}{\pi\hbar}\right)^{\frac{1}{4}} \exp\left[-\frac{\omega}{2\hbar}(x_f - q_f)^2\right] dx_f$$

$$\times \int \left(\frac{\omega}{\pi\hbar}\right)^{\frac{1}{4}} \exp\left[-\frac{\omega}{2\hbar}(x_i - q_i)^2\right] dx_i$$

$$\times \oint_{x_i}^{x_f} \exp\left\{ \frac{i}{\hbar} \int_0^T \left(\frac{1}{2}\dot{q}^2 - \omega^2 q^2 - \epsilon q Q \right) dt \right\} \mathcal{D}[q]. \qquad (15.76)$$

Now let us change the variables to the dimensionless quantities

$$\zeta(t) = \frac{\epsilon}{\sqrt{2\hbar\omega^3}}Q(t), \quad \zeta_i = \sqrt{\frac{\omega}{2\hbar}}q_i, \quad \zeta_f = \sqrt{\frac{\omega}{2\hbar}}q_f, \qquad (15.77)$$

and

$$\beta(T) = \omega \int_0^T \zeta(t)e^{-i\omega t}dt. \qquad (15.78)$$

Using these we can calculate C_q [7];

$$C_q = e^{\frac{-i\omega T}{2}} \exp\left[-\frac{1}{2}\left(\zeta_i^2 + \zeta_f^2 - 2\zeta_i\zeta_f e^{-i\omega T}\right)\right]$$

$$\times \quad \exp\left[-i\left(\zeta_i\beta + \zeta_f\beta^* e^{-i\omega T}\right) + \omega^2 \int_0^T \int_0^t \zeta(t)\zeta(s)e^{-i\omega(t-s)}dsdt\right].$$

$$(15.79)$$

We can simplify C_q by integrating $\beta(T)$ and the double integral in (15.79) by parts;

$$C_q = e^{\frac{-i\omega T}{2}} \exp\left[i\omega \int_0^T \zeta^2(t)dt - \int_0^T \int_0^t \dot{\zeta}(t)\dot{\zeta}(s)e^{-i\omega(t-s)}dsdt\right]. \qquad (15.80)$$

By substituting for $\zeta(t)$ in terms of $Q(t)$ we obtain

$$C_q = \exp\left[\frac{i}{\hbar}\int_0^T \left\{\frac{\epsilon^2}{2\omega^2}Q^2(t) - \frac{\hbar\omega}{2}\right\}dt\right]$$

$$\times \quad \exp\left[-\frac{1}{\hbar}\int_0^T \int_0^t \dot{Q}(t)\dot{Q}(s)\frac{\epsilon^2}{2\omega^3}e^{-i\omega(t-s)}dsdt\right]. \qquad (15.81)$$

This will be the contribution of a typical oscillator (e.g. the n-th) in the bath. Summing over all of the bath oscillators we find the following results:

(a) The last integral in (15.74) will be modified in that the potential term in $L(Q, q_n)$ will be replaced by the effective potential

$$V_{eff}(Q) = V(Q) - \sum_n \frac{\epsilon_n^2}{2\omega_n^2}Q^2 + \sum_n \frac{\hbar\omega_n}{2}. \qquad (15.82)$$

(b) There will be a time-retarded self interaction with the kernel

$$\Delta(t - s) = \sum_n \frac{\epsilon_n^2}{2\omega_n^3}e^{-i\omega_n(t-s)}. \qquad (15.83)$$

Having eliminated the bath degrees of freedom we extend the time integration to be from $-\infty$ to ∞ and thus (15.74) becomes

$$\langle Q_f, q_{n_f}|Q_i, q_{n_i}\rangle =$$

$$\oint \exp\left\{\frac{i}{\hbar}\int_{-\infty}^{\infty}\left[\frac{1}{2}\dot{Q}^2 - V_{eff}(Q) + i\int_{-\infty}^{t}\dot{Q}(t)\dot{Q}(s)\Delta(t-s)ds\right]dt\right\}\mathcal{D}[Q].$$

(15.84)

The effective Lagrangian, L, found in (15.84) is a complex quantity, and thus there are two degrees of freedom for the motion, Re Q and Im Q. This L is given by

$$L = \frac{1}{2}\dot{Q}^2(t) - V_{eff}(Q) + i\int_{-\infty}^{\infty}\dot{Q}(t)\dot{Q}(s)\Delta(t-s)ds.$$

(15.85)

For certain types of potential $V_{eff}(Q)$ when the particle is tunnelling through a potential barrier, it is easier to work with imaginary time, i.e. to write (15.84) as [18] [19]

$$\oint \exp\left[-\frac{1}{\hbar}\int_{\infty}^{\infty}\left\{\frac{1}{2}\dot{Q}^2 + V(Q) + \frac{1}{2}\int_{-\infty}^{\infty}\dot{Q}(\sigma)\dot{Q}(\tau)U(\tau-\sigma)d\sigma\right\}d\tau\right]\mathcal{D}[Q],$$

(15.86)

where

$$U(\tau-\sigma) = \sum_{\mathbf{k}}\frac{\epsilon_n^2}{2\omega_n^3}\exp\left[-\omega_n|\tau-\sigma|\right].$$

(15.87)

This formulation has been used extensively in the problem of macroscopic quantum tunnelling [19] (see also §8.5).

Bibliography

[1] R.P. Feynman, Rev. Mod. Phys. 20, 367 (1948).

[2] E.H. Kerner and W.G. Sutcliffe, J. Math. Phys. 11, 391 (1970).

[3] L. Cohen, J. Math. Phys. 11, 3296 (1970).

[4] L.S. Schulman, *Techniques and Applications of Path Integration*, (John Wiley & Sons, New York, 1981).

[5] B.K. Cheng, J. Phys. A 17, 2475 (1984).

[6] C-I Um, K-H Yeon and T.F. George, Phys. Rep. 362, 64 (2002).

[7] R.P. Feynman and A.R. Hibbs, *Quantum Mechanics and Path Integrals*, (McGraw-Hill, New York, N.Y. 1965).

[8] T. Dittrich, P. Hänggi, G.-L. Ingold, B. Kramer, G. Schön and W. Zwerger, *Quantum Transport and Dissipation*, (Wiley-VCH, Weinheim, 1998). Chapter 4.

[9] E.H. Kerner, Can. J. Phys. 36, 371 (1958).

[10] R.W. Hasse, J. Math. Phys. 16, 2005 (1975).

[11] G.J. Papadopoulos, J. Phys. A7, 209 (1974).

[12] A.D. Jannussis, G.N. Brodimas and A. Streclas, Phys. Lett. 74A, 6 (1979).

[13] M. Razavy, Hadronic J. 10, 7 (1987).

[14] E.R. Hanson, *A Table of Series and Products*, (Prentice-Hall, Englewood Cliffs, 1975). p. 329.

[15] I.S. Gradshetyn and I.M. Ryzhik, *Table of Integrals, Series, and Products*, (Academic Press, New York, 1965) p. 838.

[16] See for instance W. Dittrich and M. Reuter, *Classical and Quantum Dynamics*, (Springer-Verlag, 1992) Chapter 16.

[17] R.P. Feynman and F.L. Vernon, Jr. Ann. Phys. (NY) 24, 118 (1963).

[18] J.P. Sethna, Phys. Rev. B24, 698 (1981).

[19] A.O. Caldeira and A. Leggett, Ann. Phys. (NY) 149, 374 (1983).

Chapter 16

Quantization of the Motion of an Infinite Chain

In the next two chapters we want to study the problem of quantization of a conservative many-body system and the subsequent separation of the motion of the central particle from the motion of the rest of the particles. As we have seen earlier, Chapter 14, the motion of the central particle cannot be described exactly by a single particle wave function. In this chapter we study the quantum dynamics of a uniform infinite chain (Schrödinger chain) § 8.1 [1], and that of a non-uniform chain. We can also solve the problem of a finite chain of oscillators [2], but the motion of infinite chain, because of the translational symmetry, is a simpler one to solve.

16.1   Quantum Mechanics of a Uniform Chain

Our starting point is the Hamiltonian for the Schrödinger chain, which we write as

$$H = \sum_{j=-\infty}^{\infty} \frac{1}{2m} p_j^2 + \frac{1}{2} \sum_{j=-\infty}^{\infty} \sum_{k=-\infty}^{\infty} \frac{1}{4} m\nu^2 (x_j - x_k)^2 \delta_{j,k+1}. \tag{16.1}$$

The Schrödinger equation for this Hamiltonian is

$$i\hbar \frac{\partial \Psi}{\partial t} = -\frac{\hbar^2}{2m} \sum_{k=-\infty}^{\infty} \frac{\partial^2 \Psi}{\partial x_k^2} + \frac{1}{8} m\nu^2 \sum_{j,k=-\infty}^{\infty} (x_k - x_j)^2 \delta_{j,k+1} \Psi. \tag{16.2}$$

Let us ignore the correlations between the particles in the chain and consider a specific and simple solution of (16.2) which is in the form of the product of single particle wave functions;

$$\Psi(x_1, x_2 \cdots; \xi_1(t), \xi_2(t) \cdots) = \prod_{k=-\infty}^{\infty} \phi_k[x_k - \xi_1(t)] \exp\left[\frac{i}{\hbar}\left\{mx_k\dot{\xi}(t) + C_k(t)\right\}\right],$$

(16.3)

where $\phi_k[x_k - \xi_1(t)]$ is the normalized single particle wave function, $\xi_k(t)$ is the classical solution of the displacement of the k-th particle in the chain, $C_k(t)$ a function of time which is defined by

$$C_k(t) = \int^t \left[\frac{1}{8}m\nu^2 \left\{2\xi_k^2(t) - \xi_{k+1}^2(t) - \xi_{k-1}^2(t)\right\} - \frac{1}{2}\dot{\xi}_k^2(t) - \mathcal{E}_k\right] dt, \quad (16.4)$$

and $\mathcal{E}_k$ is a constant to be determined later. Substituting (16.3) in (16.2) we find the differential equation for $\phi_k[x_k - \xi_k(t)]$,

$$\sum_{k=-\infty}^{\infty} \left[-\frac{\hbar^2}{2m}\frac{\partial^2\phi_k}{\partial y_k^2} + \frac{1}{2}m\dot{\xi}_k^2\phi_k + \left(m\ddot{\xi}_k x_k + \dot{C}_k\right)\phi_k\right]$$

$$\times \exp\left[\frac{i}{\hbar}\left(mx_k\dot{\xi}_k + C_k\right)\right]\prod_{j\neq k}\phi_j(y_j)\exp\left[\frac{i}{\hbar}\left(mx_j\dot{\xi}_j + C_j\right)\right]$$

$$+ \frac{1}{8}m\nu^2\sum_{k=-\infty}^{\infty}\left[(x_{k+1} - x_k)^2 + (x_k - x_{k-1})^2\right]\prod_{j=-\infty}^{\infty}\phi_j(y_j)$$

$$\times \exp\left[\frac{i}{\hbar}\left(mx_j\dot{\xi}_j + C_j\right)\right] = 0, \tag{16.5}$$

where ϕ_k is a function of the variable y_k;

$$y_k = x_k - \xi_k(t). \tag{16.6}$$

Note that in (16.5) $\xi_k, \dot{\xi}_k$ and C_k are all functions of time. If we multiply (16.5) by

$$\prod_{j\neq k}\phi_j(y_j)\exp\left[-\frac{i}{\hbar}\left(mx_j\dot{\xi}_j + C_j\right)\right], \tag{16.7}$$

and integrate over all y_j s except y_k and use the relation

$$\int_{-\infty}^{\infty}\phi_{k+1}^*(y_{k+1})x_{k+1}\phi_{k+1}(y_{k+1})dy_{k+1} = \xi_{k+1}(t), \tag{16.8}$$

we find that ϕ_k apart from a time-dependent phase factor satisfies the wave equation for a harmonic oscillator

$$-\frac{\hbar^2}{2m}\frac{d^2\phi_k}{dy_k^2} + \left(\frac{1}{4}m\nu^2 y_k^2 - \mathcal{E}_k\right)\phi_k = 0. \tag{16.9}$$

The fact that $\phi_k(y)$ is the wave function for a harmonic oscillator justifies using Eq. (16.8) for this problem.

The eigenvalues and the eigenfunctions of (16.9) are given by

$$\mathcal{E}_k^{(n_k)} = \left(n_k + \frac{1}{2} \right) \frac{\hbar \nu}{\sqrt{2}}, \tag{16.10}$$

and

$$\phi_k^{(n_k)} = \left(\frac{\alpha}{\pi^{\frac{1}{2}} 2^{n_k} n_k!} \right)^{\frac{1}{2}} H_{n_k}(\alpha y_k) \exp\left(-\frac{1}{2}\alpha^2 y_k^2 \right), \tag{16.11}$$

respectively. Here the constant α is given by

$$\alpha^2 = \frac{m\nu}{2^{\frac{1}{2}} \hbar}. \tag{16.12}$$

We can also formulate this problem in the following way:
We write the Hamiltonian for the k-th particle as

$$H_k = \frac{1}{2m} p_k^2 + \frac{1}{8} m\nu^2 \left[(x_k - \xi_{k-1})^2 + (x_k - \xi_{k+1})^2 \right], \tag{16.13}$$

and this form of H_k implies that the k-th particle is subject to the action of two time-dependent classical forces, these forces are exerted by $(k-1)$-th and $(k+1)$-th particles on the k-th. But for the motion of the k-th particle the Hamiltonian (16.13) is not unique. The same equation of motion can be obtained from H_k', where

$$H_k' = H_k - \frac{1}{8} m\nu^2 \left(\xi_{k-1}^2(t) + \xi_{k+1}^2(t) \right). \tag{16.14}$$

In quantum mechanics the wave functions corresponding to the two Hamiltonians H_k and H_k' differ from each other by a time-dependent phase factor. The single particle wave function for the k-th particle is

$$i\hbar \frac{\partial \psi_k}{\partial t} = H_k \psi_k, \tag{16.15}$$

where H_k is defined by (16.13). The solution of (16.15) is given by

$$\psi_k^{(n_k)}(x_k, t) = \phi_k^{(n_k)}(x_k - \xi_k(t)) \exp\left\{ \frac{i}{\hbar} \left(m x_k \dot{\xi}_k + C_k(t) \right) \right\}. \tag{16.16}$$

Now using this wave function we can determine the mean energy associated with the motion

$$\begin{aligned}
\langle E_k \rangle &= \int_{\infty}^{\infty} \psi_k^{*(n_k)} \left(i\hbar \frac{\partial}{\partial t} \right) \psi_k^{(n_k)} dx_k \\
&= \frac{1}{2} \dot{\xi}_k^2 + \frac{1}{8} m\nu^2 \left[(\xi_k - \xi_{k-1})^2 + (\xi_k - \xi_{k+1})^2 \right] + \mathcal{E}_k^{(n_k)}. \tag{16.17}
\end{aligned}$$

From this relation we obtain the energy of the k-th particle

$$e_k = \langle E_k \rangle - \frac{1}{8}m\nu^2 \left(\xi_{k-1}^2 + \xi_{k+1}^2 \right), \tag{16.18}$$

and thus the total energy of the system is

$$\langle E \rangle = \sum_{k=-\infty}^{\infty} \langle E_k \rangle = \frac{1}{2} \sum_{j=-\infty}^{\infty} \left[\dot{\xi}_j^2 + \left(\sum_{k=-\infty}^{\infty} \frac{1}{4}\nu^2 (\xi_j - \xi_k)^2 \delta_{j,k+1} \right) + \mathcal{E}_j^{(n_j)} \right]. \tag{16.19}$$

If on the classical equation of motion, $\xi_k(t)$, we impose the initial conditions

$$\xi_k(0) = \dot{\xi}_k(0) = 0, \quad k \neq 0, \quad \xi_0(0) = A, \quad \dot{\xi}_0(0) = 0, \tag{16.20}$$

then as we have seen in §8.1 the total energy associated with the classical motion is $\frac{1}{4}m\nu^2 A^2$. Thus

$$\langle E \rangle = \frac{1}{4}m\nu^2 A^2 + \sum_{j=-\infty}^{\infty} \mathcal{E}_j^{(n_j)}. \tag{16.21}$$

We observe that the total momentum of the system is also a constant of motion

$$\langle P \rangle = \sum_{j=-\infty}^{\infty} \int_{-\infty}^{\infty} \psi_k^*(x_k) \left(-i\hbar \frac{\partial}{\partial x_k} \right) \psi_k(x_k) dx_k$$

$$= \sum_{k=-\infty}^{\infty} m\dot{\xi}_k(t) = \frac{1}{2}m\nu A \sum_{k=-\infty}^{\infty} (J_{2k-1} - J_{2k+1}) = 0. \tag{16.22}$$

16.2   Ground State of the Central Particle

The energy of the ground state according to (16.21) is equal to

$$\langle E \rangle = \frac{1}{4}m\nu^2 A^2 + \sum_{j=-\infty}^{\infty} \left(\frac{\hbar\nu}{2^{\frac{3}{2}}} \right), \tag{16.23}$$

which is a divergent quantity. The wave function of the system in its ground state is

$$\Psi_G = \prod_{k=-\infty}^{\infty} \left(\frac{m\nu}{2^{\frac{1}{2}}\hbar\pi} \right)^{\frac{1}{4}} \exp\left\{ \left(-\frac{m\nu}{2^{\frac{3}{2}}\hbar} \right) (x_k - \xi_k)^2 + \frac{i}{\hbar} \left(mx_k\dot{\xi}_k + C_k \right) \right\}. \tag{16.24}$$

For the central particle, $k = 0$, the wave function $\phi_0(x_0, t)$ is

$$\phi_0(x_0, t) = \left(\frac{m\nu}{2^{\frac{1}{2}}\hbar\pi} \right)^{\frac{1}{4}} \exp\left\{ \left(-\frac{m\nu}{2^{\frac{3}{2}}\hbar} \right) (x_0 - AJ_0(\nu t))^2 \right\}$$

$$= \sum_{N=0}^{\infty} \frac{\alpha^N A^N J_0^N(\nu t)}{(2^N N!)^{\frac{1}{2}}} \exp\left[-\frac{1}{4}\alpha^2 A^2 J_0^2(\nu t) \right] u_N(x_0), \tag{16.25}$$

where we have expanded the wave function using the generating function of the Hermite polynomials. In this relation $u_N(x_0)$ is the N-th eigenfunction of the harmonic oscillator and α is given by (16.12). The expansion in (16.25) shows that a large number of stationary states $u_N(x_0)$ contribute to the time-dependent wave function. However the most important contributions at the time t come from the values of N close to $N_0(t)$, where

$$N_0(t) \approx \frac{1}{2}\alpha^2 A^2 J_0^2(\nu t). \tag{16.26}$$

We can use the same argument for the expansion of $\phi_k(x_k, t)$, noting that in this case we have

$$N_k(t) \approx \frac{1}{2}\alpha^2 A^2 J_{2k}^2(\nu t). \tag{16.27}$$

This means that the probability of finding the N_k-th excited state in $\phi_k(x_k, t)$ follows a Poisson distribution with the mean value given by (16.27), and that it only becomes appreciable after a time $t = \frac{2k}{\nu}$.

The classical motion of the chain is invariant under the time reversal transformation, since the total Hamiltonian of the chain is conservative and the solution $\xi_k(t) = AJ_{2k}(\nu t)$ remains unchanged if we change t to $-t$. The single particle Hamiltonian has a time-dependent effective potential

$$V(x_k, t) = \frac{1}{8}m\nu^2 \left[\xi_{k-1}^2 + \xi_{k+1}^2 - 2x_k\left(\xi_{k-1} + \xi_{k+1}\right)\right], \tag{16.28}$$

which is also invariant under this transformation.

The time-reversed wave function for the k-th particle which we will denote by $\bar{\psi}_k(x_k, t)$ is given by

$$\bar{\psi}_k(x_k, t) = \psi_k^*(x_k, -t) = \psi_k(x_k, t). \tag{16.29}$$

We observe that in this case, in contrast with the model of a particle coupled to the heat bath (e.g. Ullersma's model), for each oscillator including the central one, the time-reversal transformed motion leads to a physically realizable "reversed" motion. But the direction of the flow of energy remains unchanged as $t \to -t$, and therefore the system is irreversible.

Similar results can be obtained when the initial conditions are:

$$\xi_k(0) = 0, \quad k = 0, \pm 1, \pm 2 \cdots \tag{16.30}$$

and

$$\dot{\xi}_0(0) = \frac{2}{\nu}B, \quad \dot{\xi}_k(0) = 0, \quad k \neq 0. \tag{16.31}$$

In this case the classical solution is $\xi_k(t) = BJ_{2k+1}(\nu t)$, where B is a constant. Here the Hamiltonian has the following invariance:

$$H(-x_k, p_k, -t) = H(x_k, p_k, t), \tag{16.32}$$

and the time-reversed wave function is related to the original wave function by the following relation [3]

$$\bar{\psi}_k(x_k, t) = \psi_k^*(x_k, -t) = (-1)^{(n_k)}\psi_k(-x_k, t). \tag{16.33}$$

16.3   Wave Equation for a Non-Uniform Chain

The problem of non-uniform chain is similar to that of the Schrödinger chain. Again we write the wave equation for the conservative Hamiltonian describing the motion of the chain, Eq. (8.21), as

$$
H = \sum_{j=1}^{\infty} \left\{ \frac{p_j^2}{2m_j} + \frac{1}{4} \sum_{k=1}^{\infty} K_{kj}(x_k - x_j)^2 \right\}, \tag{16.34}
$$

where

$$
K_{kj} = K_k \delta_{k+1,j} + K_j \delta_{j+1,k}, \tag{16.35}
$$

is a symmetric matrix.

The Hamilton canonical equations for (16.34) generate an equation for x_j which is identical to Eq. (8.32) for $\xi_k(t)$.

From the Hamiltonian we obtain the time-dependent Schrödinger equation which is given by

$$
\sum_{j=1}^{\infty} \left\{ -\frac{\hbar^2}{2m_j} \frac{\partial^2}{\partial x_j^2} + \frac{1}{4} \left[K_j(x_{j+1} - x_j)^2 + K_{j-1}(x_{j-1} - x_j)^2 \right] \right\} \Psi = i\hbar \frac{\partial \Psi}{\partial t}.
$$

$$\tag{16.36}$$

As in the case of uniform chain, we introduce a new set of variables y_j by

$$
y_j = x_j - \xi_j(t), \tag{16.37}
$$

where $\xi_j(t)$ s are defined as the solution of Eq. (8.32) with the initial conditions (8.33).

Next we write the total wave function as

$$
\Psi(x_1, x_2 \cdots, t) = \Phi(y) \exp\left[-\frac{i}{\hbar} C(y, t) \right], \tag{16.38}
$$

and choose $C(y, t)$ to be

$$
C(y, t) = \mathcal{E}t
$$
$$
+ \int \left\{ \sum_{j=1}^{\infty} \frac{1}{2} m_j \dot{\xi}_j^2 + \frac{1}{4} \sum_{j,k=1}^{\infty} \left[K_{kj}(\xi_k - \xi_j)^2 + \frac{1}{2} K_{kj}(\xi_k - \xi_j)(y_k - y_j) \right] \right\} dt.
$$

$$\tag{16.39}$$

By substituting (16.38) and (16.39) in (16.36), we find that Φ satisfies the time-dependent Schrödinger equation

$$
\sum_{j=1}^{\infty} \left(-\frac{\hbar^2}{2m_j} \frac{\partial^2 \Phi}{\partial y_j} \right) + \frac{1}{4} \sum_{k,j=1}^{\infty} K_{kj} (y_k - y_j)^2 \Phi = \mathcal{E}\Phi, \tag{16.40}
$$

provided that $\xi_k(t)$ is a solution of Eq. (8.32).

In the many-body wave function (16.38) there are two kinds of correlations between the motions of the particles. The first is the classical correlations between different mass points as can be seen from Eq. (8.32) and this enters in the expression for Φ as well as the phase $C(y,t)$. The second one is the quantum correlation as implied by the solution of (16.40). Ignoring the correlations in (16.40), we write $\Phi(y_1, y_2, \cdots)$ as a product of wave functions for each particle [4];

$$\Phi(y_1, y_2, ...) = \prod_{i=1}^{\infty} \phi_{n_k}(y_i). \tag{16.41}$$

Following the method that we used earlier for a uniform chain we find the single particle wave equation to be

$$H_k \phi_{k_n} = -\frac{\hbar^2}{2m_k} \frac{d^2}{dy_k^2} \phi_{n_k} + \frac{1}{2} \left[(K_k + K_{k-1}) y_k^2 \right] \phi_{n_k} = \mathcal{E}_{n_k} \phi_{k_n}, \tag{16.42}$$

which is the equation of motion for a simple harmonic oscillator (K_k and K_{k-1} are defined by Eq. (16.35)).

Thus ϕ_{n_k} can be written as

$$\phi_{k_n} = \left(\frac{\alpha_k}{\pi^{\frac{1}{2}} 2^{n_k} n_k!} \right)^{\frac{1}{2}} H_{n_k}(\alpha_k y_k) \exp\left(-\frac{1}{2} \alpha_k^2 y_k^2 \right), \tag{16.43}$$

where

$$\alpha_k^2 = \frac{1}{\hbar} [m_k(K_{k+1} + K_k)]^{\frac{1}{2}}. \tag{16.44}$$

To calculate the single particle phase factor we note that

$$\int \phi_{n_{k+1}}^2(y_{k+1}) \phi_{n_{k-1}}^2(y_{k-1}) dy_{k+1} dy_{k-1}$$

$$\times \quad \exp\left[-\frac{i}{\hbar} \left\{ \beta_k K_k (y_{k+1} - y_k) - \beta_{k-1} K_{k-1} (y_{k-1} - y_k) \right\} \right]$$

$$= \quad L_{n_{k+1}} \left[\frac{1}{2} \left(\frac{\beta_k K_k}{\hbar \alpha_{k+1}} \right)^2 \right] L_{n_{k-1}} \left[\frac{1}{2} \left(\frac{\beta_{k-1} K_{k-1}}{\hbar \alpha_k} \right)^2 \right]$$

$$\times \quad \exp\left[\frac{i}{\hbar} y_k (\beta_k K_k - \beta_{k-1} K_{k-1}) \right], \tag{16.45}$$

where

$$\beta_k(t) = \int^t [\xi_{k+1}(t) - \xi_k(t)] \, dt, \tag{16.46}$$

and $L_{n_{k+1}}$ is the Laguerre polynomial. We can now determine the single particle phase factor by calculating the expectation value of $\exp\left[-\frac{i}{\hbar} C \right]$;

$$\exp\left(-\frac{i}{\hbar} C_k \right) = \frac{1}{N_k(t)} \int \cdots \int \left(\prod_{j \neq k}^{\infty} \phi_{n_j}^* \right) \exp\left(-\frac{i}{\hbar} C \right) \left(\prod_{j \neq k}^{\infty} \phi_{n_j} \right) \prod_{j \neq k}^{\infty} dy_j$$

$$= \exp\left[-\frac{i}{\hbar}\left(\mathcal{E}_{n_k}t + \int\left\{\frac{1}{2}m_k\dot{\xi}_k^2 + \frac{1}{2}\sum_{j\neq k}^{\infty}K_{kj}(\xi_k - \xi_j)^2\right\}dt - y_km_k\dot{\xi}_k\right)\right],$$

(16.47)

where

$$N_k(t) = L_{n_{k+1}}\left[\frac{1}{2}\left(\frac{\beta_kK_k}{\hbar\alpha_{k+1}}\right)^2\right]L_{n_{k-1}}\left[\frac{1}{2}\left(\frac{\beta_{k-1}K_{k-1}}{\hbar\alpha_k}\right)^2\right].$$

(16.48)

With this choice of $N_k(t)$, the single particle wave function

$$\psi_k(y_k, t) = \phi_{n_k}(y_k)\exp\left[-\frac{i}{\hbar}C_k(y_k, t)\right]$$

(16.49)

becomes normalized.

16.4   Connection with Other Phenomenological Frictional Forces

We have seen that the single particle wave function can be defined by Eqs. (16.42) and (16.49). From these two relations we can derive the time-dependent Schrödinger equation

$$H_k\psi_k - i\hbar\frac{\partial\psi_k}{\partial t} = \left[y_km_k\ddot{\xi}_k - \frac{1}{2}\sum_{j\neq k}^{\infty}K_{jk}(\xi_k - \xi_j)^2\right]\psi_k.$$

(16.50)

This equation is still invariant under time-reversal transformation, i.e. $\psi_k(y_k, t)$ satisfies the same equation as $\psi_k^*(y_k, -t)$. It is convenient to work with the wave function u_k which we define as

$$u_k = \psi_k\exp\left[\frac{i}{2\hbar}\int^t\sum_{j\neq k}^{\infty}K_{jk}(\xi_k - \xi_j)^2dt\right].$$

(16.51)

Now if in (16.50) we replace $\ddot{\xi}_k(t)$ in terms of $\dot{\xi}_k(t)$ and $\xi_k(t)$, Eq. (8.48), we find that u_k is a solution of

$$i\hbar\frac{\partial u_k}{\partial t} = H_ku_k + \left[\alpha\gamma(k)m_k\dot{\xi}_k - f_k(\xi_k)\right]y_ku_k,$$

(16.52)

where α is the damping constant. We can also express u_k in terms of $\phi_k(y_k)$ using Eqs. (16.49) and (16.51);

$$u_k = \phi_k(y_k)\exp\left[-\frac{i}{\hbar}\left(\mathcal{E}_{n_k}t - m_ky_k\dot{\xi}_k - \int^t\frac{1}{2}m_k\dot{\xi}_k^2dt\right)\right].$$

(16.53)

The expectation value of the momentum of the k-th particle is given by

$$\langle p_k \rangle = \int_{-\infty}^{\infty} u_k^* \left(-i\hbar \frac{\partial}{\partial y_k} \right) u_k dy_k = m_k \dot{\xi}_k. \tag{16.54}$$

Substituting for $\dot{\xi}_k$ in Eq. (16.52) we obtain

$$i\hbar \frac{\partial u_k}{\partial t} = H_k u_k + [\alpha \langle p_k \rangle - f_k(\xi_k)] y_k u_k, \tag{16.55}$$

and this is a special case of phenomenological wave equations studied by Hasse [5] and by Stoker [6].

16.5 Fokker-Planck Equation for the Probability Density

From the wave function $\psi_k(x_k, \xi_k)$ we can find the probability density, ρ_k, of finding the k-th particle between x_k and $x_k + dx_k$;

$$\frac{\partial \rho_k}{\partial t} dx_k = |\psi_k(x_k, \xi_k)|^2 \, dx_k = |\phi_k(x_k - \xi_k)|^2 \, dx_k. \tag{16.56}$$

This probability density satisfies the Fokker-Planck equation

$$\frac{\partial \rho_k}{\partial t} + \frac{\partial}{\partial x_k}(v_k \rho_k) - \frac{\hbar}{2m_k} \frac{\partial^2 \rho_k}{\partial x_k^2} = 0, \tag{16.57}$$

provided that the velocity v_k is given by

$$v_k = \dot{\xi}_k + \frac{\hbar}{2m_k} \frac{\partial}{\partial x_k} (\ln \rho_k). \tag{16.58}$$

We can also relate the current density, $j_k(x_k, t)$, to the probability density ρ_k by noting that

$$j_k(x_k, t) = -\frac{i\hbar}{2m_k} \left(\psi_k^* \frac{\partial \psi_k}{\partial y_k} - \psi_k \frac{\partial \psi_k^*}{\partial y_k} \right), \tag{16.59}$$

and substituting for ψ_k and ψ_k^* from (16.49) to obtain

$$j_k(x_k, t) = \dot{\xi}(t) \rho_k(x_k, t). \tag{16.60}$$

The coefficients of expansion of the wave function $\psi_k(x_k, \xi_k)$ also satisfy a master equation originally derived by Bloch [7]. Let us consider the motion of the k-th particle in the chain ($k \neq 1$). Initially this particle is in its ground state, and at some later time, t, the wave function is

$$\psi_k(x_k, \xi_k) = \left(\frac{\alpha_k}{\pi^{\frac{1}{2}}} \right)^{\frac{1}{2}} \exp\left[-\frac{1}{2}\alpha_k^2(x_k - \xi_k)^2 \right] \exp\left(-\frac{i}{\hbar} C_k \right). \tag{16.61}$$

We expand $\psi_k(x_k, \xi_k)$ in terms of the complete set of harmonic oscillator states;

$$\psi(x_k, \xi_k) = \sum_{n=0}^{\infty} A_n(\xi_k)\phi_{n_k}(x_k), \qquad (16.62)$$

where

$$A_n(\xi_k) = \frac{(\alpha_k \xi_k)^n}{(2^n n!)^{\frac{1}{2}}} \exp\left[-\frac{1}{4}(\alpha_k \xi_k)^2\right]. \qquad (16.63)$$

Thus $A_n(\xi_k)$ is the time-dependent probability amplitude for the k-th particle to be in the state n_k. Since for $\alpha t > 1$, ξ_k decreases exponentially, i.e.

$$\xi_k = \left(\frac{m_1}{m_k}\right)^{\frac{1}{2}} e^{-\alpha t}, \quad t > \frac{1}{\alpha}, \qquad (16.64)$$

therefore from Eqs. (16.63) and (16.64) it follows that for $t > \frac{1}{\alpha}$, the probability amplitude $A_n(\xi_k)$ satisfies the master equation

$$\frac{1}{2\alpha}\frac{d}{dt}|A_n|^2 = (n+1)|A_{n+1}|^2 - n|A_n|^2. \qquad (16.65)$$

This equation has also been studied in detail by Bauer and Jensen [8].

Bibliography

[1] E. Schrödinger, Ann. Phys. (Leipzig) 44, 916 (1914).

[2] D.H. Zanette, Am. J. Phys. 62, 404 (1994).

[3] M. Razavy, Can. J. Phys. 57, 1731 (1979).

[4] M. Razavy, Can J. Phys. 58, 1019 (1980).

[5] R.W. Hasse, Rep. Prog. Phys. 41, 1027 (1978).

[6] W. Stoker and K. Albrecht, Ann. Phys. (NY) 117, 436 (1979).

[7] F. Bloch, Z. Phys. 29, 58 (1928).

[8] H. Bauer and J.H.D. Hansen, Z. Phys. 124, 580 (1948).

Chapter 17

The Heisenberg Equations of Motion for a Particle Coupled to a Heat Bath

The classical damped motion of a central particle coupled to a field or to a large number of oscillators forming a heat bath was discussed in Chapter 10. The corresponding quantum problem can be formulated in the same way. When the central particle is harmonically bound and the Hamiltonian is quadratic in the coordinates and momenta of the oscillators as well as the coupling, then the Heisenbergs equations have similar dependence on the initial conditions as the classical equations.

17.1 Heisenberg Equations for a Damped Harmonic Oscillator

Let us consider an oscillator of unit mass density which interacts with a scalar field, a string of length $2L$, a model similar to the one we discussed in section 9.2, but with a different type of coupling (Harris's model).

The Lagrangian for this system is given by [1] [3]

$$L_q = \int_{-L}^{L} \mathcal{L}\left(Q, \dot{Q}, \frac{\partial y}{\partial t}, \frac{\partial y}{\partial x}\right) dx, \tag{17.1}$$

where the Lagrangian density $\mathcal{L}$ is given by

$$\mathcal{L}\left(Q, \dot{Q}, \frac{\partial y}{\partial t}, \frac{\partial y}{\partial x}\right)$$

$$= \frac{1}{2}\left[\left(\frac{\partial y}{\partial t}\right)^2 - \left(\frac{\partial y}{\partial x}\right)^2\right] + \frac{1}{2}\delta(x)\left[\dot{Q}^2 - \omega_0^2 Q^2 - 2\epsilon Q\left(\frac{\partial y}{\partial t}\right)\right]. \quad (17.2)$$

We note that for this model $\mathcal{L}$ is not invariant under time reversal transformation.

To simplify the Lagrangian we expand $y(x,t)$ in a Fourier series in the interval $-L < x < L$ and assume that the string is fixed at both ends, i.e. $y(-L,t) = y(L,t) = 0$ (see also section 9.2). Thus

$$y(x,t) = \sum_k \frac{1}{\sqrt{L}} q_k(t) \cos(kx), \quad k = \left(n + \frac{1}{2}\right)\frac{\pi}{L}, \quad n = 0, 1, 2 \cdots . \quad (17.3)$$

The terms involving $\sin(kx)$ do not couple to the motion of the oscillator, and therefore are not included in the sum. By substituting (17.3) in (17.2) and simplifying the result we find

$$L_q\left(Q, \dot{Q}, q_k, \dot{q}_k\right) = \frac{1}{2}\left(\dot{Q}^2 - \omega_0^2 Q^2\right) + \frac{1}{2}\sum_k \left(\dot{q}_k^2 - k^2 q_k^2\right) - \frac{\epsilon}{\sqrt{L}}Q\sum_k \dot{q}_k. \quad (17.4)$$

The Hamiltonian obtained from the above Lagrangian has the simple form

$$H = \frac{1}{2}\left(P^2 + \omega_0^2 Q^2\right) + \frac{1}{2}\sum_k \left[\left(p_k + \frac{\epsilon Q}{\sqrt{L}}\right)^2 + k^2 q_k^2\right]. \quad (17.5)$$

Now we regard P, Q, p_k and q_k as quantum mechanical operators obeying the canonical commutation relations

$$[Q, P] = [q_k, p_k] = i\hbar. \quad (17.6)$$

Using these together with the Heisenberg equations of motion

$$\dot{Q} = -\frac{i}{\hbar}[Q, H], \quad \dot{P} = -\frac{i}{\hbar}[P, H], \quad (17.7)$$

and similar equations for $\dot{q}_k$ and $\dot{p}_k$ we find the operator equations for Q and q_k

$$\ddot{Q} + \omega_0^2 Q = -\frac{\epsilon}{\sqrt{L}}\sum_k \dot{q}_k, \quad (17.8)$$

and

$$\ddot{q}_k + k^2 q_k = \frac{\epsilon}{\sqrt{L}}\dot{Q}. \quad (17.9)$$

These are linear coupled equations for Q and q_k which we can solve exactly.
We first solve (17.9) to obtain

$$q_k(t) = \frac{\epsilon}{\sqrt{L}} \int_0^t \cos\left[k\left(t - t'\right)\right] q\left(t'\right) dt' + \left(q_k(0)\cos(kt) + p_k(0)\frac{\sin(kt)}{k}\right).$$
(17.10)

Noting that

$$\frac{1}{L} \sum_k \cos\left[k\left(t - t'\right)\right] = \delta\left(t - t'\right),$$
(17.11)

we find from (17.10) that

$$\frac{\epsilon}{\sqrt{L}} \sum_k q_k = \lambda Q + \frac{\epsilon}{\sqrt{L}} \sum_k \left(q_k(0)\cos(kt) + p_k(0)\frac{\sin(kt)}{k}\right),$$
(17.12)

where $\lambda = \frac{\epsilon^2}{2}$.

By differentiating (17.12) and substituting for $\sum_k \dot{q}_k$ in (17.8) we find the equation of motion for the oscillator [1]

$$\ddot{Q} + \lambda\dot{Q} + \omega_0^2 Q = \frac{\epsilon}{\sqrt{L}} \sum_k \left[q_k(0)k\sin(kt) - p_k(0)\cos(kt)\right].$$
(17.13)

From the commutation relation

$$[q_k(0), p_k(0)] = i\hbar,$$
(17.14)

it follows that unlike the classical case we cannot set both $q_k(0)$ and $p_k(0)$ equal to zero.

Since the equations of motion are linear in the operators Q, P, q_k and p_k we can write $Q(t)$ and $P(t)$ in terms of the initial operators [1]. These are the analogues of the equations (10.17)-(10.19) of Ullersma's model

$$Q(t) = Q(0)f_1(t) + P(0)f_2(t) + \frac{\epsilon}{\sqrt{L}} \sum_k \left\{q_k(0)f_3(k,t) - p_k(0)f_4(k,t)\right\}, \quad (17.15)$$

and

$$P(t) = Q(0)\dot{f}_1(t) + P(0)\dot{f}_2(t) + \frac{\epsilon}{\sqrt{L}} \sum_k \left\{q_k(0)\dot{f}_3(k,t) - p_k(0)\dot{f}_4(k,t)\right\}. \quad (17.16)$$

The time-dependent coefficients $f_1(t)$, $f_2(t)$, $f_3(k,t)$ and $f_4(k,t)$ will be defined later.

If we assume that initially all of the oscillators are in the ground state (or the string has no initial displacement, and no initial velocity), then we can average Q and P over the ground states of $q_k(0)$ and $p_k(0)$. Since

$$\langle 0|q_k(0)|0\rangle = \langle 0|p_k(0)|0\rangle = 0,$$
(17.17)

therefore by averaging we get

$$\frac{d}{dt}\langle 0|P(t)|0\rangle + \lambda\langle 0|P(t)|0\rangle + \omega_0^2\langle 0|Q(t)|0\rangle = 0, \qquad (17.18)$$

and

$$\frac{d}{dt}\langle 0|Q(t)|0\rangle = \langle 0|P(t)|0\rangle, \qquad (17.19)$$

where $\langle 0|Q(t)|0\rangle$ and $\langle 0|P(t)|0\rangle$ denote the averages of these operators taken over the ground state of the heat bath.

Equations (17.18) and (17.19) can be solved to yield the following results:

$$\langle 0|Q(t)|0\rangle = \langle 0|Q(0)|0\rangle f_1(t) + \langle 0|P(0)|0\rangle f_2(t), \qquad (17.20)$$

and

$$\langle 0|P(t)|0\rangle = \langle 0|Q(0)|0\rangle \dot{f}_1(t) + \langle 0|P(0)|0\rangle \dot{f}_2(t), \qquad (17.21)$$

where

$$f_1(t) = e^{\frac{-\lambda t}{2}}\left(\cos(\omega t) + \frac{\lambda}{2\omega}\sin(\omega t)\right), \qquad (17.22)$$

and

$$f_2(t) = \frac{1}{\omega}e^{\frac{-\lambda t}{2}}\sin(\omega t), \qquad (17.23)$$

i.e. $f_1(t)$ and $f_2(t)$ are the solutions of the equation of motion of a damped harmonic oscillator with the boundary conditions

$$f_1(0) = 1, \quad \dot{f}_1(0) = 0, \qquad (17.24)$$

and

$$f_2(0) = 0, \quad \dot{f}_2(0) = 1, \qquad (17.25)$$

with $\omega = \left(\omega_0^2 - \frac{\lambda^2}{4}\right)^{\frac{1}{2}}$.

From Eqs. (17.20) and (17.21) it follows that

$$[\langle 0|Q(t)|0\rangle, \langle 0|P(t)|0\rangle] = [\langle 0|Q(0)|0\rangle, \langle 0|P(0)|0\rangle]$$
$$\times \left(f_1(t)\dot{f}_2(t) - \dot{f}_1(t)f_2(t)\right) = [\langle 0|Q(0)|0\rangle, \langle 0|P(0)|0\rangle]\,e^{-\lambda t}. \qquad (17.26)$$

This relation shows that by averaging over $q_k(0)$ and $p_k(0)$, the commutator $[\langle 0|Q(t)|0\rangle, \langle 0|P(t)|0\rangle]$ will decrease as a function of time. From the solution of (17.13) we find $f_3(k,t)$ and $f_4(k,t)$ to be

$$
\begin{aligned}
f_3(k,t) =\ & \frac{k^2\exp(s_+ t)}{(s_+ - s_-)\left(s_+^2 + k^2\right)} + \frac{ik\exp(ikt)}{2(k + is_+)(k + is_-)} \\
& + \frac{k^2\exp(s_- t)}{(s_- - s_+)\left(s_-^2 + k^2\right)} - \frac{ik\exp(-ikt)}{2(k - is_-)(k - is_+)},
\end{aligned} \qquad (17.27)
$$

and

$$f_4(k,t) = \frac{s_+ \exp(s_+t)}{(s_+ - s_-)(s_+^2 + k^2)} + \frac{\exp(ikt)}{2(k + is_+)(k + is_-)}$$
$$+ \frac{s_- \exp(s_-t)}{(s_- - s_+)(s_-^2 + k^2)} + \frac{\exp(-ikt)}{2(k - is_-)(k - is_+)}. \quad (17.28)$$

In these equations $s_\pm = -\frac{\lambda}{2} \pm i\omega$ are the roots of the quadratic equation

$$s^2 + \lambda s + \omega_0^2 = 0. \quad (17.29)$$

If we calculate the commutator of $Q(t)$, and $P(t)$ rather than $[\langle 0|Q(t)|0\rangle, \langle 0|P(t)|0\rangle]$ we find that

$$[Q(t), P(t)] = [Q(0), P(0)] \left(f_1(t)\dot{f}_2(t) - \dot{f}_1(t)f_2(t) \right)$$
$$- \frac{\epsilon^2}{L} \sum_k [q_k(0), p_k(0)] \left(f_3(k,t)\dot{f}_4(k,t) - \dot{f}_3(k,t)f_4(k,t) \right). \quad (17.30)$$

But we have already shown that $(f_1(t)\dot{f}_2(t) - \dot{f}_1(t)f_2(t))$ is equal to $e^{-\lambda t}$, therefore we need to calculate the last parenthesis in (17.30) with the help of Eqs. (17.27) and (17.28);

$$\frac{\epsilon^2}{L} \sum_k \left(f_3(k,t)\dot{f}_4(k,t) - \dot{f}_3(k,t)f_4(k,t) \right) = e^{-\lambda t} - 1. \quad (17.31)$$

Substituting these in (17.30) we obtain the standard commutator for $Q(t)$ and $P(t)$;

$$[Q(t), P(t)] = i\hbar. \quad (17.32)$$

That is once all the terms contributing to $Q(t)$ and $P(t)$ are taken into account, then the commutator of $Q(t)$ and $P(t)$ becomes independent of time.

In addition to the time dependence of the expectation values of $Q(t)$ and $P(t)$ given by (17.15) and (17.16) we can calculate the time dependence of $Q^2(t)$, $P^2(t)$ and $Q(t)P(t) + P(t)Q(t)$. Again we want these expectation values over the ground state of the oscillators $q_k(0)$ and $p_k(t)$. These are given by [1]

$$\langle 0|Q^2(t)|0\rangle = \langle 0|Q^2(0)|0\rangle f_1^2(t) + \langle 0|P^2(0)|0\rangle f_2^2(t)$$
$$+ \langle 0|Q(0)P(0) + P(0)Q(0)|0\rangle f_1(t)f_2(t) + \frac{\lambda}{2\pi}g_1(t), \quad (17.33)$$

$$\langle 0|P^2(t)|0\rangle = \langle 0|Q^2(0)|0\rangle \dot{f}_1^2(t) + \langle 0|P^2(0)|0\rangle \dot{f}_2^2(t)$$
$$+ \langle 0|Q(0)P(0) + P(0)Q(0)|0\rangle \dot{f}_1(t)\dot{f}_2(t) + \frac{\lambda}{2\pi}g_2(t), \quad (17.34)$$

and

$$\langle 0|Q(t)P(t) + P(t)Q(t)|0\rangle = \frac{d}{dt}\langle 0|Q^2(t)|0\rangle, \qquad (17.35)$$

where

$$g_1(t) = \hbar \int_0^\infty k \left[\frac{1}{k^2} f_3^2(k,t) + f_4^2(k,t) \right] dk, \qquad (17.36)$$

and

$$g_2(t) = \hbar \int_0^\infty k \left[\frac{1}{k^2} \dot{f}_3^2(k,t) + \dot{f}_4^2(k,t) \right] dk. \qquad (17.37)$$

In arriving at the last two relations we have used the fact that the summation over k can be replaced by integration

$$\sum_k \rightarrow \frac{L}{2\pi} \int_{-\infty}^{+\infty} dk. \qquad (17.38)$$

The function $g_2(t)$ is logarithmically divergent, and thus $\langle p^2 \rangle$ is infinite, a defect of the present model. This can be remedied by making ϵ a function of k in such a way that the high frequency oscillators decouple from the motion of the particle. This has been achieved in Ullersma's model by assuming that ϵ_n is a decreasing function of n. The same approach can be used for the quantization of a system composed of a particle coupled to a string of finite or of infinite length (Sollfrey's model, Chapter 9), when the coupling is between Q and q_n rather than between Q and p_n.

Let us consider the interesting case of an oscillator coupled to an infinite string with the Hamiltonian given by (9.41) (Sollfrey's model). To quantize this system we replace $y(x,t)$, $\pi(x,t)$, q_0' and p_0' by operators satisfying the canonical commutation relations [2]

$$[y(x,t), \pi(x',t)] = i\hbar\delta(x - x'), \qquad (17.39)$$

$$[q_0', p_0'] = i\hbar, \qquad (17.40)$$

with all of the other commutators being equal to zero. We now define the creation and annihilation operators by

$$q'(\nu) = \left(\frac{\hbar}{2\nu} \right)^{\frac{1}{2}} \left(a^\dagger(\nu) + a(\nu) \right), \qquad (17.41)$$

$$p'(\nu) = i \left(\frac{\hbar\nu}{2} \right)^{\frac{1}{2}} \left(a^\dagger(\nu) - a(\nu) \right), \qquad (17.42)$$

$$q_0' = \left(\frac{\hbar}{2\nu_0} \right)^{\frac{1}{2}} \left(a_0^\dagger + a_0 \right), \qquad (17.43)$$

$$p_0' = i \left(\frac{\hbar\nu_0}{2} \right)^{\frac{1}{2}} \left(a_0^\dagger - a_0 \right), \qquad (17.44)$$

where these operators satisfy the commutation relations

$$[a(\nu), a^\dagger(\nu')] = \delta(\nu - \nu'), \quad [a_0, a_0^\dagger] = 1. \tag{17.45}$$

We can write the Hamiltonian in terms of $a^\dagger(\nu), a(\nu), a_0$ and $a_0^\dagger$ operators, and in this representation the Hamiltonian will not be diagonal. To find a Hamiltonian which is diagonal in the number operator, we make use of the transformations that we introduced earlier for the classical system and define $Q(\omega)$ and $P(\omega)$ operators by

$$Q(\omega) = \int_0^\infty T(\omega, \nu) q'(\nu) d\nu + T(\omega, \nu_0) q_0', \tag{17.46}$$

and

$$P(\omega) = \int_0^\infty T(\omega, \nu) p'(\nu) d\nu + T(\omega, \nu_0) p_0'. \tag{17.47}$$

These operators in turn can be replaced by creation and annihilation operators

$$Q(\omega) = \left(\frac{\hbar}{2\omega}\right)^{\frac{1}{2}} \left(A^\dagger(\omega) + A(\omega)\right), \tag{17.48}$$

and

$$P(\omega) = i \left(\frac{\hbar\omega}{2}\right)^{\frac{1}{2}} \left(A^\dagger(\omega) - A(\omega)\right). \tag{17.49}$$

From the Hamiltonian (9.41) and the transformations $T(\omega, \nu)$ and $T(\omega, \nu_0)$, Eqs. (9.42) and (9.43) it follows that H can be written as

$$H = \int_0^\infty \hbar\omega A^\dagger(\omega) A(\omega) d\omega, \tag{17.50}$$

where we have omitted the zero point energy arising from ordering of the operators.

The ground state energy of the total system, Φ_0, can be found by noting that this state is defined by

$$a(\nu)\Phi_0 = a_0\Phi_0 = 0. \tag{17.51}$$

Taking the expectation value of the Hamiltonian (17.50) with Φ_0 using (17.51) and then returning to the representation in terms of $a^\dagger(\nu), a_0^\dagger, a(\nu)$ and a_0 we obtain

$$\langle H \rangle_{\Phi_0} = \frac{\hbar}{4} \int_0^\infty \left[\int_0^\infty T^2(\omega, \nu) \frac{(\omega - \nu)^2}{\nu} d\nu + T^2(\omega, \nu_0) \frac{(\omega - \nu_0)^2}{\nu_0} \right] d\omega. \tag{17.52}$$

By substituting $T(\omega, \nu)$ and $T(\omega, \nu_0)$ from Eqs. (9.42) and (9.43) we find

$$\langle H \rangle_{\Phi_0} = \frac{\hbar}{4} \int_0^\infty \left[\frac{e^2}{\pi^2} (\omega^2 - \nu_0^2)^2 \int_0^\infty \frac{d\nu}{\nu(\nu + \omega)^2} + \frac{e^2}{\pi m} \frac{(\omega - \nu_0)^2}{\nu_0} \right] \frac{\omega^2 d\omega}{F(\omega)}. \tag{17.53}$$

This last expression shows that the integration over ν leads to a logarithmic divergence of $\langle H \rangle_{\Phi_0}$ at low frequencies.

Next let us assume that the system at $t = 0$ is in the state $\Phi_1 = a_0^\dagger \Phi_0$, we want to determine the probability that it will remain in this state at time t. This probability is given by

$$\mathcal{P}(t) = |W(t)|^2, \tag{17.54}$$

where

$$W(t) = \left\langle a_0^\dagger(t)\Phi_0, a_0^\dagger(0)\Phi_0 \right\rangle. \tag{17.55}$$

To evaluate $W(t)$ we note that $A(\omega)$ and $A^\dagger(\omega)$ have simple time-dependence

$$A(\omega, t) = A(\omega)e^{-i\omega t} \quad \text{and} \quad A^\dagger(\omega, t) = A(\omega)e^{i\omega t}, \tag{17.56}$$

Therefore starting with $A(\omega)$ and $A^\dagger(\omega)$ we expand both in terms of q_0' and p_0' using Eqs. (17.46)-(17.49), and then replace q_0' and p_0' by $a_0^\dagger$ and a_0 and let these operators act on Φ_0. This gives us the following expression for $W(t)$

$$
\begin{aligned}
W(t) &= \frac{1}{4}\int_0^\infty \int_0^\infty T(\omega, \nu_0)T(\omega', \nu_0)\left\langle \left[(\omega + \nu_0)A(\omega, t) + (\nu_0 - \omega)A^\dagger(\omega, t) \right] \right. \\
&\times \left. [(\omega' + \nu_0)A^\dagger(\omega, 0) + (\nu_0 - \omega')A(\omega', 0)] \right\rangle_{\Phi_0} \frac{d\omega' d\omega}{\nu_0 \sqrt{\omega \omega'}}.
\end{aligned}
\tag{17.57}
$$

Simplifying (17.57) we find

$$
\begin{aligned}
W(t) &= \frac{1}{4}\int_0^\infty \int_0^\infty \frac{T^2(\omega, \nu_0)T^2(\omega', \nu_0)}{\nu_0 \omega \omega'}\left[(\nu_0 + \omega)^2 e^{-i\omega t} - (\nu_0 - \omega)^2 e^{i\omega t}\right] \\
&\times \omega' d\omega' d\omega.
\end{aligned}
\tag{17.58}
$$

The integration over ω' can be carried out using Eq. (9.48) and then we are left with a simple expression for $W(t)$,

$$
\begin{aligned}
W(t) &= \frac{e^2}{4\pi m \nu_0}\int_0^\infty \left[(\nu_0 + \omega)^2 e^{-i\omega t} - (\nu_0 - \omega)^2 e^{i\omega t}\right]\frac{\omega d\omega}{F(\omega)} \\
&= -\frac{e^2}{4\pi m \nu_0}\int_0^\infty \frac{\omega(\omega - \nu_0)^2}{F(\omega)}e^{i\omega t}d\omega.
\end{aligned}
\tag{17.59}
$$

Since as $\omega \to \infty$, $F(\omega) \to \omega^6$, it follows that at $t = 0$, $W(t)$ and its first derivative are finite, however the second derivative of $W(t)$ is infinite. This corresponds to the problem of a displaced string with a corner which starts its motion with infinite acceleration. The denominator $F(\omega)$ in (17.59) can be written as the product of two cubic factors

$$F(\omega) = \left(\omega^3 - \frac{ie}{2}\omega^2 - \nu_0'^2\omega + \frac{1}{2}ie\nu_0^2\right)\left(\omega^3 + \frac{ie}{2}\omega^2 - \nu_0'^2\omega - \frac{1}{2}ie\nu_0^2\right). \tag{17.60}$$

If we calculate the roots of the first cubic factor in (17.60), then the roots of the second cubic factor will be the negatives of the roots of the first one.

Let us denote the roots of the first polynomial in ω by $\alpha + i\lambda, -\alpha + i\lambda$ and $i\gamma$ where γ is the real positive root of

$$\gamma^3 - \frac{e}{2}\gamma^2 + \nu_0'^2\gamma - \frac{1}{2}e\nu_0^2 = 0, \tag{17.61}$$

λ is the positive root of

$$8\lambda^3 - 4e\lambda^2 + \left(2\nu_0'^2 + \frac{1}{2}e^2\right)\lambda - \frac{e^2}{2m} = 0, \tag{17.62}$$

and

$$\alpha^2 = 3\lambda^2 - e\lambda + \nu_0'^2. \tag{17.63}$$

The roots of the second cubic factor in (17.60) are given by $\alpha - i\lambda, -\alpha - i\lambda$ and $-i\gamma$.

For positive values of t the integral in (17.59) can be evaluated by contour integration. Closing the contour by a large semi-circle in the upper-half of the ω-plane, we observe that the contribution to the integral comes from the residue at the poles in the upper-half plane. Thus $W(t)$ has the form

$$W(t) = -\frac{2\pi i e^2}{4\pi m \nu_0}\left[\frac{(\alpha + i\lambda - \nu_0)^2 e^{i\alpha t - \lambda t}}{8i\alpha\lambda[(\alpha + i\lambda)^2 + \gamma^2]}\right.$$
$$\left. - \frac{(\alpha - i\lambda + \nu_0)^2 e^{-i\alpha t - \lambda t}}{8i\alpha\lambda[(\alpha - i\lambda)^2 + \gamma^2]} + \frac{(i\gamma - \nu_0)^2 e^{-\gamma t}}{2[(\alpha + i\lambda)^2 + \gamma^2][(\alpha - i\lambda)^2 + \gamma^2]}\right]. \tag{17.64}$$

We can simplify (17.64) if we assume that e is small. In this case from Eqs. (17.61) and (17.62) it follows that $\alpha \approx \nu_0', \lambda \approx \mathcal{O}(e^2)$ and $\gamma \approx \mathcal{O}(e)$.

By examining the three terms in the bracket, Eq. (17.64), we find that the first and third terms are of the order of unity and the second term is of the order $\frac{1}{e^2}$. Thus $W(t)$ has the approximate form of

$$W(t) \approx \exp(-i\alpha t - \lambda t), \tag{17.65}$$

and we conclude that for the case of infinite string, when the coupling e is small, the probability $\mathcal{P}(t)$ for the system to be in its initial state is given by

$$\mathcal{P}(t) \approx \exp(-2\lambda t) \approx \exp\left(-\frac{e^2}{2m\nu_0^2}t\right), \tag{17.66}$$

i.e. the decay is exponential.

In addition to the model of Harris we want to study the quantum version of Ullersma's model. Thus we regard $Q(t), P(t), Q(0), P(0) \cdots q_j(0), p_j(0)$ as operators. Using these we can calculate the commutator

$$\left[\dot{Q}(t), Q(t)\right] = [P(t), Q(t)]$$

$$= -i\hbar \sum_{\nu=0}^{N} \sum_{\mu=0}^{N} X_{0\nu} X_{0\mu} \left[\cos(s_\nu t)\cos(s_\mu t) + \frac{s_\mu}{s_\nu}\sin(s_\nu t)\sin(s_\mu t)\right]$$

$$\times \left\{X_{0\nu} X_{0\mu} + \sum_{n=1}^{N} X_{n\nu} X_{n\mu}\right\}. \tag{17.67}$$

Then noting that

$$\left\{X_{0\nu} X_{0\mu} + \sum_{n=1}^{N} X_{n\nu} X_{n\mu}\right\} = \delta_{\mu\nu}, \tag{17.68}$$

we find that the commutator of $P(t)$ and $Q(t)$ is the same as in Eq. (17.32), whereas if we average over all of the oscillators forming the bath then we obtain (17.26).

Models similar to Harris's model have been proposed by Senitzky [4] and by Weber [5] [6]. In Senitzky's model again the central particle is coupled to a heat bath, but the detailed form of the bath is not specified. The total Hamiltonian of the system is of the form

$$H = \frac{P^2}{2M} + \frac{1}{2}M\omega^2 Q^2 + H_B + \epsilon\Gamma(t)P, \tag{17.69}$$

where H_B is the Hamiltonian of the bath, ϵ is the coupling constant and $\Gamma(t)$ is the coordinate of loss mechanism which couples the bath to the oscillator. The Heisenberg equations of motion for P and Q obtained from (17.69) are

$$\frac{dP}{dt} = -M\omega^2 Q, \tag{17.70}$$

and

$$\frac{dQ}{dt} = \frac{1}{M}P + \epsilon\Gamma(t). \tag{17.71}$$

Comparing these with Eqs. (17.8) we find that $\Gamma(t)$ plays the same role as $L^{\frac{-1}{2}}\sum_k q_k$ in Harris's model.

In Senitzky's approach it is assumed that the interaction is turned on at $t = 0$ and that in the absence of coupling, the operator $\Gamma^{(0)}(t)$ has the matrix elements

$$\Gamma_{ij}^{(0)}(t) = \Gamma_{ij}^{(0)}(0)\exp\left[\frac{i(E_i - E_j)t}{\hbar}\right] \equiv \tilde{\Gamma}_{ij}e^{i\omega_{ij}t}, \tag{17.72}$$

where E_i s are the energy states of the uncoupled loss mechanism. Now the Heisenberg equation for $\Gamma(t)$ found from the Hamiltonian (17.69) is

$$i\hbar\frac{d\Gamma(t)}{dt} = [\Gamma(t), H_B(t)], \tag{17.73}$$

where from (17.69) we can determine H_B;

$$H_B(t) = H_B^{(0)} + \frac{\epsilon}{i\hbar} \int_0^t [H_B(t_1), \Gamma(t_1)] P(t_1) dt_1. \tag{17.74}$$

By substituting (17.74) in (17.73) we obtain

$$\frac{d\Gamma(t)}{dt} = \frac{1}{i\hbar} \left[\Gamma(t), H_B^{(0)}\right] + \frac{\epsilon}{\hbar^2} \int_0^t [\Gamma(t), [\Gamma(t_1), H_B(t_1)] P(t_1)] dt_1, \tag{17.75}$$

and the integral of this first order integro-differential equation is [4]

$$\Gamma(t) = \Gamma^{(0)}(t) + \frac{\epsilon}{\hbar^2} \int_0^t dt_1 \int_0^{t_1} \exp\left[\frac{i}{\hbar} H_B^{(0)}(t - t_1)\right]$$

$$\times\ [\Gamma(t_1), [\Gamma(t_2), H_B(t_2)] P(t_2)] \exp\left[-\frac{i}{\hbar} H_B^{(0)}(t - t_1)\right] dt_2. \tag{17.76}$$

The next step is to integrate Eqs. (17.70) and (17.71);

$$P(t) = P^{(0)}(t) - M\omega^2 \epsilon \int_0^t \Gamma(t_1) \sin[\omega(t - t_1)] dt_1, \tag{17.77}$$

and

$$Q(t) = Q^{(0)}(t) + \epsilon \int_0^t \Gamma(t_1) \cos[\omega(t - t_1)] dt_1, \tag{17.78}$$

where $P^{(0)}(t)$ and $Q^{(0)}(t)$ are the solutions of the loss-free oscillator given in terms of the initial operators $P^{(0)}(0)$ and $Q^{(0)}(0)$. By substituting for $\Gamma(t)$ from (17.76) in (17.77) we find an integral equation for $P(t)$

$$\begin{aligned}P(t) =\ & P^{(0)}(t) - M\omega\epsilon \int_0^t \Gamma^{(0)}(t_1) \sin[\omega(t - t_1)] dt_1 \\ & - \frac{M\omega\epsilon^2}{\hbar^2} \int_0^t dt_1 \int_0^{t_1} dt_2 \int_0^{t_2} \sin[\omega(t - t_1)] \\ & \times\ \exp\left[\frac{i}{\hbar} H_B^{(0)}(t_1 - t_2)\right] [\Gamma(t_2), [\Gamma(t_3), H_B(t_3)] P(t_3)] \\ & \times\ \exp\left[-\frac{i}{\hbar} H_B^{(0)}(t_1 - t_2)\right] dt_3. \end{aligned} \tag{17.79}$$

The first term on the right hand side of (17.79) represents the effect of the fluctuations of the unperturbed loss mechanism on the central oscillator. If we find the average of $P(t)$ over a state of the bath which may be described by a temperature T and for which the density matrix is diagonal

$$\rho_{nm} = \frac{\delta_{mn} \exp\left(-\frac{E_n}{kT}\right)}{\sum_j \exp\left(-\frac{E_n}{kT}\right)}, \tag{17.80}$$

then the first term will not contribute. Thus the last term in (17.79) contains the interaction between the oscillator and the field and is responsible for the damping. Equation (17.79) is exact, and one can use it without specifying the details of the heat bath. This is clearly an advantage of the model, but then (17.79) can only be solved approximately. The details of the rather lengthy calculation and justifying the method of approximation are given in Senitzky's work [4]. The final result is a simple differential equation for the momentum operator $P(t)$ in the form

$$\frac{d^2P}{dt^2} + \lambda\frac{dP}{dt} + \omega^2 P = -M\omega^2\epsilon\Gamma^{(0)}(t). \tag{17.81}$$

Here the damping constant λ is given by

$$\lambda = \frac{\pi\omega\epsilon^2 M \left(1 - e^{\frac{-\hbar\omega}{kT}}\right) \int_0^\infty \rho(E + \hbar\omega)\rho(E)\tilde{\Gamma}^2(E + \hbar\omega, E)e^{\frac{-E}{kT}}\,dE}{\int_0^\infty \rho(E)e^{\frac{-E}{kT}}\,dE}. \tag{17.82}$$

In this approximate calculation of $P(t)$ we have assumed that the energy levels of the loss mechanism are closely spaced, and therefore we have replaced the summation by integration

$$\sum_{j,k} \rightarrow \int_0^\infty \rho(E_j)dE_j \int_0^\infty \rho(E_k)dE_k, \tag{17.83}$$

where $\rho(E)$ is the density of states in energy space.

17.2 Density Matrix for the Motion of a Particle Coupled to a Field

As we have seen earlier, the density matrix formulation provides an alternative way of studying the quantal problem of linearly damped harmonic oscillator. The same approach can be used for the quantum description of a particle coupled to a heat bath or to a field. For instance we can start with the Harris model with the Hamiltonian (17.5) and introduce the set of coordinates q_α and momenta p_α by

$$q_\alpha = Q, q_1, q_2 \cdots q_N \cdots, \quad p_\alpha = P, p_1, p_2 \cdots p_N \cdots, \tag{17.84}$$

and express H as a function of q_α and p_α. This H can be used to define the density matrix ρ by the operator equation

$$\frac{\partial\rho}{\partial t} = \frac{i}{\hbar}[\rho, H]. \tag{17.85}$$

For the present case it is more convenient to study the properties of the Wigner distribution function $W(q_\alpha, p_\alpha, t)$ which is directly related to ρ by the Fourier transform

$$W(q_\alpha, p_\alpha, t) = \lim_{N \to \infty} \int \rho \left(q_\alpha + \frac{x_\alpha}{2}; q_\alpha - \frac{x_\alpha}{2} \right) \exp \left(-\frac{i}{\hbar} \sum_\beta p_\beta x_\beta \right) \frac{d^N x_\alpha}{(2\pi\hbar)^N},$$

(17.86)

where in (17.86) q_α and p_α are c numbers. From Eqs. (17.85) and (17.86) we obtain a partial differential equation for $W(q_\alpha, p_\alpha, t)$

$$
\begin{aligned}
-i\hbar \frac{\partial W(q_\alpha, p_\alpha, t)}{\partial t} &= H \left[q_\alpha + \frac{\hbar}{2i} \frac{\partial}{\partial p_\alpha}, p_\alpha - \frac{\hbar}{2i} \frac{\partial}{\partial q_\alpha} \right] W(q_\alpha, p_\alpha, t) \\
&\quad - H \left[q_\alpha - \frac{\hbar}{2i} \frac{\partial}{\partial p_\alpha}, p_\alpha + \frac{\hbar}{2i} \frac{\partial}{\partial q_\alpha} \right] W(q_\alpha, p_\alpha, t).
\end{aligned}
$$

(17.87)

Using the Hamiltonian operator Eq. (17.5) we find

$$\frac{\partial W}{\partial t} + \dot{Q} \frac{\partial W}{\partial Q} + \dot{P} \frac{\partial W}{\partial P} + \sum_k \left(\dot{q}_k \frac{\partial W}{\partial q_k} + \dot{p}_k \frac{\partial W}{\partial p_k} \right) = 0,$$

(17.88)

where

$$\dot{Q} = \frac{\partial H}{\partial P} = P, \quad \dot{P} = -\frac{\partial H}{\partial Q} = -\omega_0^2 Q - \frac{\epsilon}{\sqrt{L}} \sum \left[p_k + \frac{\epsilon}{\sqrt{L}} Q \right],$$

(17.89)

and

$$\dot{q}_k = \frac{\partial H}{\partial p_k} = p_k + \frac{\epsilon}{\sqrt{L}} Q, \quad \dot{p}_k = -\frac{\partial H}{\partial q_k} = -k^2 q_k,$$

(17.90)

i.e. we have the same equation for W as the classical distribution function. This is not surprising, since for potentials of the general form $V(Q) = AQ + BQ^2$, the distribution function does not depend on $\hbar$ [7].

In order to determine the time evolution of (17.88) we can proceed in the following way [1]:
Let us assume that an acceptable initial distribution function is given by $W(q_{\alpha 0}, p_{\alpha 0}, 0) = W_0(q_{\alpha 0}, p_{\alpha 0})$ and then we integrate Eqs. (17.89)-(17.90) to obtain

$$q_\alpha(t) = q_\alpha(q_{\alpha 0}, p_{\alpha 0}, t), \quad p_\alpha(t) = p_\alpha(q_{\alpha 0}, p_{\alpha 0}, t),$$

(17.91)

where $q_{\alpha 0}$ and $p_{\alpha 0}$ are the initial values of $q_\alpha(t)$ and $p_\alpha(t)$. Solving (17.91) for $q_{\alpha 0}$ and $p_{\alpha 0}$ we have

$$q_{\alpha 0} = q_{\alpha 0}(q_\alpha(t), p_\alpha(t), t), \quad p_{\alpha 0} = p_{\alpha 0}(q_\alpha(t), p_\alpha(t), t).$$

(17.92)

Substituting these in $W_0(q_{\alpha 0}, p_{\alpha 0})$ we find

$$W(q_\alpha(t), p_\alpha(t), t) = W_0[q_{\alpha 0}(q_\alpha(t), p_\alpha(t), t), p_{\alpha 0}(q_\alpha(t), p_\alpha(t), t)].$$

(17.93)

We can express this result in terms of the functions $f_1(t), f_2(t), f_3(k,t)$ and $f_4(k,t)$ defined by (17.22)-(17.23) and (17.27)-(17.28). To this end we replace t by $-t$ in the argument of these functions and interchange q_α with $q_{\alpha 0}$ and p_α with $p_{\alpha 0}$. For instance for Q_0 we get

$$Q_0 = Q f_1(-t) + P f_2(-t) + \frac{\epsilon}{\sqrt{L}} \sum_k [q_k f_3(k,-t) - p_k f_4(k,-t)], \qquad (17.94)$$

and we find similar expressions for P_0, q_{k0} and p_{k0}.

Since we are interested in the motion of the central particle we want to find the reduced distribution function which depends on Q, P and t. This can be achieved by integrating over all q_k s and p_k s. The N-fold integrals as $N \to \infty$ can be carried out if we choose the initial distribution function to be Gaussian, i.e.

$$W_0 = C \exp\left[-\left(A_1 P^2 + A_2 Q^2 + A_3 PQ + A_4 Q + A_5 P\right) - A_6 \sum_k \left(p_k^2 + k^2 q_k^2\right) \right], \qquad (17.95)$$

where A_i s are constants. By integrating over q_k and p_k we find the reduced distribution function to be

$$W(Q, P, t) = \mathcal{N} \exp\left[-a_1^2 (\Delta P)^2 - a_2^2 (\Delta Q)^2 - a_3^2 \Delta P \Delta Q \right]. \qquad (17.96)$$

In this relation the quantities ΔP and ΔQ are defined by

$$\Delta P = P - \langle P(t) \rangle, \quad \Delta Q = Q - \langle Q(t) \rangle, \qquad (17.97)$$

where a_1, a_2, a_3 and $\mathcal{N}$ are generally functions of time. By normalizing $W(Q, P, t)$ we get

$$\mathcal{N} = \frac{1}{2\pi} \left(4 a_1^2 a_2^2 - a_3^4 \right)^{\frac{1}{2}}. \qquad (17.98)$$

Similarly we find

$$\langle \Delta P^2(t) \rangle = \frac{\partial \ln(\mathcal{N})}{\partial a_1^2}, \qquad (17.99)$$

$$\langle \Delta Q^2(t) \rangle = \frac{\partial \ln(\mathcal{N})}{\partial a_2^2}, \qquad (17.100)$$

and

$$\langle \Delta Q(t)\Delta P(t) + \Delta P(t)\Delta Q(t) \rangle = 2\frac{\partial \ln(\mathcal{N})}{\partial a_2^3}. \qquad (17.101)$$

By substituting for $\mathcal{N}$ from (17.98) in (17.99)-(17.101) we obtain the following results:

$$\langle \Delta P^2 \rangle \langle \Delta Q^2 \rangle - \frac{1}{4} \langle \Delta P \Delta Q + \Delta Q \Delta P \rangle = \frac{1}{(2\pi\mathcal{N})^2}, \qquad (17.102)$$

$$a_1^2 = \frac{(2\pi\mathcal{N})^2}{2} \langle \Delta Q^2(t) \rangle, \qquad (17.103)$$

$$a_2^2 = \frac{(2\pi \mathcal{N})^2}{2} \left\langle \Delta P^2(t) \right\rangle, \tag{17.104}$$

and

$$a_3^2 = \frac{(2\pi \mathcal{N})^2}{2} \left\langle \Delta P(t)\Delta Q(t) + \Delta Q(t)\Delta P(t) \right\rangle. \tag{17.105}$$

Thus the function $W(Q, P, t)$, Eq. (17.96), is completely determined.

17.3 Equations of Motion for the Central Particle

We have already discussed the classical equations of motion for a particle which is subject to a potential $V(Q)$ and at the same time is coupled linearly to a heat bath. The classical and the quantum Hamiltonians for the system are given by Eq. (8.49).

The Heisenberg equation of motion averaged over the ground state of the bath oscillators is a solution of

$$\frac{d}{dt}\langle 0|P(t)|0\rangle + \left\langle 0 \left| \frac{\partial V(Q)}{\partial Q} \right| 0 \right\rangle - \int_0^t K\left(t - t'\right) \langle 0|Q\left(t'\right)|0\rangle \, dt' \tag{17.106}$$

From Eqs. (17.106) and $M\dot{Q}(t) = P(t)$ we have

$$\frac{d}{dt} \left[\langle 0|P(t)|0\rangle, \langle 0|Q(t)|0\rangle \right] = \left[\frac{d}{dt} \langle 0|P(t)|0\rangle, \langle 0|Q(t)|0\rangle \right]$$

$$= \int_0^t K\left(t - t'\right) \left[\langle 0|Q(t)|0\rangle, \langle 0|Q\left(t'\right)|0\rangle \right] dt'. \tag{17.107}$$

Thus the equal time commutator of $\langle 0|Q(t)|0\rangle$ and $\langle 0|P(t)|0\rangle$ is not a constant of motion, rather, it is given by [8]

$$\left[\langle 0|P(t)|0\rangle, \langle 0|Q(t)|0\rangle \right] = -i\hbar$$

$$+ \int_0^t dt' \int_0^{t'} K\left(t - t''\right) \left[\langle 0|Q\left(t''\right)|0\rangle, \langle 0|Q\left(t'\right)|0\rangle \right] dt''. \tag{17.108}$$

In the classical formulation of this problem discussed in Chapter 8 we observed that when (a) the coupling to the bath is nonlinear or (b) the potential $V(Q)$ is not quadratic or linear in Q, the Poisson bracket of $P(t)$ and $Q(t)$ will depend on $Q(0), P(0)$ as well as t. Therefore in quantum mechanics the right hand side of (17.108) will depend not only on time, but in general, it is a q-number. Thus an essential point in Kostin's derivation of the Scrödinger-Langevin equation which is the validity of the commutation relation (17.26) is missing when the potential $V(Q)$ is not harmonic or linear.

17.4   Wave Equation for the Motion of the Central Particle

If we start with the Hamiltonian operator

$$H = \frac{P^2}{2M} + V(Q) + \sum_{n=1}^{\infty} \frac{1}{2}\left(\frac{p_n^2}{m_n} + m_n\omega_n^2 q_n^2\right) + \sum_{n} M\epsilon_n q_n Q, \qquad (17.109)$$

and eliminate $q_n(t)$ and $p_n(t)$ from the equation of motion of $Q(t)$, we find as in the classical case

$$\frac{d^2Q(t)}{dt^2} + \frac{1}{M}\frac{\partial V(Q)}{\partial Q} = -\sum_{n=1}^{\infty}\left[q_n(0)\cos(\omega_n t) + \frac{1}{\omega_n}\dot{q}_n(0)\sin(\omega_n t)\right]$$

$$+ \int_0^t K(t - t')\, Q(t')\, dt', \qquad (17.110)$$

where

$$K(t - t') = M\sum_{n=1}^{\infty}\frac{\epsilon_n^2}{m_n\omega_n}\sin[\omega_n(t - t')]. \qquad (17.111)$$

By assuming, as we did earlier, that the oscillators are initially in the ground state, i.e.

$$\langle 0\,|q_n(0)|\,0\rangle = \langle 0\,|\dot{q}_n(0)|\,0\rangle = 0, \qquad (17.112)$$

and by denoting the expectation value of $Q(t)$ over the initial states of these oscillator by $\bar{Q}(t)$, we have

$$\frac{1}{M}\frac{d\bar{P}(t)}{dt} + \frac{\partial V(\bar{Q})}{\partial \bar{Q}} = \int_0^t K(t - t')\,\bar{Q}(t')\,dt', \quad \bar{P}(t) = M\frac{d\bar{Q}(t)}{dt}. \qquad (17.113)$$

For the special case of $V(\bar{Q}) = \frac{1}{2}M\Omega_0^2\bar{Q}^2$, we can solve the operator equation in terms of two independent functions $A_1(t)$ and $A_2(t)$;

$$\bar{Q}(t) = A_1(t)\bar{Q}(0) + \frac{1}{M}A_2(t)\bar{P}(0). \qquad (17.114)$$

These functions are defined as the independent solutions of the integro-differential equations

$$\frac{d^2 A_i(t)}{dt^2} + \Omega_0^2 A_i(t) = \int_0^t K(t - t')\,A_i(t')\,dt', \quad i = 1, 2, \qquad (17.115)$$

with the boundary conditions

$$A_1(0) = 1, \quad \dot{A}_1(0) = 0, \quad A_2(0) = 0, \quad \dot{A}_2(0) = 1. \qquad (17.116)$$

Let us investigate the motion of the central particle. At the time $t = 0$, the Hamiltonian for this particle which is not coupled to the bath of oscillators is given by

$$H_P(0) = \frac{1}{2M}\bar{P}^2(0) + \frac{1}{2}M\Omega_0^2\bar{Q}^2(0). \tag{17.117}$$

But at any other time t this Hamiltonian is

$$
\begin{aligned}
H_P(t) &= \frac{1}{2M}\bar{P}^2(t) + \frac{1}{2}M\Omega_0^2\bar{Q}^2(t) \\
&= \frac{1}{2}MG_1(t)\bar{Q}^2(0) + \frac{1}{2M}G_2(t)\bar{P}^2(0) + \frac{1}{2}G_3(t)\left[\bar{Q}(0)\bar{P}(0) + \bar{P}(0)\bar{Q}(0)\right],
\end{aligned}
\tag{17.118}
$$

where $G_1(t), G_2(t)$ and $G_3(t)$ are defined by

$$G_1(t) = \dot{A}_1^2(t) + \Omega_0^2 A_1^2(t), \tag{17.119}$$

$$G_2(t) = \dot{A}_2^2(t) + \Omega_0^2 A_2^2(t), \tag{17.120}$$

and

$$G_3(t) = \dot{A}_1(t)\dot{A}_2(t) + \Omega_0^2 A_1(t)A_2(t). \tag{17.121}$$

Since the Hamiltonian for the central particle is $H_P(t)$, the wave equation for the central particle is

$$
\begin{aligned}
i\hbar\frac{\partial\psi}{\partial t} &= \frac{1}{2}M\bar{Q}^2 G_1(t)\psi - \frac{\hbar^2}{2M}G_2(t)\left(\frac{\partial^2\psi}{\partial\bar{Q}^2}\right) \\
&\quad - \frac{1}{2}i\hbar G_3(t)\left[2\bar{Q}\frac{\partial}{\partial\bar{Q}} + 1\right]\psi.
\end{aligned}
\tag{17.122}
$$

We can eliminate the term $\frac{\partial\psi}{\partial\bar{Q}}$ in (17.122) by writing

$$\psi\left(\bar{Q}, t\right) = \exp\left[-i\frac{MG_3(t)}{2\hbar G_2(t)}\bar{Q}^2\right]\phi\left(\bar{Q}, t\right), \tag{17.123}$$

and substituting for $\psi\left(\bar{Q}, t\right)$ in (17.122) to get

$$i\hbar\frac{\partial\phi}{\partial t} = -\frac{\hbar^2}{2M(t)}\left(\frac{\partial^2\phi}{\partial\bar{Q}^2}\right) + \frac{M(t)}{2}\Omega^2(t)\bar{Q}^2\phi, \tag{17.124}$$

where

$$M(t) = \frac{M}{G_2(t)}, \tag{17.125}$$

and

$$\Omega^2(t) = G_2(t)\left\{G_1(t) - \left(\frac{G_3^2(t)}{G_2(t)}\right) - \frac{d}{dt}\left(\frac{G_3(t)}{G_2(t)}\right)\right\}. \tag{17.126}$$

Wave equations for harmonic oscillators where the mass and the frequency are time-dependent have been studied extensively [9]. A very important point regarding any dissipative system is the question of the stability of its ground state. As we have already seen, in a number of models, the effect of dissipation is to cause transitions to the lower energy states. But once the system has reached its ground state, there cannot be any further decay [10].

Considering the ground state of (17.124), we note that this equation is invariant under the transformation $\bar{Q} \to -\bar{Q}$, and that the ground state, $\phi_0(\bar{Q}, t)$, is an even function of $\bar{Q}$,

$$\phi_0(\bar{Q}, t) = \exp\left(C_2(t)\bar{Q}^2 + C_0(t)\right). \tag{17.127}$$

By substituting (17.127) in (17.124) and equating different powers of $\bar{Q}$ on the two sides we find that the differential equations for $C_0(t)$ and $C_2(t)$ are:

$$i\hbar \frac{dC_2}{dt} = -\left(\frac{2\hbar^2}{M}\right) G_2(t)C_2^2(t) + \frac{1}{2}M(t)\Omega^2(t), \tag{17.128}$$

and

$$i\hbar \frac{dC_0}{dt} = -\left(\frac{\hbar^2}{M}\right) G_2(t)C_2(t). \tag{17.129}$$

We note that at $t = 0$, the oscillator is not coupled to the bath and therefore the initial values for $C_0(t)$ and $C_2(t)$ are:

$$C_0(t=0) = 0, \quad \text{and} \quad C_2(t=0) = -\frac{M\Omega_0}{2\hbar}. \tag{17.130}$$

Let us now consider the question of the transfer of energy between the central particle and the heat bath. From the Hamiltonian of the central particle we can calculate the energy eigenvalues as a function of time. Thus by calculating the diagonal elements of $H_P(t)$ we find

$$\begin{aligned}
\langle n|H_P(t)|n\rangle &= \frac{1}{2}\left[\dot{A}_1^2(t) + \Omega_0^2 A_1^2(t)\right]\langle n|\bar{Q}^2(0)|n\rangle \\
&+ \frac{1}{2}\left[\dot{A}_2^2(t) + \Omega_0^2 A_2^2(t)\right]\langle n|\bar{P}^2(0)|n\rangle.
\end{aligned} \tag{17.131}$$

The last term in Eq. (17.118) has a vanishing expectation value

$$\langle n|\bar{Q}(0)\bar{P}(0) + \bar{P}(0)\bar{Q}(0)|n\rangle = 0, \tag{17.132}$$

and does not contribute to $\langle n|H_P(t)|n\rangle$. Using the well-known relations

$$\langle n|\bar{Q}^2(0)|n\rangle = \frac{\hbar}{M\Omega_0}\left(n + \frac{1}{2}\right), \tag{17.133}$$

and

$$\langle n|\bar{P}^2(0)|n\rangle = \hbar M\Omega_0\left(n + \frac{1}{2}\right), \tag{17.134}$$

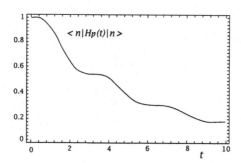

Figure 17.1: The matrix element $\langle n|H_P(t)|n\rangle$ of the energy of the central particle shown as a function of time. Since the energy is given in units of $\langle n|H_P(0)|n\rangle$, i.e. in terms of the initial energy of the system, this matrix element is independent of n.

we have

$$\frac{\langle n|H_P(t)|n\rangle}{\langle n|H_P(0)|n\rangle} = \frac{1}{2\Omega_0^2}\left[\left(\dot{A}_1^2(t) + \Omega_0^2\dot{A}_2^2(t)\right) + \Omega_0^2\left(A_1^2(t) + \Omega_0^2 A_2^2(t)\right)\right]. \quad (17.135)$$

In a similar way we can determine the Hamiltonian for the interaction between the particle and the bath in terms of $\bar{Q}(t)$ and therefore in terms of $\bar{Q}(0)$ and $\bar{P}(0)$;

$$H_I(t) = \sum_n \epsilon_n \bar{q}_n(t)\bar{Q}(t) = -\int_0^t K\left(t - t'\right)\bar{Q}(t)\bar{Q}\left(t'\right)dt'. \quad (17.136)$$

Again by substituting for $\bar{Q}(t)$ and $\bar{Q}\left(t'\right)$ in terms of $\bar{Q}(0)$ and $\bar{P}(0)$ we can write $H_I(t)$ in terms of $A_1(t)$ and $A_2(t)$ and their time derivatives. Then we find the matrix elements of $H_I(t)$ to be

$$\frac{\langle n|H_I(t)|n\rangle}{\langle n|H_P(0)|n\rangle} = -\frac{1}{2\Omega_0^2}\int_0^t K\left(t - t'\right)\left[A_1(t)A_1\left(t'\right) + \Omega_0^2 A_2(t)A_2\left(t'\right)\right]dt'. \quad (17.137)$$

Finally the energy transferred to the bath of oscillators can be obtained by noting that the initial energy of the system is given by $\langle n|H_P(0)|n\rangle$. Denoting the energy of the bath (apart from the zero point energies of the oscillators) by $\langle n|H_B(0)|n\rangle$, we have

$$\langle n|H_B(t)|n\rangle = \langle n|H_P(0)|n\rangle - \langle n|H_P(t)|n\rangle - \langle n|H_I(t)|n\rangle. \quad (17.138)$$

For the special case where $A_1(t)$ and $A_2(t)$ are given by Eqs. (17.115) and

$$K\left(t - t'\right) = 2\lambda\frac{d}{dt'}\delta\left(t' - t\right), \quad (17.139)$$

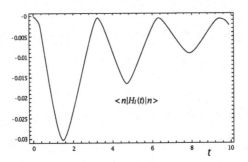

Figure 17.2: The interaction energy $\langle n|H_I(t)|n\rangle$ plotted as a function of time and expressed in units of the initial energy of the central particle.

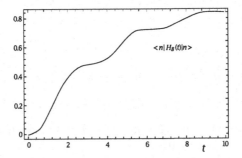

Figure 17.3: The energy transferred to the bath in units of the initial energy of the central oscillator.

we have calculated $\langle n|H_P(t)|n\rangle$, $\langle n|H_I(t)|n\rangle$ and $\langle n|H_B(t)|n\rangle$ all in units of $\langle n|H_P(0)|n\rangle$. Note that here we have used the integral

$$\int_0^t 2\lambda \frac{d}{dt}\delta\left(t - t'\right) A_i\left(t'\right) dt' = -\lambda\left(\frac{dA_i}{dt}\right) \tag{17.140}$$

since t is the end point of integration [11]. The results are shown in Figs. (17.1), (17.2) and (17.3). The rate of loss of energy by the central particle which is

$$\frac{d}{dt}\langle n|H_P(t)|n\rangle, \tag{17.141}$$

is also shown in Fig. (17.4). We observe that the rate of the energy flow is finite, unlike the situation in some other models such as the Wigner-Weisskopf model (§15.2).

For this simple case $\bar{Q}$ asymptotically goes to zero, and therefore there is a natural cut-off in the coupling between the bath and the central oscillator.

The operator equation (17.113) for any potential which is polynomial in $\bar{Q}$ can be solved numerically. Thus we write (17.113) as

$$\frac{d^2\bar{Q}(t)}{dt^2} = \frac{1}{M}f(\bar{Q}) + \int_0^t K\left(t - t'\right)\bar{Q}\left(t'\right) dt' \tag{17.142}$$

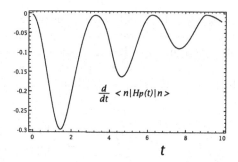

Figure 17.4: The rate of flow of energy from the particle to the bath.

where $f(\bar{Q})$ is a polynomial in $\bar{Q}$. Then we expand $\bar{Q}(t)$ as an infinite series in $\bar{Q}(0)$ and $\bar{P}(0)$;

$$
\begin{aligned}
\bar{Q}(t) \;=\; & A_0(t) + \sum_{n=1}^{\infty} \left[A_n(t)(\bar{Q}(0))^n + B_n(t)(\bar{P}(0))^n \right] \\
& + \sum_{n=0}^{\infty} \sum_{m=1}^{\infty} C_{n,m}(t) \left[(\bar{P}(0))^m (\bar{Q}(0))^{n-m+1} + ((\bar{Q}(0))^{n-m+1} \bar{P}(0))^m \right].
\end{aligned}
$$

(17.143)

By substituting (17.143) in (17.142) and ordering the operators on the right-hand side of (17.142) using the commutator

$$
[\bar{Q}(0)\bar{P}(0) - \bar{P}(0)\bar{Q}(0)] = i\hbar,
$$

(17.144)

is such a way that the right-hand side is an infinite sum of Hermitian operators,

$$
(\bar{P}(0))^m (\bar{Q}(0))^{n-m} + (\bar{Q}(0))^{n-m} (\bar{P}(0))^m,
$$

(17.145)

we find that the functions $A_0(t), A_n(t), B_n(t)$ and $C_{m,n}(t)$ all satisfy coupled second-order differential equations. Since these equations are nonlinear, the result of numerical integration is accurate only for relatively short times [12].

17.5 Motion of the Center-of-Mass in Viscous Medium

In problems such as the collision or trapping of heavy ions models have been proposed where two particles interact while moving in a viscous medium [13]-[16].

In a model where dissipation arises from the linear coupling of the motion of the two particles with the heat bath, we can separate the motion of the center of mass from the relative coordinates. For the sake of simplicity we consider the one-dimensional motion and assume that the interaction between the two particles is given by the potential $V(|Q_1 - Q_2|)$. Thus the Hamiltonian operator is

$$
\begin{aligned}
H &= \frac{P_1^2}{2M_1} + \frac{P_2^2}{2M_2} + V(|Q_1 - Q_2|) + \sum_n \frac{1}{2}\left[\frac{p_n^2}{m_n} + m_n \omega_n^2 q_n^2\right] \\
&+ \sum_n \epsilon_n q_n (M_1 Q_1 + M_2 Q_2),
\end{aligned}
\tag{17.146}
$$

where M_1 and M_2 are the masses of the two particles. Introducing the center-of-mass and the relative coordinates by

$$
(M_1 + M_2)Q = M_1 Q_1 + M_2 Q_2, \quad q = Q_1 - Q_2,
\tag{17.147}
$$

and their conjugate momenta by P and p, we can write (17.146) as

$$
\begin{aligned}
H &= \frac{P^2}{2M} + \frac{p^2}{2m} + V(q) + \sum_n \frac{1}{2}\left[\frac{p_n^2}{m_n} + m_n \omega_n^2 q_n^2\right] \\
&+ \sum_n M \epsilon_n q_n Q,
\end{aligned}
\tag{17.148}
$$

where $M = M_1 + M_2$ is the total mass and

$$
m = \left(\frac{M_1 M_2}{M_1 + M_2}\right),
\tag{17.149}
$$

is the reduced mass. It can be seen from (17.148) that the relative motion is not coupled to the oscillators, whereas the center-of-mass is.

The equation of motion for the center-of-mass averaged over the ground states of the oscillators q_n is

$$
\frac{d^2 \bar{Q}(t)}{dt^2} = \int_0^t K(t - t') \bar{Q}(t') dt',
\tag{17.150}
$$

and for $\bar{q}$ we have

$$
m \frac{d^2 \bar{q}}{dt^2} = -\frac{\partial V(\bar{q})}{\partial \bar{q}}.
\tag{17.151}
$$

Since the motion of the center-of-mass operator $\bar{Q}$ is independent of $V(q)$, we can find the wave function for the motion by writing $\bar{Q}(t)$ in terms of $\bar{Q}(0)$ and $\bar{P}(0)$, Eq. (17.115), where

$$\frac{d^2 A_i(t)}{dt^2} = \int_0^t K\left(t - t'\right) A_i\left(t'\right) dt', \quad i = 1, 2, \tag{17.152}$$

with the same initial conditions as those given in (17.116).

The Hamiltonian for the center-of-mass motion is given by

$$H_{cm} = \frac{1}{2M} \bar{P}^2(t) + \mathcal{F}(t)\bar{Q}, \quad \bar{P} = M\frac{d\bar{Q}(t)}{dt}, \tag{17.153}$$

where

$$\mathcal{F}(t) = M \sum_n \epsilon_n q_n(t), \tag{17.154}$$

is the time-dependent driving force. Thus we have the Schrödinger equation

$$
\begin{aligned}
i\hbar \left(\frac{\partial \Psi}{\partial t}\right) &= \frac{1}{2} M \dot{A}_1^2 \bar{Q}^2 \Psi - \frac{i\hbar}{2M} \dot{A}_1 \dot{A}_2 \Psi \\
&\quad - \frac{\hbar^2}{2M} \dot{A}_2^2 \left[\frac{\partial \Psi}{\partial \bar{Q}^2} + \left(\frac{2iM\dot{A}_1}{\hbar \dot{A}_2}\right) \bar{Q} \left(\frac{\partial \Psi}{\partial \bar{Q}}\right)\right] + \mathcal{F}(t)\bar{Q}\Psi.
\end{aligned}
\tag{17.155}
$$

Again if we eliminate the first order derivative $\frac{\partial \Psi}{\partial \bar{Q}}$ in (17.155) by changing Ψ to Φ, where

$$\Psi = \Phi \exp\left[-\frac{iM\dot{A}_1 \bar{Q}^2}{2\hbar \dot{A}_2}\right], \tag{17.156}$$

we obtain a partial differential equation for Φ,

$$i\hbar \left(\frac{\partial \Phi}{\partial t}\right) = -\frac{\hbar^2}{2M} \dot{A}_2^2 \left(\frac{\partial \Phi}{\partial \bar{Q}^2}\right) - \frac{M}{2}\left[\frac{d}{dt}\left(\frac{\dot{A}_1}{\dot{A}_2}\right)\right] \bar{Q}^2 \Phi + \mathcal{F}(t)\bar{Q}\Phi. \tag{17.157}$$

For the special case where $K\left(t - t'\right) = 2\lambda \frac{d}{dt'}\delta\left(t' - t\right)$, Eq. (17.152) can be solved exactly;

$$A_1(t) = 1, \quad A_2(t) = \frac{1 - e^{-\lambda t}}{\lambda}. \tag{17.158}$$

Thus for this form of the kernel $K\left(t - t'\right)$, the equation of of motion of the center-of-mass becomes;

$$i\hbar \frac{\partial \Psi}{\partial t} = -\frac{\hbar^2}{2M} e^{-\lambda t} \frac{\partial^2 \Psi}{\partial \bar{Q}^2} + \mathcal{F}(t)\bar{Q}\Psi, \tag{17.159}$$

i.e. the wave function in this case is identical with the one that we found by quantizing Kanai-Caldirola Hamiltonian. This simple result follows from the special form of the coupling that we have assumed. If we replace

$$\sum_n \epsilon_n q_n \left(M_1 Q_1 + M_2 Q_2\right) \tag{17.160}$$

by

$$\sum_n q_n \left(\epsilon_n M_1 Q_1 + \varepsilon_n M_2 Q_2 \right) \tag{17.161}$$

then the damping force acting on the two particles will be different.

17.6   Invariance Under Galilean Transformation

In the microscopic derivation of dissipative forces from a many-particle system, we have seen that the time-reversal invariance and the constancy of the commutation relation $\left[\bar{P}(t), \bar{Q}(t) \right]$ are violated in the process of eliminating the degrees of freedom of the oscillators forming the bath. Now let us consider the question of the Galilean invariance of a dissipative system derived from a many-particle system which is assumed to be invariant under this transformation.

Let us start with a modified version of the Hamiltonian (17.109) which we write as

$$H = \frac{P^2}{2M} + \sum_j \frac{p_j^2}{2m_j} + \sum_j \frac{1}{2} m_j \omega_j^2 \left(Q - q_j \right)^2. \tag{17.162}$$

Here we assume that the frequencies ω_j and masses m_j satisfy the relation

$$\sum_{j=1}^{\infty} m_j \omega_j^2 = M\Omega^2, \tag{17.163}$$

and therefore the force $\frac{1}{2} M\Omega^2 Q^2$ on the particle is a finite. The equations of motion in this case are

$$M\ddot{Q} + \sum_j m_j \omega_j^2 \left(Q - q_j \right) = 0, \tag{17.164}$$

and

$$m_j \ddot{q}_j - m_j \omega_j^2 \left(Q - q_j \right) = 0. \tag{17.165}$$

These equations remain invariant under the Galilean transformation

$$P' = P + Mv, \quad Q' = Q + vt, \quad q_j' = q_j + vt. \tag{17.166}$$

As before by eliminating q_j between (17.164) and (17.165) we find the equation of motion for Q to be

$$M\ddot{Q} \;+\; M\Omega^2 Q - \sum_j m_j \omega_j^2 \left\{ q_j(0) \cos(\omega_j t) + \frac{1}{\omega_j} \dot{q}_j(0) \sin(\omega_j t) \right\}$$

$$+ \; \int_0^t \Gamma \left(t - t' \right) Q \left(t' \right) dt' = 0,$$

$$\tag{17.167}$$

where

$$\Gamma\left(t-t'\right)=\sum_j m_j\omega_j^3 \sin\left[\omega_j\left(t-t'\right)\right].\tag{17.168}$$

Let us consider the equation of motion for Q' i.e. the motion of the central particle referred to a frame S' which is moving with constant velocity with respect to the initial frame of reference S;

$$
M\ddot{Q}' \; + \; M\Omega^2 Q' - \sum_j m_j\omega_j^2 \left\{ q_j(0)\cos(\omega_j t) + \frac{1}{\omega_j}[\dot{q}_j(0) + v]\sin(\omega_j t) \right\}
$$
$$
+ \; \int_0^t \Gamma\left(t-t'\right) Q'\left(t'\right) dt' = 0,
$$

$$\tag{17.169}$$

If we average Q' over the initial distributions of $q_j(0)$ and $\dot{q}_j(0)$, the presence of the additional term proportional to v destroys the Galilean invariance of the original equations (17.164) and (17.165).

17.7   Velocity Coupling and Coordinate Coupling

In §8.2 we briefly mentioned two types of classical coupling between the central particle and the bath oscillators; either we can couple the coordinate of the central particle Q to the coordinates of the oscillators of the bath q_n s or we can couple its momentum P to q_n s . Here we want to show that these two models of interaction are related to each other. This connection can be made either at the classical level or at the quantum level.

Let us start with the momentum coupling model for which the total Hamiltonian operator is given by [17]

$$H_P = \frac{1}{2M}\left[P + \sum_j m_j\omega_j^2 q_j\right]^2 + V(Q) + \sum_j \left[\frac{p_j^2}{2m_j} + \frac{1}{2}m_j\omega_j^2 q_j^2\right].\tag{17.170}$$

Next consider the unitary transformation

$$U_1 = \exp\left(-\frac{i}{\hbar}Q\sum_j m_j\omega_j q_j\right),\tag{17.171}$$

which transforms P, Q, p_j and q_j to

$$P \rightarrow U_1^\dagger P U_1 = P - \sum_j m_j\omega_j q_j, \quad Q \rightarrow Q,\tag{17.172}$$

$$p_j \rightarrow U_1^\dagger p_j U_1 = p_j - m_j \omega_j Q, \quad q_j \rightarrow q_j, \tag{17.173}$$

and thus H_P transforms to H_I,

$$H_P \rightarrow H_I = U_1^\dagger H_P U_1 = \frac{P^2}{2M} + V(Q) + \sum_j \left[\frac{(p_j - m_j \omega_j Q)^2}{2m_j} + \frac{1}{2} m_j \omega_j^2 q_j^2 \right],$$
$$\tag{17.174}$$

This Hamiltonian is not yet of the form where Q couples to q_j s. Therefore we make a second unitary transformation

$$U_2 = \exp \left[\frac{i\pi}{2\hbar} \sum_j \left(\frac{p_j^2}{2m_j \omega_j} + \frac{1}{2} m_j \omega_j q_j^2 \right) \right]. \tag{17.175}$$

Under this transformation we change q_j s, p_j s and H_I;

$$q_j \rightarrow U_2^\dagger q_j U_2 = -\frac{p_j}{m_j \omega_j}, \quad p_j \rightarrow m_j \omega_j q_j, \tag{17.176}$$

$$H_I \rightarrow H_Q = U_2^\dagger H_I U_2 = \frac{P^2}{2M} + V(Q) + \sum_j \left[\frac{p_j^2}{2m_j} + \frac{1}{2} m_j \omega_j^2 \left(q_j - Q \right)^2 \right]. \tag{17.177}$$

Thus after the second transformation we get a Hamiltonian where Q is coupled to all q_j s.

17.8   Equation of Motion for a Harmonically Bound Radiating Electron

From the Heisenberg equations of motion for a harmonically bound electron coupled to electromagnetic field we can derive an operator equation for the position of the electron which is the quantum analogue of the Abraham-Lorentz equation [18] [19].

We start with the quantum mechanical Hamiltonian for one-dimensional motion of the electron coupled to the radiation field in the dipole approximation

$$H = \frac{1}{2m_0} \left[p_x - \frac{e}{c} A_x \right]^2 + \frac{1}{2} K x^2 + \frac{1}{2} \sum_{\mathbf{k},\sigma} \left(p_{\mathbf{k},\sigma}^2 + \omega_k^2 q_{\mathbf{k},\sigma}^2 \right), \tag{17.178}$$

where we have written the Hamiltonian for the radiation field in terms of $p_{\mathbf{k},\sigma}$ and $q_{\mathbf{k},\sigma}$, and these are the components of $\mathbf{p_k}$ and $\mathbf{q_k}$ (see §9.1). In Eq. (17.178) $\omega_k = ck$, and the x-component of the vector potential A_x is given by

$$A_x = \sum_{\mathbf{k},\sigma} \left(\frac{2\pi \hbar c}{kV} \right)^{\frac{1}{2}} \left(\delta_k^* a_{\mathbf{k},\sigma} \mathbf{e}_{\mathbf{k},\sigma} \cdot \hat{\mathbf{x}} + \delta_k a_{\mathbf{k},\sigma}^\dagger \mathbf{e}_{\mathbf{k},\sigma}^* \cdot \hat{\mathbf{x}} \right). \tag{17.179}$$

Here δ_k is the electron form factor, $\mathbf{e}_{\mathbf{k},\sigma}$ is the polarization unit vector, $\hat{\mathbf{x}}$ is the unit vector in the direction $\mathbf{x}$ and V is the volume. The creation and annihilation operators $a^\dagger_{\mathbf{k},\sigma}$ and $a_{\mathbf{k},\sigma}$ for the electromagnetic field are related to $\mathbf{q}_{\mathbf{k},\sigma}$ and $\mathbf{p}_{\mathbf{k},\sigma}$ in (17.178) by the relations

$$a_{\mathbf{k},\sigma} = \frac{1}{\sqrt{2\omega}}\left(\omega_k q_{\mathbf{k},\sigma} + i p_{\mathbf{k},\sigma}\right), \tag{17.180}$$

and

$$a^\dagger_{\mathbf{k},\sigma} = \frac{1}{\sqrt{2\omega}}\left(\omega_k q_{\mathbf{k},\sigma} - i p_{\mathbf{k},\sigma}\right) \tag{17.181}$$

and satisfy the commutation relations

$$\left[a_{\mathbf{k},\sigma}, a^\dagger_{\mathbf{k}',\sigma'}\right] = \delta_{\mathbf{k},\mathbf{k}'}\delta_{\sigma,\sigma'}, \tag{17.182}$$

and

$$\left[a^\dagger_{\mathbf{k},\sigma}, a^\dagger_{\mathbf{k}',\sigma'}\right] = [a_{\mathbf{k},\sigma}, a_{\mathbf{k}',\sigma'}] = 0. \tag{17.183}$$

Substituting for $p_{\mathbf{k},\sigma}$ and $q_{\mathbf{k},\sigma}$ in terms of $a^\dagger_{\mathbf{k},\sigma}$ and $a_{\mathbf{k},\sigma}$ we can write H as

$$H = \frac{1}{2m_0}\left[p_x - \frac{e}{c}A_x\right]^2 + \frac{1}{2}Kx^2 + \frac{1}{2}\sum_{\mathbf{k},\sigma}\hbar kc\, a^\dagger_{\mathbf{k},\sigma}a_{\mathbf{k},\sigma}, \tag{17.184}$$

where we have omitted the zero point energy term. From this Hamiltonian operator we find the Heisenberg equation of motion for x, p_x and $a_{\mathbf{k},\sigma}$;

$$\dot{x} = \frac{1}{m_0}\left(p_x - \frac{e}{c}A_x\right), \quad \dot{p}_x = -Kx, \tag{17.185}$$

and

$$\dot{a}_{\mathbf{k},\sigma} = -icka_{\mathbf{k},\sigma} + i\left(\frac{2\pi e^2}{\hbar ckV}\right)^{\frac{1}{2}}(\delta_k\hat{\mathbf{x}}.\mathbf{e}_{\mathbf{k},\sigma})\,\dot{x}, \tag{17.186}$$

with a similar relation for $\dot{a}^\dagger_{\mathbf{k},\sigma}$. Now we integrate (17.186) and assume that the coupling is turned off at $t = -\infty$, i.e.

$$a_{\mathbf{k},\sigma}(t) = a^f_{\mathbf{k},\sigma}(t) + i\left(\frac{2\pi e^2}{\hbar ckV}\right)^{\frac{1}{2}}\delta_k\hat{\mathbf{x}}.\mathbf{e}_{\mathbf{k},\sigma}\int_{-\infty}^t e^{-i\omega(t-t')}\dot{x}(t')\,dt'. \tag{17.187}$$

The operator $a^f_{\mathbf{k},\sigma}(t)$ is the free-field Heisenberg annihilation operator. We can also find a similar result for $a^\dagger_{\mathbf{k},\sigma}(t)$. Now we substitute for $a^\dagger_{\mathbf{k},\sigma}(t)$ and $a_{\mathbf{k},\sigma}(t)$ in A_x and eliminate p_x between the two equations in (17.185) and obtain the Heisenberg equation for the position operator x;

$$m_0\ddot{x} + \int_{-\infty}^t J(t-t')\,\dot{x}(t')\,dt' + Kx = F(t), \tag{17.188}$$

where

$$J\left(t - t'\right) = \left(\frac{8\pi e^2}{3V}\right) \sum_{\mathbf{k}} |\delta_k|^2 \cos\left[ck\left(t - t'\right)\right],\qquad(17.189)$$

and

$$F(t) = -\frac{e}{c}\left(\frac{\partial A_x^f(t)}{\partial t}\right).\qquad(17.190)$$

Once the electron form factor δ_k is given, then $J\left(t - t'\right)$ can be determined and the integro-differential equation (17.188) can be solved. For instance if we choose the electron form factor to be [17]

$$|\delta_k|^2 = \frac{\Omega^2}{\Omega^2 + c^2 k^2},\qquad(17.191)$$

where Ω is a large cutoff frequency, and with this choice of $|\delta_k|^2$ we have

$$J\left(t - t'\right) = \left(\frac{8\pi e^2}{3V}\right) \sum_{\mathbf{k}} \frac{\Omega^2}{\Omega^2 + c^2 k^2} \cos\left[ck\left(t - t'\right)\right].\qquad(17.192)$$

We replace the summation over $\mathbf{k}$ by integration

$$J\left(T\right) = \left(\frac{8\pi e^2}{3(2\pi)^3}\right) \int_0^\infty \frac{4\pi k^2 \Omega^2}{\Omega^2 + c^2 k^2} \cos\left(ckT\right) dk,\qquad(17.193)$$

and then we can evaluate the integral in (17.193) with the result that

$$J\left(T\right) = m\Omega^2 \tau \left[2\delta(T) - \Omega \exp(-\Omega T)\right],\qquad(17.194)$$

where $\tau = \frac{2e^2}{3mc^2}$. Here m is the renormalized mass of the electron (see e.g. Eq. (9.11) of the van Kampen model). For the present model this renormalized mass is given by [17] [20]

$$m = m_0 + \frac{2e^2\Omega}{3c^3}.\qquad(17.195)$$

Substituting for $J\left(t - t'\right)$ from (17.194) in (17.188) we find

$$m_0\ddot{x}(t) + m\Omega^2\tau\dot{x}(t) - m\Omega^3\tau \int_{-\infty}^t \exp\left[-\Omega\left(t - t'\right)\right]\dot{x}\left(t'\right) dt' + Kx(t) = F(t).\qquad(17.196)$$

Now we multiply this equation by $e^{-\Omega t}\frac{d}{dt}e^{\Omega t}$ to get an operator differential equation for x;

$$\left(\frac{m_0}{\Omega}\right)\frac{d^3 x}{dt^3} + m\frac{d^2 x}{dt^2} + K\left(x + \frac{1}{\Omega}\dot{x}\right) = F(t) + \frac{1}{\Omega}\dot{F}(t).\qquad(17.197)$$

This is the quantum mechanical analogue of the classical Abraham-Lorentz equation [21].

Bibliography

[1] E.G. Harris, Phys. Rev. A42, 3685 (1990).

[2] W. Sollfrey, Ph.D. Dissertation, New York University (1950)

[3] W.G. Unruh and W.H. Zurek, Phys. Rev. D40, 1071 (1989).

[4] I.R. Senitzky, Phys. Rev. 119, 1280 (1960).

[5] J. Weber, Phys. Rev. 101, 1619 (1956).

[6] J. Weber, Phys. Rev. 90, 977 (1953).

[7] C.W. Gardiner, *Quantum Noise*, (Springer-Verlag, Berlin 1991).

[8] M. Razavy, Phys. Rev. A41, 1211 (1990).

[9] V.V. Dodonov and V.I Man'ko in *Invariants and the Evolution of Non-stationary Quantum Systems*, Preceeding of the Lebedev Physics Institute vol. 183, Edited by M.A. Markov. (Nova Scientific, Commack, 1989) p. 103.

[10] M. Razavy, Can. J. Phys. 69, 1235 (1991).

[11] B. Friedman *Principles and Techniques of Applied Mathematics*, (John Wiley & Sons, New York, 1957) p. 154.

[12] M. Razavy, Phys. Rev. A41, 6668 (1990).

[13] W. Nöremberg and H.A. Weidemüler, *Introduction to the Theory of Heavy-Ion Collision*, (Springer-Verlag, Berlin, 1976).

[14] R.W. Hasse, J. Math. Phys. 16, 2005 (1976).

[15] R.W. Hasse, Rep. Prog. Phys. 41, 1027 (1978).

[16] K. Albrecht and R.W. Hasse, Physica 4D, 244 (1982).

[17] G.W. Ford, J.T. Lewis and R.F. O'Connell, Phys. Rev. A37, 4419 (1998).

[18] G.W. Ford, J.T. Lewis and R.F. O'Connell, Phys. Rev. Lett. 55, 2273 (1985).

[19] G.W. Ford, and R.F. O'Connell, Phys. Lett. 157, 217 (1991).

[20] G.W. Ford, and R.F. O'Connell, J. Stat. Phys. 57, 803 (1989).

[21] G.W. Ford and R.F. O'Connell, Appl. Phys. B 60, 301 (1995).

Chapter 18

Quantum Mechanical Models of Dissipative Systems

In our studies of the quantum dissipative systems, up to this point, we relied on the quantization of classically damped motions. We observed that this classical damping can be introduced phenomenologically in the equation of motion, or can be derived from a conservative many-particle system. But there are damped or decaying systems in quantum mechanics with no classical analogues. Perhaps the most striking example of a decaying system with no classical counterpart is the escape of a particle originally trapped behind a barrier by the mechanism of quantum tunneling. Here we want to consider few examples were the starting point is quantum mechanical, and these systems may or may not correspond to the classical damped motion.

18.1 Forced Vibration with Damping

Let us start with the Schrödinger equation [1]

$$-\frac{\hbar^2}{2m}\frac{\partial^2\psi(x,t)}{\partial x^2} + \frac{1}{2}Kx^2\psi - g(t)x\psi(x,t)$$
$$- i\hbar\lambda x\psi(x)\int \psi^*(y,t)\frac{\partial\psi(y,t)}{\partial y}dy = i\hbar\frac{\partial\psi(x,t)}{\partial t}, \qquad (18.1)$$

where $g(t)$ is the external force and

$$i\hbar\lambda \int \psi^*(y,t)\frac{\partial\psi(y,t)}{\partial y}dy = -\lambda\langle p\rangle \tag{18.2}$$

is the damping force. We observe that the effect of damping is zero if $\psi(x,t)$ is an eigenstate of the free lattice. This suggests, as in the case of Schrödinger-Langevine equation, that we change x to z, where

$$z = x - \xi(t), \tag{18.3}$$

and assume a solution of the form

$$\psi(x,t) = u(z) \exp\left[-\frac{i}{\hbar}\Gamma(x,t)\right]. \tag{18.4}$$

Here $u(z)$ is the solution of the Schrödinger equation for a free undamped oscillator

$$-\frac{\hbar^2}{2m}\frac{d^2u(z)}{dz^2} + \frac{1}{2}Kz^2u(z) = Eu(z). \tag{18.5}$$

By substituting (18.4) in (18.1) and making use of (18.5) we find that $\Gamma(x,t)$ must satisfy the relation

$$\Gamma(x,t) = Et + \int \left[\frac{1}{2}m\dot{\xi}^2 - \frac{1}{2}K\xi^2 + \left(K\xi - g(t) + m\lambda\dot{\xi}\right)x\right]dt, \tag{18.6}$$

where ξ is a solution of the classical equation of motion

$$m\ddot{\xi}(t) = -K\xi(t) + g(t) - m\lambda\dot{\xi}(t). \tag{18.7}$$

By calculating $|\psi(x,t)|^2$ we find that

$$|\psi(x,t)|^2 = |u(z)|^2 = |u(x - \xi(t)|^2; \tag{18.8}$$

therefore the center of the wave packet oscillates exactly as that of a damped and forced harmonic oscillator, i.e. $\xi(t)$.

In this model the expectation value of the energy of the system is given by

$$\langle E\rangle = i\hbar \int \psi^*(x,t)\frac{\partial\psi(x,t)}{\partial t}dx = E + \frac{1}{2}\dot{\xi}^2 + \frac{1}{2}K\xi^2 - g(t)\xi + m\lambda\xi\dot{\xi}, \tag{18.9}$$

which is the sum of the quantum energy and the classical energy associated with the oscillatory motion of $\xi(t)$.

There are different ways of generalizing this model:

(a) We can change it to a form that can be used for dissipative forces proportional to any power of velocity.

(b) We can modify it so that it can be applied to N interacting particles with pair interaction.

First let us consider the following modification of (18.1) where we replace the term

$$-i\hbar\lambda x \int \psi^*(y,t)\frac{\partial\psi(y,t)}{\partial y}dy \qquad (18.10)$$

by

$$\gamma x \left[\int \psi^*(y,t)\left(-i\hbar\frac{\partial\psi(y,t)}{\partial y}\right)dy\right]^2. \qquad (18.11)$$

The new $\Gamma(x,t)$ and $\xi(t)$ must satisfy the following relations

$$\Gamma(x,t) = Et + \int \left[\frac{1}{2}m\dot{\xi}^2 - \frac{1}{2}K\xi^2 + \left(K\xi - g(t) + m\gamma\dot{\xi}^2\right)x\right]dt \qquad (18.12)$$

and

$$m\ddot{\xi}(t) = -K\xi(t) + g(t) - m\gamma\dot{\xi}^2(t). \qquad (18.13)$$

Similar results can be found for dissipative forces proportional to $\dot{\xi}^\alpha, (\alpha > 0)$.

The second possible generalization is when there are N interacting particles with $3N$ degrees of freedom. Here the simple harmonic pair interaction is of the form

$$V(x_1, x_2, \ldots x_{3N}) = \frac{1}{4}\sum_{j,k=1}^{3N} K_{jk}(x_k - x_j)^2, \qquad (18.14)$$

where both j and k run over all of the allowed values. For an externally applied force we assume the potential W to be

$$W(x_1, x_2, \ldots x_{3N}, t) = -\sum_{j=1}^{3N} g_j(t)x_j. \qquad (18.15)$$

With these two potentials the Schrödinger equation becomes

$$-\frac{\hbar^2}{2m}\sum_j \left(\frac{\partial^2\psi}{\partial x_j^2}\right) + V(x_1, x_2, \ldots x_{3N})\psi(x_1 \cdots x_{3N}, t) - \sum_j g_j(t)x_j\psi$$
$$-i\hbar\lambda\sum_j \left(\int \psi^*\frac{\partial\psi}{\partial y_j}dy_j\right)x_j\psi(x_1 \cdots x_{3N}, t) = i\hbar\frac{\partial}{\partial t}\psi(x_1 \cdots x_{3N}, t). \qquad (18.16)$$

Next we change the variable x_k to z_k where

$$z_k = x_k - \xi_k(t), \qquad (18.17)$$

and again seek a solution of the form

$$\psi(x_1 \cdots x_{3N}, t) = u(z_1, z_2, \ldots z_{3N})\exp\left[-\frac{i}{\hbar}\Gamma(x_1, x_2, \ldots x_{3N}, t)\right]. \qquad (18.18)$$

Here $u(z_1, z_2, \ldots z_{3N})$ is the solution of the Schrödinger equation for the lattice with no damping force

$$\left[-\frac{\hbar^2}{2m} \sum_j \left(\frac{\partial^2}{\partial z_j^2} \right) + V(z_1, z_2, \ldots z_{3N}) \right] u(z_1, \ldots z_{3N}) = Eu(z_1, \ldots z_{3N}).$$

(18.19)

Just as before, Eq. (18.18) will satisfy (18.16) provided that $\Gamma(x_1, x_2, \ldots x_{3N}, t)$ is given by

$$\Gamma(x_1, x_2, \ldots x_{3N}, t) = Et + \int \left[\frac{m}{2} \sum_{j=1}^{3N} \dot{\xi}_j^2 - \frac{1}{4} \sum_{j,k=1}^{3N} K_{jk} \left(\xi_k(t) - \xi_j(t) \right)^2 \right.$$

$$\left. + \frac{1}{2} \sum_{j,k=1}^{3N} K_{jk} (x_k - x_j) (\xi_k(t) - \xi_j(t)) - \sum_{j=1}^{3N} g_j(t) x_j + m\lambda \sum_{j=1}^{3N} \dot{\xi}_j x_j \right] dt,$$

(18.20)

and where $\xi_j(t)$ satisfies the classical equation of motion

$$m\ddot{\xi}_j = -\sum_k K_{jk} [\xi_j(t) - \xi_k(t)] + g_j(t) - m\lambda\dot{\xi}_j.$$

(18.21)

From Eq. (18.18) it follows that the centers at $x_j = \xi_j(t)$ of the quantum wave packets follow the classical motion of the lattice.

The expression for the energy can be found in the same way as we obtained Eq. (18.9). The result shows that the total energy is the sum of quantum energy of the lattice, E, plus the energy of the classical motion

$$\langle E \rangle = E + \sum_j \left(\frac{m}{2} \right) \dot{\xi}_j^2 + \frac{1}{4} \sum_{jk} K_{jk} [\xi_j(t) - \xi_k(t)]^2$$

$$- \sum_j g_j(t) \xi_j(t) + m\lambda \sum_j \xi_j(t) \dot{\xi}_j(t).$$

(18.22)

18.2 The Wigner-Weisskopf Model

This model describes an unstable system which consists of a central particle with wave function $\psi(\mathbf{r}, t)$, and this particle interacts with a group of motionless objects with the amplitude $\chi_n(t), n = 1, 2, \cdots$ at the origin [2]-[6]. In the following discussion we use the units where $\hbar = 1$ and $m = \frac{1}{2}$. The Schrödinger equations for the coupled systems are:

$$i \frac{\partial \psi(\mathbf{r}, t)}{\partial t} = -\nabla^2 \psi(\mathbf{r}, t) + \sum_n g_n \delta(\mathbf{r}) \chi_n(t)$$

(18.23)

and

$$i\frac{d\chi_n}{dt} = \mathcal{E}_n\chi_n + g_n\psi(0,t), \quad n = 1, 2\cdots, \tag{18.24}$$

where $\psi(0,t)$ is defined as the finite part of $\psi(\mathbf{r},t)$ at the origin [4]. Since the interaction takes place at $r = 0$, therefore only S wave will be affected by the coupling. In this model one can also introduce a form factor $\Lambda(\mathbf{r})$ and write the interaction term as $\sum_n g_n\Lambda(\mathbf{r})\chi_n(t)$ in Eq. (18.23) and $g_n\Lambda(\mathbf{r})\psi(\mathbf{r},t)$ in (18.24) [5].

We can use the time Fourier transform and write the solution for the wave number k as

$$\begin{cases} \psi_k(r) = \frac{1}{r}\sin(kr + \eta(k)) \\ \chi_{nk} \end{cases}, \quad n = 1, 2\cdots. \tag{18.25}$$

Substituting (18.25) in (18.23) and equating the coefficients of $\delta(\mathbf{r})$ on the two sides yields the condition

$$4\pi\sin\eta(k) + \sum_n g_n\chi_{nk} = 0. \tag{18.26}$$

The same substitution in (18.24) gives us

$$\left(k^2 - \mathcal{E}_n\right)\chi_{nk} = g_n k \cos\eta(k), \tag{18.27}$$

where the finite part of $\psi(0,r)$ from (18.25) is $k\cos\eta(k)$. Denoting the eigenvalues by $\mathcal{E} = \mathcal{E}_n$, we note that if all $\mathcal{E}_n$ s are positive then we have no bound state, however if some $\mathcal{E}_n$ s are negative then we have bound states. Here we assume that all $\mathcal{E}_n$ s are positive. The completeness relation for this system is

$$\langle k'|k\rangle = \int \psi^*(\mathbf{r},k')\,\psi(\mathbf{r},k)\,d^3r + \sum_n \chi_{nk'}\chi_{nk} = 2\pi^2\delta(k-k'). \tag{18.28}$$

By eliminating χ_{nk} between (18.26) and (18.27) we find

$$\tan\eta(k) = -k\sum_n \frac{g_n^2}{4\pi\left(k^2 - \mathcal{E}_n\right)}. \tag{18.29}$$

Equation (18.29) shows that there are resonances at $k^2 = \mathcal{E}_n$ when $\mathcal{E}_n > 0$.

Now let us impose the initial conditions where one of the motionless particles, say $\chi_0(t = 0)$ has the maximum amplitude of one, while all the others have zero amplitudes, i.e.

$$|\Psi_0\rangle = \begin{cases} \psi(\mathbf{r},0) = 0 \\ \chi_0 = 1, \quad \text{and} \quad \chi_i = 0, \quad i \neq 0 \end{cases}. \tag{18.30}$$

We expand this initial state $|\Psi_0\rangle$ in terms of the complete set of states

$$|\Psi_0\rangle = \int_0^\infty |k\rangle\,\rho_0(k)dk, \tag{18.31}$$

where $\rho_0(k)$ can be determined from Eqs. (18.28)-(18.30)

$$\rho_0(k) = \frac{1}{2\pi^2} \langle k | \Psi_0 \rangle = \frac{g_0}{2\pi^2} \left(\frac{k \cos \eta(k)}{k^2 - \mathcal{E}_0} \right). \tag{18.32}$$

While we do have the orthogonality of states, these states are not physically realizable. The reason for this is that the mean energy of the state $|\Psi_0\rangle$ is infinite

$$\langle \Psi_0 | k^2 | \Psi_0 \rangle = \frac{1}{2\pi^2} \int_0^\infty \frac{k^4 \cos^2 \eta(k)}{k^2 - \mathcal{E}_0} dk \to \infty. \tag{18.33}$$

This problem can be remedied either by introducing an appropriate form factor $\Lambda(\mathbf{r})$ that we mentioned earlier, or by replacing the interaction by a nonlocal one in time, i.e. by considering the following Lagrangian for the system [6]

$$\begin{aligned}
L = \sum_n \chi_n^*(t) \left(i\frac{d}{dt} - \mathcal{E}_n \right) \chi_n(t) + \int \Bigg[i\psi^* \frac{\partial \psi}{\partial t} - \nabla \psi^* \cdot \nabla \psi \\
- \Lambda(\mathbf{r}) \sum_n \int_{-\infty}^\infty K(t - t') \left\{ \chi_n(t') \psi^*(\mathbf{r}, t') + \chi_n^*(t') \psi(\mathbf{r}, t') \right\} dt' \Bigg] d^3r,
\end{aligned} \tag{18.34}$$

where $K(t - t')$ is the solution of the differential equation

$$\left[P\left(\frac{d^2}{dt^2} \right) - \beta^2 \right] K(t - t') = -\beta^2 \delta(t - t'), \tag{18.35}$$

and P is a polynomial of its argument. The boundary conditions for (18.35) is that the function $K(t - t')$ should tend to zero as $t - t'$ goes to infinity. This Lagrangian generates the equations of motion (18.23) and (18.24) provided that

$$\Lambda(\mathbf{r}) \to \delta(\mathbf{r}), \quad \text{and} \quad K(t - t') \to \delta(t - t'). \tag{18.36}$$

Thus while $|\Psi_0\rangle$ is not a realizable state, we can find realizable states arbitrarily close to $|\Psi_0\rangle$ with finite mean energy.

Now let us study the way that $|\Psi_0\rangle$ decays in time. For this we need to examine the overlap between the initial state of the system $|\Psi_0(0)\rangle$ and the state of the system at the time t, i.e. $|\Psi_0(t)\rangle$. We note that $|\Psi_0(0)\rangle$ is expressible as a superposition of the energy eigenstates, Eq. (18.31), and hence

$$|\Psi_0(t)\rangle = \int_0^\infty |k\rangle \rho_0(k) \exp\left(-ik^2 t\right) dk. \tag{18.37}$$

By calculating $\langle \Psi_0(0) | \Psi_0(t) \rangle$ we find that

$$\langle \Psi_0(0) | \Psi_0(t) \rangle = 2\pi^2 \int_0^\infty |\rho_0(k)|^2 \exp\left(-ik^2 t\right) dk, \tag{18.38}$$

and by substituting for ρ_0 from (18.32) we have

$$
\begin{aligned}
\langle \Psi_0(0)|\Psi_0(t)\rangle &= \frac{g_0^2}{2\pi^2} \int_0^\infty \frac{k^2 \cos^2 \eta(k) e^{-ik^2 t} dk}{(k^2 - \mathcal{E}_0)^2} \\
&= \frac{g_0^2}{2\pi^2} \int_0^\infty \frac{k^2 e^{-ik^2 t} dk}{\left[(k^2 - \mathcal{E}_0)^2 + k^2 g_0^4 \right]}.
\end{aligned}
\tag{18.39}
$$

The integral in (18.39) can be evaluated in terms of the error function. Let us assume that g_0^2 is not very large, then the integrand in (18.39) has four poles, one in each quadrant. If the pole in the first quadrant is located at $k = \kappa$, then the other poles are at $k = -\kappa$ and at $k = \pm\kappa^*$. When this is the case the integral in (18.39) can be calculated analytically:

$$
\begin{aligned}
\langle \Psi_0(0)|\Psi_0(t)\rangle &= \frac{ig_0^2}{4\pi \left(\kappa^2 - \kappa^{*2} \right)} \left[\kappa e^{-i\kappa^2 t} \mathrm{erfn} \left\{ \exp\left(\frac{-i\pi}{4} \right) \kappa \sqrt{t} \right\} \right. \\
&\quad \left. - \kappa^* e^{-i\kappa^{*2} t} \mathrm{erfn} \left\{ \exp\left(\frac{-i\pi}{4} \right) \kappa^* \sqrt{t} \right\} + 2\kappa^* e^{-i\kappa^{*2} t} \right],
\end{aligned}
\tag{18.40}
$$

where the error function, $\mathrm{erfn}(z)$, is defined by

$$
\mathrm{erfn}(z) = \frac{2}{\sqrt{\pi}} \int_z^\infty e^{-x^2} dx.
\tag{18.41}
$$

When g_0^2 is small and t is not too large, the largest term in (18.40) which dominates the decay is the last term, i.e.

$$
\langle \Psi_0(0)|\Psi_0(t)\rangle \approx \frac{ig_0^2 \kappa^* e^{-i\kappa^{*2} t}}{2\pi \left(\kappa^2 - \kappa^{*2} \right)},
\tag{18.42}
$$

and from this relation it follows that $|\langle \Psi_0(0)|\Psi_0(t)\rangle|^2$ decreases exponentially in time. This exponential decay is for the initial state $\chi_0(0)$ which decays into $\psi(\mathbf{r}, t)$ and $\chi_i(t), (i \neq 0)$. Thus $\chi_0(t)$ plays the role of the central particle which by its absorption by the particle $\psi(\mathbf{r}, t)$ and re-emission into other $\chi_i(t)$ s loses energy to the other parts of the system.

From Eq. (18.40) we also deduce that the initial decay rate of $|\langle \Psi_0(0)|\Psi_0(t)\rangle|^2$ is infinite at $t = 0$ and hence this probability as a function of t has a cusp at $t = 0$. This result should be compared with the result obtained for the model discussed in §17.4 (Figure 17.4).

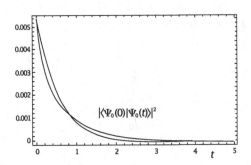

Figure 18.1: A plot of $|\langle \Psi_0(0)|\Psi_0(t)\rangle|^2$, Eq. (18.40), as a function of time t shows that the decay is very close to an exponential decay. For comparison the exponential decay curve which is higher in the range $0 < t < 1$ is also shown.

18.3   Quantum Theory of Line Width

The Wigner-Weisskopf model was one of the earliest models proposed to explain the quantum nature of the line width. Later the quantum theory of radiation offered a perturbative approach to the problem of the line width for the general case of a bound electron which interacts with the electromagnetic field [7]. Here we want to study a simple solvable model of a two level system where the two levels are coupled together by the transverse electromagnetic field $\mathbf{E}_\perp(\mathbf{r}, t)$ [8]. The total Hamiltonian of the system is given by

$$H = H_0 + H', \tag{18.43}$$

where

$$
\begin{aligned}
H_0 &= \sum_{\alpha=i,f} \int \left[\frac{1}{2m} \nabla \psi_\alpha^*(\mathbf{r}, t) \cdot \nabla \psi_\alpha(\mathbf{r}, t) + V(\mathbf{r}) \psi_\alpha^*(\mathbf{r}, t) \psi_\alpha(\mathbf{r}, t) \right] d^3r \\
&\quad + \frac{1}{8\pi} \int \left\{ \mathbf{E}_\perp^2(\mathbf{r}, t) + [\nabla \wedge \mathbf{A}(\mathbf{r}, t)]^2 \right\} d^3r,
\end{aligned}
\tag{18.44}
$$

and

$$
\begin{aligned}
H' &= \frac{ie}{2m} \int \mathbf{A}^+(\mathbf{r}, t) \cdot [\psi_i^*(\mathbf{r}, t) \nabla \psi_f(\mathbf{r}, t) - \nabla \psi_i^*(\mathbf{r}, t) \psi_f(\mathbf{r}, t)] \, d^3r \\
&\quad + \frac{ie}{2m} \int \mathbf{A}^-(\mathbf{r}, t) \cdot [\psi_f^*(\mathbf{r}, t) \nabla \psi_i(\mathbf{r}, t) - \nabla \psi_f^*(\mathbf{r}, t) \psi_i(\mathbf{r}, t)] \, d^3r.
\end{aligned}
\tag{18.45}
$$

In these equations we have set $\hbar = c = 1$, i and f refer to the initial and the final states of the system respectively, and $\mathbf{A}^+$ and $\mathbf{A}^-$ are the positive and negative frequency parts of the magnetic potential $\mathbf{A}$. The binding potential $V(\mathbf{r})$ can be short-range or long-range. Denoting the eigenfunctions of the Schrödinger equation with the potential $V(\mathbf{r})$ by $u_\alpha(\mathbf{r})$, i.e.

$$\left[-\frac{1}{2m}\nabla^2 + V(\mathbf{r}) \right] u_\alpha(\mathbf{r}) = E_\alpha u_\alpha(\mathbf{r}), \quad \alpha = i, f, \tag{18.46}$$

we can expand ψ_α and ψ_α^* as

$$\psi_\alpha(\mathbf{r}, t) = b_\alpha(t) u_\alpha(\mathbf{r}), \tag{18.47}$$

and

$$\psi_\alpha^*(\mathbf{r}, t) = b_\alpha^\dagger(t) u_\alpha^*(\mathbf{r}). \tag{18.48}$$

Here $b_\alpha^\dagger(t)$ and $b_\alpha(t)$ denote the creation and annihilation operators for the electron satisfying the anti-commutation relation

$$\left[b_\alpha(t), b_{\alpha'}^\dagger(t) \right]_+ = \delta_{\alpha,\alpha'}. \tag{18.49}$$

By choosing the periodic boundary conditions and unit normalization volume we can expand $\mathbf{A}^\pm(\mathbf{r}, t)$ in terms of plane waves:

$$\mathbf{A}^\pm(\mathbf{r}, t) = \sum_{\mathbf{k},\sigma} \left(\frac{2\pi}{k} \right)^{\frac{1}{2}} \mathbf{e}_\sigma(\mathbf{k}) \left(\begin{array}{c} a_{\mathbf{k},\sigma}(t) \\ a_{\mathbf{k},\sigma}^\dagger(t) \end{array} \right) \exp\left(\pm i \mathbf{k}.\mathbf{r}\right), \tag{18.50}$$

where the photon energy is $k = |\mathbf{k}|$, and $\mathbf{e}_\sigma(\mathbf{k})$ is the unit polarization vector which is orthogonal to $\mathbf{k}$. The sum over $\mathbf{k}$ extends over all plane waves in the unit normalization volume and the sum over σ is for the two allowed directions of polarization. In Eq. (18.50) the operators $a_{\mathbf{k},\sigma}^\dagger(t)$ and $a_{\mathbf{k},\sigma}(t)$ satisfy the commutation relation

$$\left[a_{\mathbf{k},\sigma}(t), a_{\mathbf{k}',\sigma'}^\dagger(t) \right] = \delta_{\mathbf{k},\mathbf{k}'}\delta_{\sigma,\sigma'}. \tag{18.51}$$

By substituting from (18.46) and (18.50) in H, Eqs. (18.43)-(18.45) and carrying out the integration over $\mathbf{r}$, we find H in the second quantized form:

$$\begin{aligned} H = {} & E_f b_f^\dagger b_f + E_i b_i^\dagger b_i + \sum_{\mathbf{k}} k \left[a_{\mathbf{k}}^\dagger a_{\mathbf{k}} + \frac{1}{2} \right] \\ & + e \sum_{\mathbf{k}} \left(\frac{1}{2k} \right)^{\frac{1}{2}} \left[\beta(k) b_i^\dagger b_f a_{\mathbf{k}} + \beta(k)^* b_f^\dagger b_i a_{\mathbf{k}}^\dagger \right], \end{aligned} \tag{18.52}$$

where we have suppressed the sum over σ. In the above equation the coefficient $\beta(k)$ is defined by

$$\beta(k) = \frac{2i}{m}\sqrt{\pi} \int e^{-i\mathbf{k}\cdot\mathbf{r}} u_i(\mathbf{r}) \mathbf{e}_\sigma(\mathbf{k}).\nabla u_f(\mathbf{r}) d^3 r. \tag{18.53}$$

Since $u_i(\mathbf{r})$ is the bound state wave function, the integral in (18.53) is convergent and $\beta(k)$ goes to zero as $k \to \infty$.

We assume that at $t = 0$, the particle is in the state i and no photon is present. Thus we can write the wave function as

$$|\Psi(0)\rangle = |0, 1_i\rangle,\qquad(18.54)$$

where 0 refers to the number of photons, and 1_i shows that the electron is in the state i. Because of the structure of the Hamiltonian, Eq. (18.52), after a time t we can have a photon, and an electron. The total wave function now takes the form

$$|\Psi(t)\rangle = C_0(t)\,|0, 1_i\rangle + \sum_{k} C_k(t)\,|1_k, 1_f\rangle.\qquad(18.55)$$

Here $|1_k, 1_f\rangle$ indicates a state which includes the electron in the state f plus one photon with the energy k and polarization $\mathbf{e}_\sigma(\mathbf{k})$. The coefficients $C_0(t)$ and $C_k(t)$ measure the probability amplitudes for the no photon and a single photon states respectively.

By comparing (18.54) and (18.55), we find the initial conditions for $C_0(t)$ and $C_k(t)$;

$$C_0(t = 0) = 1, \quad C_k(t = 0) = 0.\qquad(18.56)$$

To determine these amplitudes we write the time-dependent Schrödinger equation for this case as

$$i\frac{\partial}{\partial t}|\Psi(t)\rangle = H\,|\Psi(t)\rangle,\qquad(18.57)$$

and then by substituting for H and $|\Psi(t)\rangle$ using Eqs. (18.52) and (18.55) and equating the coefficients of the state vectors $|0, 1_i\rangle$ and $|1_k, 1_f\rangle$ we obtain

$$i\frac{d}{dt}C_0(t) = E_i C_0(t) + \sum_{k}\frac{e\beta(k)}{\sqrt{2k}}C_k(t),\qquad(18.58)$$

and

$$i\frac{d}{dt}C_k(t) = (E_f + k)\,C_k(t) + \frac{e\beta^*(k)}{\sqrt{2k}}C_0(t).\qquad(18.59)$$

We can find the normal modes for this coupled system by expressing $C_0(t)$ and $C_k(t)$ in terms of their Fourier transforms [9]:

$$C_\mathbf{q}(t) = \sum_{\omega} p(\omega)C_\mathbf{q}(\omega)\exp\left[-i(\omega + E_i)t\right],\quad \mathbf{q} = 0, \mathbf{k},\qquad(18.60)$$

where

$$p(\omega) = C_0^*(\omega)C_0(t = 0) + \sum_{k} C_k^*(\omega)C_k(t = 0) = C_0^*(\omega).\qquad(18.61)$$

By substituting from (18.60) in (18.58) and (18.59) we obtain

$$\omega C_0(\omega) = e\sum_{k}\frac{\beta(k)}{\sqrt{2k}}C_k(\omega),\qquad(18.62)$$

and

$$w C_{\mathbf{k}}(\omega) = (E_f + k - E_i)\, C_{\mathbf{k}}(\omega) + \frac{e\beta^*(k)}{\sqrt{2k}} C_0(\omega). \tag{18.63}$$

Now if we eliminate $C_{\mathbf{k}}(\omega)$ between (18.62) and (18.63) and replace the summation over $\mathbf{k}$ by integration

$$\sum_{\mathbf{k}} \rightarrow \frac{2}{(2\pi)^3} \int d^3 k, \tag{18.64}$$

we find the eigenvalue equation for ω

$$\omega + \frac{e^2}{2\pi^2} \int_0^\infty \frac{|\beta(k)|^2 k dk}{(E_f + k - E_i - \omega)} = 0. \tag{18.65}$$

In order to calculate $C_0(\omega)$, we first normalize $C_0(\omega)$ and $C_{\mathbf{k}}(\omega)$ by requiring

$$C_0^*(\omega) C_0(\nu) + \sum_{\mathbf{k}} C_{\mathbf{k}}^*(\omega) C_{\mathbf{k}}(\nu) = \delta_{\omega,\nu}, \tag{18.66}$$

and then we solve (18.63) for $C_{\mathbf{k}}(\omega)$ and substitute it in (18.66) and set $\omega = \nu$. This gives us

$$C_0(\omega) = \left(\frac{dG(\omega)}{d\omega} \right)^{-\frac{1}{2}}, \tag{18.67}$$

where now $G(\omega)$ is given by

$$G(\omega) = \omega + \frac{e^2}{2\pi^2} \int_0^\infty \frac{|\beta(k)|^2 k dk}{(E_f + k - E_i - \omega)}. \tag{18.68}$$

Similarly for $C_{\mathbf{k}}(\omega)$ we find

$$C_{\mathbf{k}}(\omega) = \frac{e\beta^*(k)}{(\omega - E_f - k + E_i)\sqrt{2k}} \left(\frac{dG(\omega)}{d\omega} \right)^{-\frac{1}{2}}. \tag{18.69}$$

Now from Eqs. (18.60), (18.61), (18.67) and (18.68) we have

$$C_0(t) = \sum_\omega \exp\left[-i(\omega + E_i)t\right] \left(\frac{dG(\omega)}{d\omega} \right)^{-1}. \tag{18.70}$$

Changing the variable ω to z, where

$$z = \omega + E_i - E_f, \tag{18.71}$$

we can write (18.70) as a contour integral

$$C_0(t) = \frac{1}{2\pi i} e^{-iE_f t} \oint_C \frac{e^{-izt}}{G(z)} dz, \tag{18.72}$$

where the contour C contains all the roots of $G(z)$. We expect the amplitude $C_0(t)$ to decay more or less exponentially, and thus we want to consider the roots of $G(z)$ with negative imaginary part. For this we must look onto the second Reimann sheet for the desired root. If in the first Reimann sheet $G^I(z)$ represents the analytic continuation of $G(z)$, then

$$G^I(z) = z - E_i + E_f + \frac{e^2}{2\pi^2} \int_0^\infty \frac{|\beta(k)|^2 k \, dk}{k - z}, \qquad (18.73)$$

where $0 < \arg z < 2\pi$. By analytic continuation of $G(z)$ in the second Reimann sheet we find $G^{II}(z)$, and for $0 > \arg z > -2\pi$, we have

$$G^{II}(x - i\epsilon) = G^I(x + i\epsilon). \qquad (18.74)$$

Thus

$$G^{II}(z) = G^I(z) + \frac{ie^2}{\pi} z |\beta(z)|^2. \qquad (18.75)$$

If we assume that only one root of $G^{II}(z)$ at $z = z_0$ makes the major contribution to the integral (18.72) [10], i.e.

$$G^{II}(z_0) = 0, \quad \text{for} \quad z_0 = E_R - E_f - \frac{i}{2}\Gamma, \qquad (18.76)$$

then from Eqs. (18.68), (18.75) and (18.76) it follows that

$$E_R - E_i \;+\; \left(\frac{e^2}{2\pi^2}\right) \int_0^\infty \frac{(k + E_f - E_R)|\beta(k)|^2 k \, dk}{\left[(k + E_f - E_R)^2 + \frac{1}{4}\Gamma^2\right]}$$

$$+\; \left(\frac{e^2 \Gamma}{2\pi}\right) \left|\beta\left(E_R - E_f - \frac{i}{2}\Gamma\right)\right|^2 = 0, \qquad (18.77)$$

and

$$\frac{\Gamma}{2}\left\{1 + \frac{e^2}{2\pi^2} \int_0^\infty \frac{|\beta(k)|^2 k \, dk}{\left[(k + E_f - E_R)^2 + \frac{1}{4}\Gamma^2\right]}\right\}$$

$$-\; \left(\frac{e^2}{\pi}\right)(E_R - E_f)\left|\beta\left(E_R - E_f - \frac{i}{2}\Gamma\right)\right|^2 = 0. \qquad (18.78)$$

These solutions are obtained by equating the real and imaginary parts of $G^{II}(z)$. The quantity $E_R - E_i$ presents the shift in the energy of the electron and $\Gamma/2$ is the line width of the emitted electromagnetic wave.

A very good approximation to the coupled nonlinear integral equations for E_R and Γ can be found by noting that

$$E_R \approx E_i + \mathcal{O}\left(e^2\right), \quad \text{and} \quad \Gamma \approx \mathcal{O}\left(e^2\right), \qquad (18.79)$$

where $\mathcal{O}\left(e^2\right)$ means of the order e^2, which in proper units is $e^2/(\hbar c) \approx (1/137)$. Using these we can find the approximate solution to (18.77) and (18.78);

$$E_R - E_i \approx -\left(\frac{e^2}{2\pi^2}\right) \mathcal{P} \int_0^\infty \frac{|\beta(k)|^2 k \, dk}{k + E_f - E_i}, \qquad (18.80)$$

and

$$\Gamma \approx \frac{e^2}{\pi}(E_i - E_f)|\beta(E_i - E_f)|^2, \tag{18.81}$$

where $\mathcal{P}$ stands for the principal value of the integral.

These results agree with the results obtained from the first-order perturbation theory [7]. Evaluating the contour integral (18.72) following the same technique used in solving Ullersma's model, Chapter 10, we find

$$C_0(t) = \frac{\exp\left(-iE_R t - \frac{1}{2}\Gamma t\right)}{\left(\frac{dG^{II}(z)}{dz}\right)_{z_0}} - \left(\frac{e^2}{2\pi^2}\right)e^{-iE_f t}\int_{-\infty}^{0}\frac{|\beta(x)|^2 e^{-ixt}x\,dx}{G^I(x)G^{II}(x)}. \tag{18.82}$$

Again if we ignore terms of the order $\mathcal{O}\left(e^2\right)$ we have

$$\frac{dG^{II}(z)}{dz} = 1 + \left(\frac{e^2}{2\pi^2}\right)\int_{0}^{\infty}\frac{|\beta(k)|^2 k\,dk}{(k-z)^2} + \frac{ie^2}{\pi}\frac{d}{dz}[z|\beta(z)|^2] \approx 1 + \mathcal{O}\left(e^2\right). \tag{18.83}$$

Therefore

$$C_0(t) \approx \exp\left[-iE_R t - \frac{1}{2}\Gamma t\right] + \mathcal{O}\left(e^2\right), \tag{18.84}$$

and

$$C_{\mathbf{k}}(t) \approx -e\beta^*(k)\frac{\left\{\exp\left(-iE_R t - \frac{1}{2}\Gamma t\right) - \exp\left[-i(k+E_f)t\right]\right\}}{\sqrt{2k}\left(E_f + k - E_R + \frac{i}{2}\Gamma\right)} + \mathcal{O}\left(e^2\right). \tag{18.85}$$

In this, the first order approximation, we have the approximate result

$$|C_0(t)|^2 \approx e^{-\Gamma t}, \tag{18.86}$$

or the decay is purely exponential. However a more careful analysis of the contour integral (18.72) shows that both the initial form of the decay and the decay after a long time $t >> \Gamma^{-1}$ are not exponential [10].

A two-level system similar to the one that we have studied in this section has been discussed by Morse and Feshbach [11].

18.4   The Optical Potential

In Chapter 4 we observed that one of the ways of describing the damped harmonic motion in classical dynamics was by introducing complex potentials. The phenomenological optical (or complex) potential can also be utilized in quantum theory of dissipative system as was done in §15.2. In this section we want to show that such a potential can also be derived from a conservative many-body system S by isolating the motion of a part of the system S_1, from the rest of the system S_2 and study its dependence on the coupling between different particles in the system. However in this case, unlike the coupling of a particle to a heat bath the inter-particle forces are assumed to be of short range, i.e. have an exponential dependence on inter-particle distances. As we will see the imaginary part of the complex potential derived in this way has to satisfy certain conditions.

Since this type of complex potential has been extensively used in nuclear reaction theory, the simplest model which we will study is the description of the scattering of a nucleon which is the subsystem S_1 from the target nucleus, S_2, which contains a number of nucleons. Thus we are concerned with a system composed of $A + 1$ nucleons [12] [13].

Let us denote all of the degrees of the j-th nucleon including its spin and isospin by $\mathbf{r}_j$, and the j-th state of the target by $\psi_j (\mathbf{r}_1, \ldots \mathbf{r}_A)$. We write the Hamiltonian of the total system as

$$H\Psi = E\Psi, \tag{18.87}$$

where

$$H = H_A (\mathbf{r}_1, \ldots \mathbf{r}_A) - \frac{\hbar^2}{2m}\nabla_0^2 + V (\mathbf{r}_0, \mathbf{r}_1, \ldots \mathbf{r}_A). \tag{18.88}$$

In Eqs. (18.87) and (18.88) E is the total energy and H_A is the Hamiltonian operator of the target nucleus. The operator $\left(-\frac{\hbar^2}{2m}\nabla_0^2\right)$ is the kinetic energy operator of the incoming nucleon and $V (\mathbf{r}_0, \mathbf{r}_1, \ldots \mathbf{r}_A)$ is the interaction potential between the incoming particle and the nucleons in the target.

Now we expand the total wave function Ψ in terms of the complete set of states, j, of the target

$$\Psi = \sum_j u_j (\mathbf{r}_0) \psi_j (\mathbf{r}_1, \ldots \mathbf{r}_A). \tag{18.89}$$

By substituting (18.89) in (18.88) and using the orthogonality of ψ_j s we find

$$\left(-\frac{\hbar^2}{2m}\nabla_0^2 + V_{jj} + \epsilon_j - E\right) u_j (\mathbf{r}_0) = -\sum_{k \neq j} V_{jk} (\mathbf{r}_0) u_k (\mathbf{r}_0), \tag{18.90}$$

where

$$V_{jk} (\mathbf{r}_0) = \int \psi_j^* V \psi_k d^3 r_1 \cdots d^3 r_A, \tag{18.91}$$

and

$$H_A \psi_j = \epsilon_j \psi_j. \tag{18.92}$$

Let $j = 0$ term in (18.89) denote the state where the target is in its ground state and a nucleon of energy E is in the incident channel and has a wave function $u_0 (\mathbf{r}_0)$. We want to drive an uncoupled Schrödinger equation for $u_0 (\mathbf{r}_0)$ by eliminating all of the other channels. To do this we introduce the column matrix $\mathbf{\Phi}$ by

$$\mathbf{\Phi} = \begin{bmatrix} u_1 \\ u_2 \\ \vdots \end{bmatrix}, \tag{18.93}$$

and a matrix operator $\hat{H}$ with the matrix elements

$$H_{jk} = -\frac{\hbar^2}{2m}\nabla_0^2\delta_{jk} + V_{jk} + \epsilon_j\delta_{jk}, \quad i,j \neq 0. \tag{18.94}$$

For the coupling between $\boldsymbol{\Phi}$ and u_0 we define the matrix potentials

$$\mathbf{V}_0 = (V_{01}, V_{02}\cdots), \tag{18.95}$$

and

$$\mathbf{V}_0^\dagger = \begin{bmatrix} V_{01}^* \\ V_{02}^* \\ \vdots \end{bmatrix}. \tag{18.96}$$

With the introduction of these potentials we can write (18.90) in the simpler form of

$$\left(-\frac{\hbar^2}{2m}\nabla_0^2 + V_{00} - E\right)u_0 = -\mathbf{V}_0\boldsymbol{\Phi}, \tag{18.97}$$

and

$$\left(\hat{H} - E\right)\boldsymbol{\Phi} = -\mathbf{V}_0^\dagger u_0. \tag{18.98}$$

We can solve (18.98) for $\boldsymbol{\Phi}$ to get

$$\boldsymbol{\Phi} = \left(\frac{1}{E^+ - \hat{H}}\right)\mathbf{V}_0^\dagger u_0, \tag{18.99}$$

where

$$E^+ = E + i\varepsilon, \quad \varepsilon \to 0^+. \tag{18.100}$$

The positive small number ε in (18.100) guarantees that we have only outgoing waves in the exit channels u_i ($i \geq 1$). Now by inserting (18.99) in (18.97) we obtain

$$\left[-\frac{\hbar^2}{2m}\nabla_0^2 + V_{00} + \mathbf{V}_0\left(\frac{1}{E^+ - \hat{H}}\right)\mathbf{V}_0^\dagger - E\right]u_0(\mathbf{r}_0) = 0. \tag{18.101}$$

From Eq. (18.101) we find the effective potential $\mathcal{V}$ to be

$$\mathcal{V} = V_{00} + \mathbf{V}_0\left(\frac{1}{E^+ - \hat{H}}\right)\mathbf{V}_0^\dagger. \tag{18.102}$$

Note that because of the coupling potential, $\mathbf{V}_0^\dagger$, the incoming particle leaves the channel u_0 and is then emitted in the exit channel u_i which is a part of $\boldsymbol{\Phi}$ in accordance with the Eq. (18.97), and that is why $\mathcal{V}$ is complex. This does not happen unless the energy E is larger than the lowest eigenvalue of the unperturbed Schrödinger equation (see below) ($E > \epsilon_1$).

In general the spectrum of $\hat{H}$ consists of a discrete part and a continuum.

Denoting the wave function for the discrete spectrum by Φ_n and for the continuum states by $\Phi(\mathcal{E}, \alpha)$, where α labels various states having a common $\mathcal{E}$, we have

$$\hat{H}\Phi_n = \mathcal{E}_n \Phi_n, \tag{18.103}$$

and

$$\hat{H}\Phi(\mathcal{E}, \alpha) = \mathcal{E}\Phi(\mathcal{E}, \alpha). \tag{18.104}$$

If we want to express $\mathcal{V}$ in terms of Φ_n and $\Phi(\mathcal{E}, \alpha)$, we introduce a complete set of states and write

$$\begin{aligned}
\mathcal{V} = V_{00} &+ \sum_n \frac{|V_0 \Phi_n\rangle \langle \Phi_n V_0^\dagger|}{(E - \mathcal{E}_n)} \\
&+ \int d\alpha \int_{\epsilon_1}^{\infty} \frac{|V_0 \Phi(\mathcal{E}', \alpha)\rangle \langle \Phi(\mathcal{E}', \alpha) V_0^\dagger|}{(E^+ - \mathcal{E}')} d\mathcal{E}'.
\end{aligned} \tag{18.105}$$

In order to show that $\mathcal{V}$ is nonlocal, we let $\mathcal{V}$ to operate on u_0. Consider the operator sum in (18.105) when it operates on u_0;

$$\sum_n \frac{|V_0 \Phi_n\rangle \langle \Phi_n | V_0^\dagger | u_0 \rangle}{(E - \mathcal{E}_n)} = \int K(\mathbf{r}_0, \mathbf{r}') u_0(\mathbf{r}') d^3 r', \tag{18.106}$$

then the kernel $K(\mathbf{r}_0, \mathbf{r}')$ which is defined by

$$K(\mathbf{r}_0, \mathbf{r}') = \sum_n \sum_{j,k \neq 0} \frac{V_{0j}(\mathbf{r}_0) u_j^{(n)}(\mathbf{r}_0) \left[u_k^{(n)}(\mathbf{r}')\right]^* V_{k0}(\mathbf{r}')}{E - \mathcal{E}_n}, \tag{18.107}$$

shows that $\mathcal{V}$ is nonlocal.

In addition to nonlocality the optical potential $\mathcal{V}$ has the following general properties:

(1) The numerators of the two terms in the expansion in (18.105) are positive definite, i.e. for any arbitrary function w we have

$$\langle w V_0 \Phi_n \rangle \langle \Phi_n V_0^\dagger w \rangle = |\langle w V_0 \Phi_n \rangle|^2 \geq 0 \tag{18.108}$$

and

$$|\langle w V_0 \Phi(\mathcal{E}', \alpha) \rangle|^2 \geq 0. \tag{18.109}$$

From (18.105) it readily follows that

$$\text{Im}\,\mathcal{V} = -\pi \int |V_0 \Phi(E, \alpha)\rangle \langle \Phi(E, \alpha) V_0^\dagger|\theta(E - \epsilon_1)\, d\alpha, \tag{18.110}$$

where $\theta(x)$ is a step function

$$\theta(x) = \begin{cases} 1 & \text{for } x > 0 \\ 0 & \text{for } x < 0 \end{cases}, \tag{18.111}$$

and where ϵ_1 is the eigenvalue of the uncoupled Schrödinger equation

$$\left(-\frac{\hbar^2}{2m}\nabla_0^2 + V_{11}\right)u_1 = -(\epsilon_1 - \mathcal{E})u_1. \tag{18.112}$$

We observe that $\text{Im}\mathcal{V} \le 0$ as it should be since $\mathcal{V}$ is an absorptive potential.
(2) Equation (18.105) shows that $\text{Re}\mathcal{V}$ has simple poles at $E = \mathcal{E}_n$ and a branch line for $E > \epsilon_1$. By taking the derivative of $\text{Re}\mathcal{V}$ with respect to E we find that

$$\frac{\partial}{\partial E}(\text{Re}\mathcal{V}) \le 0. \tag{18.113}$$

(3) The real and the imaginary parts of $\mathcal{V}$ are related to each other. Thus if we substitute (18.110) in (18.105) we find

$$\text{Re}\mathcal{V} = V_{00} + \sum_n \frac{|\mathbf{V}_0\mathbf{\Phi}_n\rangle\langle\mathbf{\Phi}_n\mathbf{V}_0^\dagger|}{E - \mathcal{E}_n} - \frac{\mathcal{P}}{\pi}\int_{\epsilon_1}^\infty \frac{\text{Im}\mathcal{V}(\mathcal{E}')}{E - \mathcal{E}'}d\mathcal{E}', \tag{18.114}$$

where $\mathcal{P}$ denotes Cauchy principal value. For nuclear matter the sum will not appear, therefore (18.114) will reduce to [12]

$$\text{Re}\mathcal{V} = V_{00} - \frac{\mathcal{P}}{\pi}\int_0^\infty \frac{\text{Im}\mathcal{V}(\mathcal{E}')}{E - \mathcal{E}'}d\mathcal{E}'. \tag{18.115}$$

18.5   Gisin's Nonlinear Wave Equation

A phenomenological nonlinear wave equation with complex interaction has been proposed by Gisin to account for decaying states in wave mechanics [14]-[16]. This model is completely quantum mechanical and does not correspond to a simple classical dissipative system of the types that we have seen in the earlier chapters. The fact that a complex potential with negative imaginary part can account for the damped motion of a particle suggests that a more general complex interaction may also be used as a model of dissipation. Gisin's model is described by the wave equation

$$i\hbar\frac{\partial\psi}{\partial t} = \left(1 - i\frac{\kappa}{2}\right)H\psi + i\frac{\kappa}{2}\langle\psi|H|\psi\rangle\psi, \tag{18.116}$$

where H is the Hamiltonian of the system and κ is a dimensionless positive real damping constant. This nonlinear wave equation has the following properties:
(1) The norm of the wave function ψ is independent of time, and therefore once it is normalized it will remain so at later times.
(2) When ψ is an eigenfunction of H then $(\langle\psi|H|\psi\rangle - H)\psi$ is zero and (18.116) reduces to the Schrödinger equation.
(3) The rate of change of $\langle\psi|H|\psi\rangle$ is negative definite, i.e.

$$\frac{d}{dt}\langle\psi|H|\psi\rangle = \left\langle\frac{\partial\psi}{\partial t}|H|\psi\right\rangle - \left\langle\psi|H|\frac{\partial\psi}{\partial t}\right\rangle = -\frac{\kappa}{\hbar}(\Delta H)^2 \le 0. \tag{18.117}$$

In this relation $(\Delta H)^2$ is defined as

$$(\Delta H)^2 = \left(\langle \psi | H | \psi \rangle^2 - \langle \psi | H^2 | \psi \rangle \right). \tag{18.118}$$

The inequality in (18.117) holds only when ψ is an eigenfunction of H. This result is obtained by substituting for $\frac{\partial \psi}{\partial t}$ and $\frac{\partial \psi^*}{\partial t}$ from the nonlinear equation (18.116) in (18.117).

(4) For a Hamiltonian with discrete spectrum the solution of the nonlinear equation (18.116) can be expressed in terms of the eigenfunctions $\psi_{j,\alpha}(\mathbf{r})$ of H.

Consider the eigenvalue equation

$$H\psi_{j,\alpha}(\mathbf{r}) = E_j \psi_{j,\alpha}(\mathbf{r}), \tag{18.119}$$

where α labels various states with common E_j and $\psi_{j,\alpha}(\mathbf{r})$ s form an orthonormal set

$$\langle \psi_{j,\alpha} | \psi_{k,\beta} \rangle = \delta_{jk} \delta_{\alpha\beta}. \tag{18.120}$$

By expanding $\psi(\mathbf{r}, t)$ in terms of the complete set of states $\psi_{j,\alpha}$, i.e.

$$\psi(\mathbf{r}, t) = \sum_{j,\alpha} C_{j,\alpha}(t) \psi_{j,\alpha}(\mathbf{r}), \tag{18.121}$$

and substituting in (18.116) we obtain the time-dependent coefficients of expansion

$$C_{j,\alpha}(t) = \frac{C_{j,\alpha}(0)}{\sqrt{N(t)}} \exp\left[-\frac{(2i + \kappa) E_j t}{2\hbar} \right], \tag{18.122}$$

where

$$N(t) = \sum_{j,\alpha} |C_{j,\alpha}(0)|^2 \exp\left[-\frac{\kappa E_j t}{\hbar} \right]. \tag{18.123}$$

The solution given by Eq. (18.122) is interesting since it shows that all $|C_{j,\lambda}|^2$ s tend to zero as $t \to \infty$, except for the ones corresponding to the lowest energy state (when the ground state is non-degenerate), i.e.

$$\psi(\mathbf{r}, t) \to \exp\left(-\frac{iE_0 t}{\hbar} \right), \quad \text{as} \quad t \to \infty. \tag{18.124}$$

For the general case of H we have the formal solution of (18.116) which is

$$\psi(\mathbf{r}, t) = \frac{\exp\left[-\frac{(2i+\kappa)}{2\hbar} Ht \right] \psi_0(\mathbf{r})}{\langle \psi_0 | \exp\left(-\frac{\kappa}{\hbar} Et \right) | \psi_0 \rangle^{\frac{1}{2}}}. \tag{18.125}$$

As an example let us consider the motion of a damped harmonic oscillator in Gisin's formulation. Here we assume that the initial wave function is of the form of a displaced normalized Gaussian [17]

$$\psi(x, 0) = \frac{\sqrt{\alpha}}{\pi^{\frac{1}{4}}} \exp\left[-\frac{\alpha^2}{2} (x - \xi_0)^2 \right], \tag{18.126}$$

where $\alpha = \sqrt{\frac{m\omega}{\hbar}}$. We expand (18.126) in terms of the complete set of eigenfunctions of the harmonic oscillator $\psi_n(x)$

$$\psi(x,0) = \sum_{n=0}^{\infty} C_n(0)\psi_n(x). \tag{18.127}$$

From Eqs. (18.126) and (18.127) we calculate $C_n(0)$;

$$C_n(0) = \int_{-\infty}^{\infty} \psi(x,0)\psi_n(x)dx = \frac{1}{\sqrt{2^n n!}} \left\{ (\alpha\xi_0)^2 \exp\left[-\left(\frac{\alpha\xi_0}{2}\right)^2\right] \right\}. \tag{18.128}$$

The time dependent normalization can be found from Eqs. (18.123) and (18.128);

$$
\begin{aligned}
N(t) &= \sum_{0}^{\infty} |C_n|^2 \exp\left[-\kappa\left(n+\frac{1}{2}\right)\omega t\right] \exp\left[-\frac{1}{2}\alpha^2\xi_0^2\right] \\
&= \exp\left(\frac{-\lambda t}{2}\right) \exp\left[-\frac{1}{2}\alpha^2\xi_0^2\left(1-e^{\frac{-\lambda t}{2}}\right)\right],
\end{aligned} \tag{18.129}
$$

where we have introduced a new damping constant $\lambda = \kappa\omega$.

Having found $N(t)$, we can determine $\psi(x,t)$ by utilizing Eqs. (18.121) and (18.122);

$$
\begin{aligned}
\psi(x,t) &= \sum_{n=0}^{\infty} C_n \exp\left[\left(i\omega + \frac{\lambda}{2}\right)t\right] \psi_n(x) \\
&= \frac{\sqrt{\alpha}}{\pi^{\frac{1}{4}}} \sum_{n=0}^{\infty} \frac{C_n(0)}{\sqrt{2^n n!}} \exp\left[-\left(i\omega + \frac{\lambda}{2}\right)t\right] H_n(\alpha x) \exp\left[-\frac{\alpha^2 x^2}{2}\right].
\end{aligned} \tag{18.130}
$$

The last sum can be obtained in closed form which simplifies the result

$$
\begin{aligned}
\psi(x,t) &= \frac{\sqrt{\alpha}}{\pi^{\frac{1}{4}}} \exp\left[-\frac{1}{2}\alpha^2\left(x - \xi_0 e^{-\frac{\lambda t}{2}}\cos(\omega t)\right)^2\right. \\
&\quad \left. - i\left(\frac{1}{2}\omega t + \alpha^2 x\xi_0 \sin(\omega t)e^{-\frac{\lambda t}{2}} - \frac{1}{4}x\xi_0^2 e^{-\lambda t}\sin(2\omega t)\right)\right].
\end{aligned} \tag{18.131}
$$

The absolute square of $\psi(x,t)$ which is the probability density is given by

$$|\psi(x,t)|^2 = \frac{\alpha}{\pi^{\frac{1}{2}}} \exp\left[-\alpha^2\left(x - \xi_0 e^{\frac{-\lambda t}{2}}\cos(\omega t)\right)^2\right], \tag{18.132}$$

and this relation shows that the center of the packet oscillates as a damped harmonic oscillator;

$$\langle x\rangle = \xi(t) = \xi_0 \exp\left[-\frac{\lambda t}{2}\right]\cos(\omega t). \tag{18.133}$$

In the same way the momentum of the center of the wave packet can be obtained from the wave function (18.131)

$$\langle p \rangle = \int_{-\infty}^{\infty} \psi^*(x,t) \left(-i\hbar \frac{\partial \psi(x,t)}{\partial x} \right) dx = -m\omega\xi_0 \exp\left[-\frac{\lambda t}{2} \right] \sin(\omega t). \quad (18.134)$$

Thus $\langle x \rangle$ and $\langle p \rangle$ satisfy the equations of motion [14]

$$\frac{d\langle x \rangle}{dt} = -\frac{\lambda}{2}\langle x \rangle + \frac{1}{m}\langle p \rangle, \quad (18.135)$$

and

$$\frac{d\langle p \rangle}{dt} = -\frac{\lambda}{2}\langle p \rangle - m\omega^2\langle x \rangle. \quad (18.136)$$

By eliminating $\langle p \rangle$ between (18.135) and (18.136) we find the equation of motion for $\langle x \rangle$,

$$m\frac{d^2\langle x \rangle}{dt^2} + m\lambda\frac{d\langle x \rangle}{dt} + m\left(\omega^2 + \frac{\lambda^2}{4} \right)\langle x \rangle = 0, \quad (18.137)$$

and a similar equation for $\langle p \rangle$.

18.6 Nonlinear Generalization of the Wave Equation

By quantizing the modified Hamilton-Jacobi equation for a linearly damped system using the Schrödinger method, Chapter 13, we obtained the nonlinear Schrödinger-Langevin equation. Earlier in this chapter we also found that certain generalizations of the quantum mechanical systems correspond to classically damped motions. In this section we consider the following nonlinear extension of the wave equation which also describes a dissipative system.

We start with the nonlinear wave equation:

$$i\hbar\frac{\partial \psi}{\partial t} = \hat{H}\psi + iD\hbar G(\psi), \quad (18.138)$$

where $\hat{H} = -\frac{\hbar^2}{2m}\nabla^2 + V$, D is the diffusion constant and $G(\psi)$ is given by

$$G(\psi) = \nabla^2\psi + \frac{|\nabla\psi|^2}{|\psi|^2}\psi. \quad (18.139)$$

Defining the probability density ρ and the current $\mathbf{j}$ in the usual way by

$$\rho = |\psi|^2, \quad (18.140)$$

and

$$\mathbf{j} = \frac{i\hbar}{2m}\left(\psi\nabla\psi^* - \psi^*\nabla\psi \right), \quad (18.141)$$

from (18.138) and its complex conjugate we find the continuity equation to be

$$\frac{\partial \rho}{\partial t} + \nabla \cdot \mathbf{j} = D\nabla^2 \rho. \tag{18.142}$$

Thus the continuity equation in this case is given by the Fokker-Planck equation, and this is why this particular form of $G(\psi)$ is chosen [18] [19]. Equation (18.142) shows that $\int \rho d^3 r$ remains constant in time and also from (18.139) it follows that

$$\int \psi^* G(\psi)\psi d^3 r = 0. \tag{18.143}$$

From the wave equation we can determine $\langle \mathbf{r} \rangle$ and $\langle \mathbf{p} \rangle$ in the same way as in Ehrenfest's theorem

$$\frac{d}{dt}\langle \mathbf{r} \rangle = \frac{1}{m}\langle \mathbf{p} \rangle, \tag{18.144}$$

and

$$\frac{d}{dt}\langle \mathbf{p} \rangle = -\langle \nabla V \rangle + 2D\mathrm{Re}\int \psi^* \left(-i\hbar\nabla\right) G(\psi)d^3 r. \tag{18.145}$$

The last term in Eq. (18.145) is related in a complicated way to the expectation value of the momentum operator $(-i\hbar\nabla)$ showing the presence of a damping term proportional to the diffusion coefficient D.

Now let us consider the one-dimensional motion where $V(x)$ is

$$V(x) = \frac{1}{2}m\omega_0 x^2 - f(t)x, \tag{18.146}$$

and a Gaussian wave function ψ given by

$$\psi(x,t) = N(t)\exp\left[-a(t)x^2 + b(t)x\right]. \tag{18.147}$$

Substituting (18.147) in (18.138) we find that $a(t)$ satisfies the nonlinear differential equation

$$\frac{da(t)}{dt} = \frac{2\hbar}{im}a^2(t) + \frac{im\omega_0^2}{2\hbar} - 8Da(t)\mathrm{Re}\, a(t). \tag{18.148}$$

Equation (18.148) shows that $a(t)$ is a complex function. By writing it as

$$a(t) = \alpha(t) + i\beta(t), \tag{18.149}$$

we can separate the real and imaginary parts in (18.148) and obtain two first order coupled nonlinear equations for $\alpha(t)$ and $\beta(t)$;

$$\frac{d\alpha(t)}{dt} = \frac{4\hbar}{m}\alpha(t)\beta(t) - 8D\alpha^2(t), \tag{18.150}$$

and

$$\frac{d\beta(t)}{dt} = \frac{m\omega_0^2}{2\hbar} - \frac{2\hbar}{m}\left(\alpha^2(t) - \beta^2(t)\right) - 8D\alpha(t)\beta(t). \tag{18.151}$$

We can also write for $\alpha(t)$ and $\beta(t)$ in terms of variances of the coordinate and momentum of this Gaussian wave function.

Here we define $\sigma_{xx}(t), \sigma_{xp}(t)$ and $\sigma_{pp}(t)$ by

$$\sigma_{xx}(t) = \frac{1}{4\alpha(t)}, \quad \sigma_{pp}(t) = \hbar^2 \frac{\alpha^2(t) + \beta^2(t)}{\alpha(t)} \quad \text{and} \quad \sigma_{xp}(t) = -\frac{\hbar\beta(t)}{2\alpha(t)}, \quad (18.152)$$

and substitute these for $\alpha(t)$ and $\beta(t)$ to obtain

$$\frac{d\sigma_{xx}(t)}{dt} = \frac{2}{m}\sigma_{xp}(t) + 2D, \quad (18.153)$$

$$\frac{d\sigma_{xp}(t)}{dt} = \frac{1}{m}\sigma_{pp}(t) - m\omega_0^2\sigma_{xx}(t), \quad (18.154)$$

and

$$\frac{d\sigma_{pp}(t)}{dt} = -2m\omega_0^2\sigma_{xp}(t) - 2D\frac{\sigma_{pp}(t)}{\sigma_{xx}(t)}. \quad (18.155)$$

Of course these three functions $\sigma_{xx}(t), \sigma_{xp}(t)$ and $\sigma_{pp}(t)$ are not independent and from their definitions it is easy to show that

$$\sigma_{xx}(t)\sigma_{pp}(t) - \sigma_{xp}^2(t) = \frac{\hbar^2}{2m}. \quad (18.156)$$

By eliminating $\sigma_{xp}(t)$ and $\sigma_{pp}(t)$ between Eqs. (18.153)-(18.155) we find that $\sigma_{xx}(t)$ satisfies a third order nonlinear differential equation

$$\frac{d^3\sigma_{xx}}{dt^3} + \frac{2D}{\sigma_{xx}}\frac{d^2\sigma_{xx}}{dt^2} + 4\omega_0^2\frac{d\sigma_{xx}}{dt} = 0. \quad (18.157)$$

We want to find a solution of (18.157) which deviates slightly from the stationary solution. For this we note that if we ignore the time dependence of $a(t)$, Eq. (18.148), as a first approximation we have

$$a(t) = \frac{m\omega_0}{2\hbar}\frac{1 - i\Gamma}{\sqrt{1 + \Gamma^2}}, \quad (18.158)$$

where $\Gamma = 2mD/\hbar$ is a dimensionless number. Thus to this order, from (18.148), (18.149) and (18.152) we find that σ_{xx} is given by

$$\sigma_{xx} = \frac{1}{\alpha} = \frac{\hbar\left(1 + \Gamma^2\right)^{\frac{1}{2}}}{2m\omega_0}. \quad (18.159)$$

Substituting for σ_{xx} in the coefficient $(2D/\sigma_{xx})$ in Eq. (18.157), we linearize this equation. From the solution of the linearized equation we find that

$$\frac{d\sigma_{xx}(t)}{dt} \approx \left(\frac{d\sigma_{xx}(t)}{dt}\right)_{t=0} \exp\left[\frac{-\omega_0 t}{\sqrt{1 + \Gamma^2}}\left\{\Gamma \pm i\left(4 + 3\Gamma^2\right)^{\frac{1}{2}}\right\}\right], \quad (18.160)$$

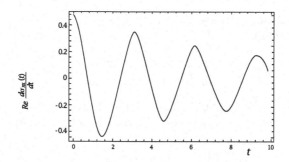

Figure 18.2: For the case of weak damping $D = 0.1$, the exact solution of the differential equation (18.157) for $\text{Re}[(d\sigma_{xx}/dt)]$ is plotted as a function of time.

i.e. the damping constant in this approximation is given by

$$\lambda = \frac{\omega_0 \Gamma}{\sqrt{1 + \Gamma^2}}. \tag{18.161}$$

In Fig. (18.2) the exact solution of the nonlinear differential equation (18.157) for $\text{Re}[\frac{d\sigma_{xx}}{dt}]$ is plotted as a function of time. This graph shows that for weak damping, $\text{Re}[\frac{d\sigma_{xx}}{dt}]$ follows the motion of a damped oscillator. A plot of the approximate solution given by (18.160) overlaps with the exact solution for the range of t shown in the figure.

18.7 Dissipation Arising from the Motion of the Boundaries

In quantum mechanics a change in the boundaries of a system has an effect similar to the action of an external force. There are many examples where the motion of the boundaries play an important role in the dynamical behavior of the system. These include problems from relativity [20] [21], quantum physics [22] statistical mechanics [23], and condensed matter physics [24].

The simplest model of this type of decay is the one where there are N noninteracting bosons occupying different energy levels of a cavity of length L_0 at $t = 0$. Now if the length of the cavity increases (or decreases) as a function of time according to the relation

$$L(t) = \zeta(t)L_0, \tag{18.162}$$

where $\zeta(t)$ is a continuous and twice differentiable function of time, then the particles will make transition from the original levels to lower or higher levels by exchanging energy with the moving walls of the cavity. Thus in this problem the coupling is between each individual particle and the walls of the cavity. The

Hamiltonian for these noninteracting particles which we assume to satisfy the Schrödinger equation is given by

$$H_0 = -\frac{i\hbar}{2m} \int_0^{L(t)} \frac{\partial \pi(x,t)}{\partial x} \frac{\partial \psi(x,t)}{\partial x} dx, \tag{18.163}$$

where $\psi(x,t)$ is the field amplitude and $\pi(x,t)$ is its conjugate momentum density. These operators must satisfy the boundary conditions at the walls of the cavity

$$\psi(0,t) = \psi(L(t),t) = 0, \tag{18.164}$$

and

$$\pi(0,t) = \pi(L(t),t) = 0, \tag{18.165}$$

as well as the canonical commutation relation [17]

$$[\psi(x',t), \pi(x'',t)] = i\hbar\delta(x' - x''). \tag{18.166}$$

The Hamiltonian (18.163) is defined over a length $L(t)$ which depends on time. To remove the time-dependence from the boundary $x = L(t)$ and show that the motion of the boundary is equivalent to the action of a force, we use two sets of unitary transformations and we denote these by $U_1(t)$ and $U_2(2)$ respectively. The first operator acting on the Schrödinger equation

$$H_0\psi = i\hbar\frac{\partial\psi}{\partial t}, \tag{18.167}$$

changes the Hamiltonian to H_1 where

$$H_1 = \exp(iU_1(t))\left(H_0 - i\hbar\frac{\partial}{\partial t}\right)\exp(-iU_1(t)). \tag{18.168}$$

We choose $U_1(t)$ to be given by

$$U_1(t) = \frac{1}{2\hbar}\ln\zeta(t)\int_0^{L(t)} 2x'\frac{\partial\pi(x',t)}{\partial x'}\psi(x')\,dx'. \tag{18.169}$$

Using the Baker-Hausdorff theorem, Eq. (12.103), we find

$$[U_1(t), \pi(x)] = i\ln(\zeta(t))x\frac{\partial\pi}{\partial x}, \tag{18.170}$$

and therefore from (18.169) it follows that

$$\exp(iU_1(t))\pi(x)\exp(-iU_1(t)) = \exp\left[-(\ln\zeta(t))x\frac{\partial}{\partial x}\right]\pi(x) = \pi\left(\frac{x}{\zeta(t)}\right). \tag{18.171}$$

Similarly for the $\psi(x)$ we have

$$\exp(iU_1(t))\psi(x)\exp(-iU_1(t)) = \frac{1}{\zeta(t)}\psi\left(\frac{x}{\zeta(t)}\right). \tag{18.172}$$

The factor $\frac{1}{\zeta(t)}$ in front of $\frac{1}{\zeta(t)}\psi\left(\frac{x}{\zeta(t)}\right)$ in (18.172) introduces an asymmetry between $\psi(x,t)$ and $\pi(x,t)$ fields. We can make the transformed ψ and π symmetrical by means of a second unitary transformation $\exp(iU_2(t))$, where

$$U_2(t) = \frac{1}{2\hbar}\ln\zeta(t)\int_0^{L(t)}\pi\left(\frac{x'}{\zeta(t)}\right)\psi\left(\frac{x'}{\zeta(t)}\right)\frac{dx'}{\zeta(t)}. \tag{18.173}$$

Evaluating the transformed field operators using (12.103) we get

$$\exp(iU_2(t))\frac{1}{\zeta(t)}\psi\left(\frac{x}{\zeta(t)}\right)\exp(-iU_2(t)) = \frac{1}{\sqrt{\zeta(t)}}\psi\left(\frac{x}{\zeta(t)}\right), \tag{18.174}$$

and

$$\exp(iU_2(t))\pi\left(\frac{x}{\zeta(t)}\right)\exp(-iU_2(t)) = \frac{1}{\sqrt{\zeta(t)}}\pi\left(\frac{x}{\zeta(t)}\right). \tag{18.175}$$

Thus the transformation $\exp(iU_2(t))\exp(iU_1(t))$ is a unitary transformation which leaves the commutation relation (18.166) invariant:

$$\left[\frac{1}{\sqrt{\zeta(t)}}\psi\left(\frac{x'}{\zeta(t)},t\right),\frac{1}{\sqrt{\zeta(t)}}\pi\left(\frac{x''}{\zeta(t)},t\right)\right] = i\hbar\delta\left(x'-x''\right). \tag{18.176}$$

Next let us consider the effect of the transformation $\exp(iU_1(t))$ on the operator $\left[H_0 - i\hbar\frac{\partial}{\partial t}\right]$;

$$
\begin{aligned}
H_1 &= e^{iU_1(t)}\left[H_0 - i\hbar\left(\frac{\partial}{\partial t}\right)\right]e^{-iU_1(t)} \\
&= \int_0^{L(t)}\left[\frac{-i\hbar}{2m\zeta(t)}\left\{\frac{\partial}{\partial x}\pi\left(\frac{x}{\zeta(t)}\right)\right\}\left\{\frac{\partial}{\partial x}\psi\left(\frac{x}{\zeta(t)}\right)\right\}\right. \\
&\quad \left. - \frac{\dot\zeta(t)}{\zeta^2(t)}x\left\{\frac{\partial}{\partial x}\pi\left(\frac{x}{\zeta(t)}\right)\psi\left(\frac{x}{\zeta(t)}\right)\right\}\right]dx.
\end{aligned}
\tag{18.177}
$$

By changing x to $\xi = \frac{x}{\zeta(t)}$ we note that the integrand in (18.177) can be written in terms of ξ alone, and then the upper limit becomes L_0 independent of time. The result of the second transformation is

$$
\begin{aligned}
H_2 &= e^{iU_2(t)}\left[H_1 - i\hbar\left(\frac{\partial}{\partial t}\right)\right]e^{-iU_2(t)} \\
&= \int_0^{L_0}\left[\frac{-i\hbar}{2m\zeta^2(t)}\left(\frac{\partial\pi}{\partial\xi}\frac{\partial\psi}{\partial\xi}\right) - \frac{\dot\zeta(t)}{\zeta(t)}\left\{\xi\frac{\partial\pi(\xi)}{\partial\xi}\psi(\xi) + \frac{1}{2}\pi(\xi)\psi(\xi)\right\}\right]d\xi.
\end{aligned}
\tag{18.178}
$$

Finally to simplify the result we consider a third unitary transformation of the form $\exp(iS(t))$;

$$S(t) = -\frac{i}{\hbar}\int_0^{L_0}\beta(t)\eta^2\pi(\eta)\psi(\eta)d\eta, \tag{18.179}$$

where $\beta(t)$ is a function of time which will be determined later. The final result, a Hamiltonian which we denote by H_{eff}, is obtained by applying this transformation:

$$
\begin{aligned}
H_{eff} &= e^{iS(t)} \left[H_2 - i\hbar \left(\frac{\partial}{\partial t} \right) \right] e^{-iS(t)} = \\
&= \int_0^{L_0} \left[\frac{-i\hbar}{2m\zeta^2(t)} \left(\frac{\partial \pi(\xi)}{\partial \xi} \frac{\partial \psi(\xi)}{\partial \xi} \right) - \left(\frac{\hbar\beta(t)}{m\zeta^2(t)} + \frac{\dot\zeta(t)}{2\zeta(t)} \right) \right. \\
&\quad \times \left\{ 2\xi \frac{\partial \pi(\xi)}{\partial \xi} \psi(\xi) + \pi(\xi)\psi(\xi) \right\} \\
&\quad \left. - i \left(\frac{2\hbar\beta^2(t)}{m\zeta^2(t)} + \frac{2\beta(t)\dot\zeta(t)}{\zeta(t)} - \frac{d\beta(t)}{dt} \right) \xi^2 \pi(\xi)\psi(\xi) \right] d\xi. \quad (18.180)
\end{aligned}
$$

Since $\beta(t)$ is an arbitrary function of time we choose it is such a way that H_{eff} takes a simple form. Thus we set the coefficient of $\left(\frac{\partial \pi(\xi)}{\partial \xi} \right) \psi$ equal to zero to get

$$
\beta(t) = -\frac{m}{2\hbar} \zeta(t)\dot\zeta(t), \quad (18.181)
$$

and with this choice of $\beta(t)$, H_{eff} becomes [25]

$$
H_{eff} = \int_0^{L_0} \left[\frac{-i\hbar}{2m\zeta^2(t)} \left(\frac{\partial \pi(\xi)}{\partial \xi} \frac{\partial \psi(\xi)}{\partial \xi} \right) - \frac{im}{2\hbar} \zeta(t)\ddot\zeta(t)\xi^2 \pi(\xi)\psi(\xi) \right] d\xi. \quad (18.182)
$$

Now we write this Hamiltonian in terms of the creation and annihilation operators. To this end we first define a complete set of functions by the differential equations

$$
-\frac{\hbar^2}{2m} \frac{d^2\phi_j}{d\xi^2} + \frac{1}{2}\xi^2\Omega^2\phi_j = \varepsilon_j\phi_j, \quad (18.183)
$$

with the boundary conditions

$$
\phi_j(0) = \phi_j(L_0) = 0. \quad (18.184)
$$

In the differential equation (18.183), Ω^2 is a nonnegative constant. Since $\{\phi_j\}$ s form a complete set we expand $\psi(\xi,t)$ and $\pi(\xi,t)$ in terms of $\phi_j(\xi)$;

$$
\psi(\xi,t) = \sum_j a_j(t)\phi_j(\xi), \quad (18.185)
$$

and

$$
\pi(\xi,t) = \sum_j i\hbar a_j^\dagger(t)\phi_j^*(\xi). \quad (18.186)
$$

By substituting (18.185) and (18.186) in (18.182), we find H_{eff} in terms of $a_j^\dagger(t)$ and $a_j(t)$;

$$
H_{eff} = \frac{1}{\zeta^2(t)} \sum_j \varepsilon_j a_j^\dagger(t)a_j(t) + \frac{m}{2\zeta^2(t)} \left[\zeta^3(t)\ddot\zeta(t) - \Omega^2 \right] \sum_{j,k} I_{kj} a_k^\dagger(t)a_j(t),
$$

$$
(18.187)
$$

where

$$I_{kj} = I_{jk}^* = \int_0^{L_0} \xi^2 \phi_k^*(\xi)\phi_j(\xi)d\xi. \tag{18.188}$$

The set of eigenfunctions $\{\phi_j(t)\}$ are given by

$$\phi_j(\xi) = N_j \xi \exp\left[-\frac{m\Omega}{2\hbar^2}\xi^2\right] {}_1F_1\left(\frac{3}{4} - \frac{\varepsilon_j}{2\hbar\Omega}, \frac{3}{2}, \frac{m\Omega}{\hbar}\xi^2\right), \tag{18.189}$$

and the eigenvalues are the roots of

$${}_1F_1\left(\frac{3}{4} - \frac{\varepsilon}{2\hbar\Omega}, \frac{3}{2}, \frac{m\Omega}{\hbar}L_0^2\right) = 0. \tag{18.190}$$

In Eq. (18.189) N_j is the normalization constant. In this model for arbitrary $\zeta(t)$, the number of particles in each state will change, but the total number of particles in the cavity remains constant, i.e.

$$i\hbar\frac{dN}{dt} = \sum_j \left[a_j^\dagger(t)a_j(t), H_{eff}\right] = 0. \tag{18.191}$$

But for the special case of

$$\zeta^3(t)\ddot{\zeta}(t) = \Omega^2, \tag{18.192}$$

or

$$\zeta(t) = \left[\left(\frac{c^2}{L_0^2} + \Omega^2\right)t^2 + \frac{2c}{L_0}t + 1\right]^{\frac{1}{2}}, \tag{18.193}$$

where c is an arbitrary constant, the particles in each level stay in that level, and only the energy of the system decreases (or increases). Thus the total energy of the particles when $\zeta(t)$ is given by (18.193) is

$$E(t) = \sum_j \frac{\varepsilon_j n_j}{\zeta(t)}. \tag{18.194}$$

As a special case of this model let us consider the decay of a single particle at the i-th level of a cavity to the lower states. We write the time-dependent wave function for this particle as

$$|\Psi(t)\rangle = C_i(t)|1,0,0\cdots\rangle + C_{i-1}(t)|0,1,0\cdots\rangle + C_{i-2}|0,0,1,\cdots\rangle + \cdots, \tag{18.195}$$

where $|C_{i-k}(t)|^2$ is the probability of the particle to be at the $(i-k)$-th level at the time t. This wave function together with the Hamiltonian (18.187) gives us the set of coupled equations

$$i\hbar\frac{dC_j(t)}{dt} = \frac{1}{\zeta^2(t)}C_j(t) + \frac{m}{2\zeta^2(t)}\left[\zeta^3(t)\ddot{\zeta}(t) - \Omega^2\right]\sum_{k=1}^i I_{jk}C_k(t), \quad j = 1,2\cdots i. \tag{18.196}$$

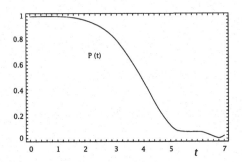

Figure 18.3: The probability of finding a particle in its initial state when the particle is confined to a cavity with a moving wall.

These equations are subject to the boundary conditions

$$C_i(0) = 1, \quad C_{i-k}(0) = 0, \quad k = 1, \cdots i - 1. \tag{18.197}$$

For instance if we choose $\Omega = 0$ and $L_0 = 1$, then

$$I_{jk} = 2 \int_0^1 \xi^2 \sin(j\pi\xi) \sin(k\pi\xi) d\xi, \tag{18.198}$$

and if we choose $\zeta(t)$ to be given by

$$\zeta(t) = \left[1 + \frac{a^4 t^4}{1 + a^2 t^2} \right], \tag{18.199}$$

then we can integrate Eq. (18.196) and calculate $|C_j(t)|^2$. For $a = 0.5$ and $i = 4$, there are four $C_j(t)$'s and they satisfy the initial conditions (18.197). In Fig. (18.4) the probability $P(t) = |C_4(t)|^2$ is plotted as a function of time.

We can formulate this problem for massive particles satisfying the Klein-Gordon equation [26], or for photons [27] [28], i.e. Maxwell's equations.

For the particles satisfying the Klein-Gordon equation we start with the Hamiltonian

$$H = \frac{1}{2} \int_0^{L(t)} \left[\pi^2(x,t) + \left(\frac{\partial \psi(x,t)}{\partial x} \right)^2 + m^2 \psi^2(x,t) \right] dx, \tag{18.200}$$

where we have set $\hbar = c = 1$. The fields $\psi(x,t)$ and $\pi(x,t)$ satisfy the boundary conditions (18.164) and (18.165) and the canonical commutation relation (18.166). Using the unitary transformations $\exp(iU_1(t))$ and $\exp(iU_2(t))$ on the operator $H - i\frac{\partial}{\partial t}$ we obtain the effective Hamiltonian for this field

$$\begin{aligned} H_{eff} &= \int_0^{L_0} \left[\frac{1}{2} \left\{ \pi^2 + \frac{1}{\zeta^2(t)} \left(\frac{\partial \psi}{\partial \xi} \right)^2 + m^2 \psi^2 \right\} \right. \\ &\quad \left. - \frac{\dot{\zeta}(t)}{\zeta(t)} \left\{ \xi \frac{\partial \pi}{\partial \xi} \psi + \frac{1}{2} \pi \psi \right\} \right] d\xi. \end{aligned} \tag{18.201}$$

From the effective Hamiltonian density, i.e. the integrand in (18.201) we obtain the equations for $\psi(\xi, t)$ and $\pi(\xi, t)$. The above unitary transformation preserves the reciprocal symmetry of the field, viz, the equations for $\psi(\xi, t)$ and $\pi(\xi, t)$ are identical in form. Thus for $\psi(\xi, t)$ we find

$$
\frac{1}{\zeta^2(t)} \left(1 - \dot{\zeta}^2(t)\xi^2\right) \frac{\partial^2 \psi}{\partial \xi^2} - \frac{\partial^2 \psi}{\partial t^2} + \frac{2\dot{\zeta}(t)}{\zeta(t)} \xi \frac{\partial^2 \psi}{\partial t \partial \xi} + \frac{\dot{\zeta}(t)}{\zeta(t)} \frac{\partial \psi}{\partial t}
$$
$$
+ \frac{1}{\zeta^2(t)} \left(\ddot{\zeta}(t)\zeta(t) - 3\dot{\zeta}^2(t)\right) \xi \frac{\partial \psi}{\partial \xi} + \left[\frac{\ddot{\zeta}(t)}{2\zeta(t)} - \frac{3}{4} \left(\frac{\dot{\zeta}(t)}{\zeta(t)}\right)^2 - m^2\right] \psi = 0.
$$

$$(18.202)$$

We can also write H_{eff} in terms of creation and annihilation operators. For the present case it is convenient to take $L_0 = \pi$ and expand both $\psi(\xi, t)$ and $\pi(\xi, t)$ as

$$
\psi(\xi, t) = \left(\frac{\zeta(t)}{\pi}\right)^{\frac{1}{2}} \sum_{k=1}^{\infty} \frac{1}{\sqrt{\omega_k(t)}} \left(a_k^\dagger + a_k\right) \sin(k\xi), \qquad (18.203)
$$

and

$$
\pi(\xi, t) = \frac{i}{(\pi\zeta(t))^{\frac{1}{2}}} \sum_{k=1}^{\infty} \sqrt{\omega_k(t)} \left(a_k^\dagger - a_k\right) \sin(k\xi), \qquad (18.204)
$$

where in these relations

$$
\omega_k(t) = \left[k^2 + m^2\zeta^2(t)\right]^{\frac{1}{2}}. \qquad (18.205)
$$

Substituting (18.203) and (18.204) in (18.201) and carrying out the integration over ξ, we obtain

$$
H_{eff} = \frac{1}{\zeta(t)} \sum_k \omega_k(t) \left(a_k^\dagger a_k + \frac{1}{2}\right) + \frac{i\dot{\zeta}(t)}{\zeta(t)} \sum_k \sum_{j \neq k} (-1)^{k+j}
$$
$$
\times \frac{jk}{j^2 - k^2} \left(\frac{\omega_k(t)}{\omega_j(t)}\right)^{\frac{1}{2}} \left(a_k^\dagger a_j^\dagger - a_k^\dagger a_j + a_k a_j^\dagger - a_k a_j\right). \quad (18.206)
$$

As in the non-relativistic problem we can find the equation of motion for a_k and $a_k^\dagger$ from (18.206). Thus for da_k/dt we have

$$
i\frac{da_k}{dt} = \frac{\omega_k(t)}{\zeta(t)} a_k + iR_k(t), \qquad (18.207)
$$

where

$$
R_k(t) = \frac{\dot{\zeta}(t)}{\zeta(t)} \sum_{j \neq k} \frac{(-1)^{k+j} jk}{j^2 - k^2} \left[\left(\frac{\omega_j(t)}{\omega_k(t)}\right)^{\frac{1}{2}} \left(a_j - a_j^\dagger\right) + \left(\frac{\omega_k(t)}{\omega_j(t)}\right)^{\frac{1}{2}} \left(a_j + a_j^\dagger\right)\right].
$$

$$(18.208)$$

If at $t = 0$ there are no particles in the state k, i.e. if

$$a_k^\dagger(0)a_k|0\rangle = 0, \qquad (18.209)$$

then the number of particles created in this state between $t = 0$ and $t = \infty$ is given by

$$\langle 0|N_k|0\rangle = \int_0^\infty \langle 0|a_k^\dagger R_k + R_k^\dagger a_k|0\rangle dt. \qquad (18.210)$$

The asymptotic form of the Hamiltonian as $t \to \infty$ can be expressed in terms of $\langle 0|N_k|0\rangle$;

$$H_{eff}(\infty) \to \sum_{k=1}^\infty \left(m^2 + \frac{k^2}{\zeta^2(\infty)}\right)^{\frac{1}{2}} \left(\langle 0|N_k|0\rangle + \frac{1}{2}\right), \qquad (18.211)$$

and, therefore, the change in the energy of the system is

$$E_f - E_i = \sum_0^\infty \left[\left(m^2 + \frac{k^2}{\zeta^2(\infty)}\right)^{\frac{1}{2}} \left(\langle 0|N_k|0\rangle + \frac{1}{2}\right) - \frac{1}{2}\left(m^2 + k^2\right)\right]. \qquad (18.212)$$

This change in the energy can be finite or infinite depending on how $\zeta(t)$ evolves in time [26].

18.8   Decaying States in a Many-Boson System

A solvable many-boson system with nonlinear coupling is the Bassichis-Foldy model [29]-[31]. This model consists of three interacting states with the Hamiltonian

$$H = a_1^\dagger a_1 + a_3^\dagger a_3 + g\left[a_2^\dagger a_2\left(a_1^\dagger a_1 + a_3^\dagger a_3\right) + a_2^2 a_1^\dagger a_3^\dagger + a_2^{\dagger 2} a_1 a_3\right] - Fg a_1^\dagger a_1 a_3^\dagger a_3, \qquad (18.213)$$

where g and Fg are the coupling constants, and where we have set $\hbar = 1$. This Hamiltonian admits two constant of motion:
(a) The number operator N

$$N = a_1^\dagger a_1 + a_2^\dagger a_2 + a_3^\dagger a_3, \qquad (18.214)$$

and
(b) The difference between the number of particles in the states 1 and 3, i.e.

$$\Delta = a_1^\dagger a_1 - a_3^\dagger a_3. \qquad (18.215)$$

The time evolution of this many-body system is given by the time-dependent Schrödinger equation

$$i\frac{\partial \Psi}{\partial t} = H\Psi. \qquad (18.216)$$

This wave function depends on the number of particles in three different states. Let n be the number of particles in the state 3, i.e.

$$a_3^\dagger a_3 \Psi = n\Psi, \tag{18.217}$$

then the wave function which depends on the number of particles in each of the three states can be written as

$$|\Psi\rangle = \sum_{n=0}^{\frac{1}{2}(N-\Delta)} C_n |N, \Delta, n\rangle. \tag{18.218}$$

Substituting for H from Eq. (18.213) in (18.216) we find that the Schrödinger equation (18.216) can be transformed into a linear differential-difference equation,

$$g\left[(N - \Delta - 2n + 2)(N - \Delta - 2n + 1)(\Delta + n)n\right]^{\frac{1}{2}} C_{n-1}$$
$$+ \left[\Delta + 2n + g(N - \Delta - 2n)(\Delta + 2n) - Fgn(n + \Delta) - i\frac{d}{dt}\right] C_n$$
$$+ g\left[(N - \Delta - 2n - 1)(N - \Delta - 2n)(\Delta + n + 1)(n + 1)\right]^{\frac{1}{2}} C_{n+1} = 0. \tag{18.219}$$

This equation for fixed Δ and N can be solved numerically.

Here let us study the following simple case:

(a) We set $\Delta = 0$, i.e. when the number of particles in the states 1 and 3 are equal.

(b) We take the limit of $N \to \infty$ and $g \to 0$ in such a way that gN remains finite.

Then Eq. (18.219) will reduce to

$$i\frac{dC_n}{dt} = 2n(1 + gN)C_n + gN\left[nC_{n-1} + (n + 1)C_{n+1}\right]. \tag{18.220}$$

We solve this equation with the initial conditions

$$C_0(t = 0) = 1, \quad \text{and} \quad C_n(t = 0) = 0, \quad \text{for} \quad n \neq 0. \tag{18.221}$$

These initial conditions imply that at $t = 0$ all of the N particles are in the state 2. Because of the interaction, n particles move to the state 3 with equal number moving to the state 1. The probability of having $N - 2n$ particles left in 2 at t is given by $|C_n(t)|^2$.

In order to solve (18.220) we first obtain the solution of the first order partial differential equation

$$\frac{\partial G}{\partial t} + \left[\frac{1}{L} + \frac{2L}{\pi}gN\sin\left(\frac{\pi x}{L}\right)\right]\frac{\partial G}{\partial x} + gN\exp\left(\frac{i\pi x}{L}\right)G = 0, \tag{18.222}$$

subject to the boundary conditions

$$G(L,t) = G(-L,t), \quad G(x,t=0) = (2L)^{-\frac{1}{2}}. \tag{18.223}$$

The partial differential equation (18.222) is found by the method of generating function which was used in Chapter 8.

If ω^2, which is defined by

$$\omega^2 = \frac{\pi^2}{4L^2} - N^2 g^2, \tag{18.224}$$

is positive, then $G(x,t)$ will be a periodic function of time and is given by

$$G(x,t) = \frac{(2L)^{-\frac{1}{2}} \exp\left(\frac{i\pi t}{2L^2}\right)}{\left\{\cos\omega t + \frac{gN}{\omega}\left[\exp\left(\frac{i\pi x}{L}\right) + \frac{i\pi}{2L^2 gN}\right]\sin\omega t\right\}}. \tag{18.225}$$

By expanding $G(x,t)$ in terms of the orthogonal set $(2L)^{-\frac{1}{2}} \exp\left(\frac{in\pi x}{L}\right)$, i.e.

$$G(x,t) = (2L)^{-\frac{1}{2}} \sum_{n=0}^{\infty} D_n(t) \exp\left(\frac{in\pi x}{L}\right), \tag{18.226}$$

and substituting in (18.222) we find that D_n must satisfy the differential-difference equation;

$$\frac{dD_n}{dt} = -\frac{in\pi}{L^2} D_n + gN\left[(n+1)D_{n+1} - nD_{n-1}\right], \tag{18.227}$$

thus showing that $G(x,t)$ is the generating function for the differential-difference equation (18.227). At the same time we observe that if we expand (18.225) in powers of $\exp\left(\frac{i\pi t}{2L^2}\right)$ and compare the result with (18.227) we obtain D_n for the case when $\frac{\pi}{2L^2} > gN$,

$$D_n(t) = \left(-\frac{gN}{\omega}\right)^n \frac{\exp\left(\frac{i\pi t}{2L^2}\right)\sin^n\omega t}{\left\{\cos\omega t + \frac{i\pi}{2\omega L^2}\sin\omega t\right\}^{n+1}}. \tag{18.228}$$

Now we return to our original set of equations for $C_n(t)$ and compare (18.220) and (18.227) to find $C_n(t)$

$$C_n(t) = i^n D_n(t) = \left(-\frac{igN}{\omega}\right)^n \frac{\exp\left(i(1+gN)t\right)\sin^n\omega t}{\left\{\cos\omega t + \frac{i(1+gN)}{\omega}\sin\omega t\right\}^{n+1}}, \tag{18.229}$$

where

$$\omega^2 = (1+gN)^2 - g^2 N^2 = 1 + 2gN. \tag{18.230}$$

For g negative and less than $\left[-(2N)^{-1}\right]$, ω^2 in (18.230) becomes negative

$$-\omega^2 \to \nu^2 = 2|g|N - 1, \tag{18.231}$$

where ν^2 is a positive number. For this case $C_n(t)$ satisfies the equation

$$i\frac{dC_n}{dt} = -2\nu^2 n C_n + |g|N\left[(n+1)C_{n+1} + nC_{n-1}\right].\qquad(18.232)$$

We can solve (18.232) as before with the result that

$$C_n(t) = \left(-\frac{|g|N}{\nu}\right)^n \frac{\exp\left(i(|g|N-1)t\right)\sinh^n \nu t}{\left\{\cosh\nu t + \frac{i(|g|N-1)}{\nu}\sinh\nu t\right\}^{n+1}}.\qquad(18.233)$$

A simpler version of this model where the Hamiltonian is given by

$$H = ig\left(a_2^{\dagger 2}a_3 a_1 - a_3^\dagger a_1^\dagger a_2^2\right)\qquad(18.234)$$

is also exactly solvable. Here we have the same constants of motion N and Δ as in (18.219). Again for $\Delta = 0$ and in the limit of large N, and gN finite the time-dependent Schrödinger equation reduces to

$$\frac{dC_n}{dt} = gN\left[(n+1)C_{n+1} - nC_{n-1}\right].\qquad(18.235)$$

This equation is a special case of (18.227) and for this model the initial state $C_0(t)$ decays with the probability

$$|C_0(t)|^2 = \frac{1}{\cosh^2(gNt)}.\qquad(18.236)$$

An interesting feature of this model is that for finite but large N it has a degenerate ground state, and this corresponds to the value of F for which the classical limit of the system possesses a special symmetry [31]. The quantal Hamiltonian has an approximate symmetry for the same value of F. Thus if we choose $\Delta = 0$, set F equal to $\frac{4}{gN}$, and use the approximations

$$[(N-2n+2)(N-2n+1)]^{\frac{1}{2}} \approx (N-2n+2),\qquad(18.237)$$

and

$$[(N-2n+-1)(N-2n)]^{\frac{1}{2}} \approx (N-2n),\qquad(18.238)$$

in Eq. (18.219), then the resulting equation which is

$$gn(N-2n+2)C_{n-1} + \left[2n(N-2n)\left(g+\frac{1}{N}\right) - i\frac{d}{dt}\right]C_n$$
$$+ g(n+1)(N-2n)C_{n+1} = 0,\qquad(18.239)$$

will remain invariant under the transformation

$$n \to \frac{1}{2}N - j, \quad C_{n\pm 1} \to C_{\frac{1}{2}N-(j\mp 1)} = C_{j\mp 1}.\qquad(18.240)$$

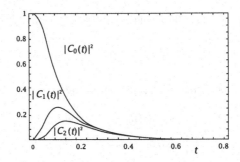

Figure 18.4: The probability of decay of the initial state $|C_0(t)|^2$ is plotted as a function of time. For comparison $|C_1(t)|^2$ and $|C_2(t)|^2$ are also shown.

Therefore if we find the solution of (18.219) with $\Delta = 0$ and $F = \frac{4}{gN}$ and with the initial conditions (18.221) the result would be the same as that of solving (18.219) for the same Δ and F but with the initial conditions

$$C_{\frac{N}{2}}(t = 0) = 1, \quad C_n(t = 0) = 0, \quad \text{for} \quad n \neq 0. \tag{18.241}$$

For the general case with N finite and large, the system will return to its original state but after a very long time. The decay except for very short time is exponential. Let us calculate the probability of finding the many-boson system in its original state where all of the particles are in the state 2, or zero momentum state, this is given by $|C_0(t)|^2$. The curves shown in Fig. (18.3) are calculated for $N = 200$, $g = -0.05$ and $F = 0$. For these values the numerical solution of (18.219) is very close to the analytic but approximate solution given by (18.233). In addition to $|C_0(t)|^2$, the time-dependence of $|C_1(t)|^2$ and $|C_2(t)|^2$ are also displayed.

The equations of motion derived from the Bassichis-Foldy model are very similar to the classical equations describing the modulational stability of a Langmuir condensate [32].

Bibliography

[1] W. Band, Am. J. Phys. 30, 646 (1962).

[2] V. Weisskopf and E. Wigner Z. Phys. 63, 54 (1930).

[3] The English translation of the paper of Weisskopf and Wigner is reprinted in W.R. Hindmarch, *Atomic Spectra*, (Pergamon Press, Oxford, 1967).

[4] J.L. Martin, Proc. Camb. Phil. Soc. 60, 587 (1964).

[5] R.J. Newton, *Scattering Theory of Waves and Particles*, Second Edition, (Springer-Verlag, New York, 1982) §17.3.3.

[6] M. Razavy, Can. J. Phys. 45, 1469 (1967).

[7] W. Heitler, *The Quantum Theory of Radiation*, Third Edition (Oxford University Press, London 1954) p. 181.

[8] M. Razavy and E.A. Henley Jr., Can. J. Phys. 48, 2439 (1970).

[9] F. Haake and W. Weidlich, Z. Phys. 213, 445 (1968).

[10] See for instance, M.L. Goldberger and K.M. Watson, *Collision Theory*, (John Wlley & Sons, New York, N.Y.) Chapter 8.

[11] P.M. Morse and H. Feshbach, *Methods of Theoretical Physics*, vol. II (McGraw-Hill, New York, 1953) p. 1754.

[12] H. Feshbach, Ann. Phys. (NY) 5, 357 (1958).

[13] N.F. Mott and H.S.W. Massey, *The Theory of Atomic Collisions*, Third Edition (Oxford University Press, London, 1971) p. 404.

[14] N.Gisin, J. Phys. A 14, 2259 (1981).

[15] N. Gisin, Physica A 111, 364 (1982).

[16] Y. Huang, S-I. Chu and J.O. Hirshfelder, Phys. Rev. A 40, 4171 (1989).

[17] See for example L.I. Schiff, *Quantum Mechanics*, Third Edition (McGraw-Hill, New York, 1968).

[18] H.-D. Doebner and G.A. Goldin, Phys. Lett. A162, 397 (1992).

[19] V.V. Dodonov, J. Korean Phys. Soc. 26, 111 (1993).

[20] S.A. Fulling and P.C.W. Davies, Proc. R. Soc. London, A 348, 393 (1976).

[21] B.S. De Witt, Phys. Rep. C19, 295 (1975).

[22] M. Razavy, Phys. Rev. A48, 3486 (1993).

[23] Y. Takahashi and H. Umezawa, Nuovo Cimento 6, 1324 (1957).

[24] E.P. Gross, in *Mathematical Methods in Solid States and Superfluid Theory*, edited by R.C. Clark and G.H. Derrick, (Edinburgh 1969).

[25] M. Razavy, Lett. Al Nuovo Cimento, 37, 449 (1983).

[26] M. Razavy and J. Terning, Lett. Al Nuovo Cimento, 41, 561 (1984).

[27] G.T. Moore, J. Math. Phys. 11, 2679 (1970).

[28] M. Razavy and J. Terning, Phys. Rev. D31, 307 (1985).

[29] W.H. Bassichis and L.L. Foldy, Phys. Rev. A133, 935 (1964).

[30] M. Razavy Int. J. Theor. Phys. 13, 237 (1975).

[31] M. Razavy and R.B. Ludwig, Phys. Rev. A33, 1519 (1985).

[32] R.O. Dendy and D. ter Haar, J. Plasma Phys. 31, 67 (1984).

Chapter 19

More on the Concept of Optical Potential

The Feshbach theory of optical potential has been very successful in explaining nucleon-nucleus and nucleus-nucleus scattering [1]-[4]. Since the wave function is in general complex, the presence of a complex (optical potential) in the Schrödinger equation arises naturally from the interaction of a particle (or particles) from a system of particles. But the idea of a complex force law is not compatible with the real position, real velocity, and the real trajectory of the particle in classical mechanics. The nonlocality of the interaction also adds to the problem of finding the classical limit of the optical potentials.

Similarly the connection between the scattering of the two complex nuclear systems interacting via optical potential and the classical (or quantal) description of heavy-ion scattering based on velocity-dependent damping forces is not, at present, clear. In the following section we study the classical limit of the wave equation with the optical potential assuming that this potential is local and independent of the energy of the particle. We will first consider the question of the nonlocality.

19.1 The Classical Analogue of the Nonlocal Interaction

As we have seen in Feshbach's theory, the optical potential is not only a complex function of $\mathbf{r}$, but it is also nonlocal. Here we will investigate the classical analogue of a symmetric and real nonlocal potential, and then consider how the complex nature of the potential manifests itself in the classical limit [5].

We write the Schrödinger equation in the simple form of

$$\left(\nabla^2 + k^2\right) \psi(\mathbf{r}) = \int K\left(\mathbf{r}, \mathbf{r}'\right) \psi\left(\mathbf{r}'\right) d^3 r', \tag{19.1}$$

where $K\left(\mathbf{r}, \mathbf{r}'\right)$ is a symmetric kernel

$$K\left(\mathbf{r}, \mathbf{r}'\right) = K\left(\mathbf{r}', \mathbf{r}\right). \tag{19.2}$$

Now we expand $K\left(\mathbf{r}, \mathbf{r}'\right)$ in the following way [5]:

$$\int K\left(\mathbf{r}, \mathbf{r}'\right) d^3 r' = U_0(r), \tag{19.3}$$

$$\int K\left(\mathbf{r}, \mathbf{r}'\right) \left(x - x'\right) d^3 r' = -x U_1(r), \tag{19.4}$$

$$\cdots\cdots\cdots\cdots$$

$$\int K\left(\mathbf{r}, \mathbf{r}'\right) \left(x - x'\right) \left(y - y'\right) d^3 r' = x y U_2(r), \tag{19.5}$$

$$\cdots\cdots\cdots\cdots$$

Thus the right hand side of (19.1) can be written as

$$\int K\left(\mathbf{r}, \mathbf{r}'\right) \psi\left(\mathbf{r}'\right) d^3 r = U_0 \psi(\mathbf{r}) + U_1 \left(\sum x \frac{\partial}{\partial x}\right) \psi(\mathbf{r})$$

$$+ \frac{1}{2} U_2 \sum \sum x y \frac{\partial^2}{\partial x \partial y} \psi\left(\mathbf{r}'\right) + \cdots, \tag{19.6}$$

where the sum is over the components x, y and z. Rearranging the terms on the right hand side of (19.6) we get

$$V(\mathbf{r}, \nabla) \psi(\mathbf{r}) = \left[U_0 + \frac{1}{2} U_1 \left(\mathbf{r}.\nabla + \nabla.\mathbf{r}\right) + \frac{1}{4} U_2 \left(\mathbf{r}.\nabla + \nabla.\mathbf{r}\right)^2 + \cdots\right] \psi(\mathbf{r}), \tag{19.7}$$

where the potential $V(\mathbf{r}, \nabla)$ is an infinite differential operator.

We can write $V(\mathbf{r}, \nabla)$ as a momentum (or velocity) dependent potential by noting that $\mathbf{p} = -i\hbar\nabla$, and therefore

$$V(\mathbf{r}, \mathbf{p})$$

$$= \left[U_0 + \frac{i}{2\hbar} U_1 \left(\mathbf{r} \cdot \mathbf{p} + \mathbf{p} \cdot \mathbf{r}\right) - \frac{1}{4\hbar^2} U_2 \left\{\mathbf{r}^2 \mathbf{p}^2 + 2\left(\mathbf{r} \cdot \mathbf{p}\right)\left(\mathbf{p} \cdot \mathbf{r}\right) + \mathbf{p}^2 \mathbf{r}^2\right\}\right] + \cdots. \tag{19.8}$$

Thus the classical momentum-dependent potential is

$$V(\mathbf{r}, \mathbf{p}) = \left[U_0 + \frac{i}{\hbar} U_1 (\mathbf{r} \cdot \mathbf{p}) - \frac{1}{\hbar^2} U_2 (\mathbf{r} \cdot \mathbf{p})^2 + \cdots \right]. \tag{19.9}$$

If we impose the requirement of invariance under rotation, then $V(\mathbf{r}, \mathbf{p})$ must be a function of r^2, p^2, $(\mathbf{r} \times \mathbf{p})^2$, and $(\mathbf{r} \cdot \mathbf{p} + \mathbf{p} \cdot \mathbf{r})$. The time reversal invariance implies that only even powers of $(\mathbf{r} \cdot \mathbf{p} + \mathbf{p} \cdot \mathbf{r})$ appear in $V(\mathbf{r}, \mathbf{p})$. Thus the most general velocity-dependent local potentials satisfying these requirements will have the general form of $U\left(r^2, p^2, L^2\right)$, where $\mathbf{L} = \mathbf{r} \times \mathbf{p}$ is the angular momentum.

A potential of the form

$$V(\mathbf{r}, \mathbf{p}) = U_0(r) + \frac{1}{2m} \mathbf{p} \cdot W(r) \mathbf{p}, \tag{19.10}$$

which is a special case of (19.8), has been used to describe the interaction between two nucleons [6]. Note the similarity between (19.10) and the Hamiltonian for the dissipative motion with quadratic dependence on velocity [7].

Next we will study the classical limit of a local optical potential. We assume that the potential is independent of the energy of the particle and is of the form

$$V_0(r) = v(r) + i\mathcal{V}(r). \tag{19.11}$$

Now according to Feshbach's theory $\mathcal{V}$ must be negative definite, i.e. $\mathcal{V}(r) \leq 0$ for all r. One of the possible ways of studying the classical limit of the motion in a complex potential field is by examining the time evolution of the expectation values of the radial coordinate and momentum of the particle. To this end we start with the Schrödinger equation with the complex potential

$$i\hbar \frac{\partial \psi}{\partial t} = \left[-\frac{\hbar^2}{2m} \nabla^2 + v(r) + i\mathcal{V}(r) \right] \psi. \tag{19.12}$$

Noting that for any operator $\hat{O}(\mathbf{r}, -i\hbar\nabla)$, the rate of change of the expectation value is given by

$$\frac{d}{dt} \left\langle \hat{O} \right\rangle = \frac{i}{\hbar} \left(\int H^* \psi^* \hat{O} \psi d^3 r - \int \psi^* \hat{O} H \psi d^3 r \right), \tag{19.13}$$

where H is the non-Hermitian operator given in the square bracket in (19.12), we want to determine the expectation values of r and p_r.

The last equation can be written in terms of a Hermitian and an anti-Hermitian operator [8]. Let us write

$$H = H_H + H_A = \frac{1}{2} \left(H + H^* \right) + \frac{1}{2} \left(H_H - H_A^* \right), \tag{19.14}$$

where the first term $(1/2)\left(H + H^*\right)$ is the Hermitian part. Then (19.13) becomes

$$\frac{d}{dt} \left\langle \hat{O} \right\rangle = \int \psi^* \left(\frac{i}{\hbar} \left[H_H, \hat{O} \right] - \frac{i}{\hbar} \left[H_A, \hat{O} \right]_+ \right) \psi \, d^3 r, \tag{19.15}$$

where $[\,,]$ and $[\,,]_+$ denote the commutator and the anti-commutator respectively. Substituting for $\hat{O}$ first r and then p_r in Eq. (19.15), we find

$$\frac{d}{dt}\langle r\rangle = \frac{\langle p_r\rangle}{m} + \frac{\langle 2r\mathcal{V}(r)\rangle}{\hbar}, \tag{19.16}$$

and

$$\frac{d}{dt}\langle p_r\rangle = -\left\langle \frac{\partial v(r)}{\partial r} + i\frac{\partial \mathcal{V}}{\partial r}\right\rangle + 2\frac{\langle \mathcal{V}(r)p_r\rangle}{\hbar}. \tag{19.17}$$

These relations reduce to the usual form of the Ehrenfest's theorem when $\mathcal{V}(r) = 0$ [9] [10].

From these relations we draw the following conclusions [11]:

(1) For the existence of a well-defined classical limit $\mathcal{V}(r)$ has to be proportional to $\hbar^\alpha, (\alpha \geq 1)$. We can also reach the same result by considering the hydrodynamical formulation of the Schrödinger equation [12] [13]. In the latter formulation we start with the wave function $\psi(\mathbf{r}, t)$ and write it as

$$\psi(\mathbf{r}, t) = \sqrt{\rho}\exp\left(\frac{iS}{\hbar}\right). \tag{19.18}$$

By substituting (19.18) in (19.12) and separating the real and imaginary parts we find two equations for ρ and S;

$$\frac{\partial \rho}{\partial t} + \nabla \cdot \left(\frac{\rho \nabla S}{m}\right) - \frac{2\rho \mathcal{V}}{\hbar} = 0, \tag{19.19}$$

and

$$\frac{\partial S}{\partial t} + \frac{1}{2m}(\nabla S)^2 - \hbar^2\left[\frac{\nabla^2\rho}{\rho} - \left(\frac{\nabla\rho}{\rho}\right)^2\right] + v(r) = 0. \tag{19.20}$$

If we integrate (19.19) over all space and observe that $\rho\nabla S \to 0$ as $r \to \infty$, we obtain

$$\frac{d}{dt}\int \rho d^3r = \frac{2}{\hbar}\int \rho\mathcal{V}(r)d^3r \leq 0. \tag{19.21}$$

The negative-definiteness in (19.21) follows from the fact that $\rho(r)$ is positive or zero and $\mathcal{V}(r)$ is negative or zero for all r. This relation shows that $\mathcal{V}(r)$ has to go to zero as $\hbar^a (a \geq 1)$ in order to have a well-defined classical limit.

(2) Equation (19.16) shows that the canonical momentum is not the same as the mechanical momentum. As we have seen before this creates a problem with using the minimal coupling rule when the particle is charged.

(3) The time evolution equations for $\langle x\rangle$ and $\langle p\rangle$ show that both of these are complex quantities. Thus the three-dimensional motion of a particle corresponds to six degrees of freedom. This confirms the result of §4.4 where we noticed that the motion of a spring with a complex spring constant can be decomposed into the motion of two real damped harmonic oscillators.

19.2   Minimal and/or Maximal Coupling

In the case of the optical potential found in the previous section there is no general rule for the coupling of a charged particle to the electromagnetic field which is compatible with the gauge transformation.

First let us write Eq. (18.101) as

$$\left(\frac{\hbar^2}{2m}\nabla^2 + E\right)\psi(\mathbf{r}) = \int \mathcal{V}(\mathbf{r},\mathbf{r}')\,\psi(\mathbf{r}')\,d^3r', \tag{19.22}$$

with $\mathcal{V}(\mathbf{r},\mathbf{r}')$ the potential acting between, say, a proton and a neutron. If in addition we assume that this potential is separable then we can write it as

$$\mathcal{V}(\mathbf{r},\mathbf{r}') \rightarrow v(\mathbf{r}_p - \mathbf{r}_n)\,v(\mathbf{r}'_p - \mathbf{r}'_n). \tag{19.23}$$

In this relation $\mathbf{r}_p$ and $\mathbf{r}_n$ are the coordinates of the proton and neutron respectively, and $\mathbf{r} = \mathbf{r}_p - \mathbf{r}_n$ and $\mathbf{r}' = \mathbf{r}'_p - \mathbf{r}'_n$. Now if $\mathbf{R}$ denotes the center of mass coordinate of this two-body problem

$$\mathbf{R} = \frac{1}{2}(\mathbf{r}_p + \mathbf{r}_n) = \frac{1}{2}(\mathbf{r}'_p + \mathbf{r}'_n), \tag{19.24}$$

then in the presence of the electromagnetic field $\mathbf{A}$, the potential $\mathcal{V}(\mathbf{r},\mathbf{r}')$ is replaced by

$$\mathcal{V}(\mathbf{r},\mathbf{r}',\mathbf{A}) = v(\mathbf{r}_p - \mathbf{r}_n)\,v(\mathbf{r}'_p - \mathbf{r}'_n)$$
$$\times\ \exp\left\{ie\int_{\mathbf{R}}^{\mathbf{r}_p}\mathbf{A}\cdot d\mathbf{s} + ie\int_{\mathbf{r}_{p'}}^{\mathbf{R}}\mathbf{A}\cdot d\mathbf{s}\right\}. \tag{19.25}$$

The line integral is taken along a straight line connecting the lower to the upper limit [14].

It should be pointed out that (19.25) is not the most general form of the coupling which is possible. Thus if F and G are two gauge invariant quantities such that as the electromagnetic field tends to zero, $F \rightarrow 1$ and $G \rightarrow 0$, then (19.25) can be replaced by [15] [16]

$$\mathcal{V}(\mathbf{r},\mathbf{r}',\mathbf{A})\,F + G. \tag{19.26}$$

Now let us consider the Schrödinger equation for two particles where only one of the particles (e.g. particle 1) is charged. For this case the charge and current densities are given by

$$\langle \Psi_f\,|\rho_1(\mathbf{r})|\,\Psi_i\rangle = e\int \Psi_f^*(\mathbf{r},\mathbf{r}_2)\Psi_i(\mathbf{r},\mathbf{r}_2)d^3r_2, \tag{19.27}$$

and

$$\langle \Psi_f\,|\mathbf{j}_1(\mathbf{r})|\,\Psi_i\rangle = \frac{-ie}{2m}\int \left[\Psi_f^*(\mathbf{r},\mathbf{r}_2)\nabla\Psi_i(\mathbf{r},\mathbf{r}_2) - \Psi_i(\mathbf{r},\mathbf{r}_2)\nabla\Psi_f^*(\mathbf{r},\mathbf{r}_2)\right]d^3r_2, \tag{19.28}$$

respectively. To find the conservation law for the charge and current density we start with the Schrödinger equation with the nonlocal potential

$$-\frac{\hbar^2}{2m}\left(\nabla_1^2 + \nabla_2^2\right)\Psi\left(\mathbf{r}_1, \mathbf{r}_2, t\right) \; + \; \int \mathcal{V}\left(\mathbf{r}_1, \mathbf{r}_2; \mathbf{r}_1', \mathbf{r}_2'\right)\Psi\left(\mathbf{r}_1', \mathbf{r}_2', t\right) d^3 r_1' d^2 r_2'$$

$$= \; i\hbar\frac{\partial}{\partial t}\Psi\left(\mathbf{r}_1, \mathbf{r}_2, t\right), \tag{19.29}$$

and a similar equation for $\Psi^*\left(\mathbf{r}_1, \mathbf{r}_2, t\right)$. The potential $\mathcal{V}\left(\mathbf{r}_1, \mathbf{r}_2; \mathbf{r}_1', \mathbf{r}_2'\right)$ depends on the relative coordinates of the two particles, therefore

$$\mathcal{V}\left(\mathbf{r}_1, \mathbf{r}_2; \mathbf{r}_1', \mathbf{r}_2'\right) = \delta(\mathbf{R} - \mathbf{R}')\left\langle \mathbf{r}\left|\mathcal{V}\right|\mathbf{r}'\right\rangle, \tag{19.30}$$

where $\mathbf{R}$ is the center of mass coordinate and $\mathbf{r}$ is the relative coordinate of the two particles.

From (19.29) and its complex conjugate we find

$$\frac{\partial}{\partial t}\left(e\Psi^*\Psi\right) = -\left(\frac{ie\hbar}{2m}\right)\nabla_1\cdot\left(\Psi\nabla_1\Psi^* - \Psi^*\nabla_1\Psi\right) - Y\left(\mathbf{r}_1, \mathbf{r}_2, t\right), \tag{19.31}$$

where

$$Y\left(\mathbf{r}_1, \mathbf{r}_2, t\right) = \left(\frac{ie}{\hbar}\right)\int \mathcal{V}\left(\mathbf{r}_1, \mathbf{r}_2; \mathbf{r}_3, \mathbf{r}_4\right)$$

$$\times \quad \left\{\Psi\left(\mathbf{r}_3, \mathbf{r}_4, t\right)\Psi^*\left(\mathbf{r}_1, \mathbf{r}_2, t\right) - \Psi^*\left(\mathbf{r}_3, \mathbf{r}_4, t\right)\Psi\left(\mathbf{r}_1, \mathbf{r}_2, t\right)\right\} d^3 r_3 d^3 r_4. \tag{19.32}$$

The charge density in this case is expressible as an integral

$$\rho(\mathbf{r}_1, t) = e\int \Psi^*\left(\mathbf{r}_1, \mathbf{r}_2, t\right)\Psi\left(\mathbf{r}_1, \mathbf{r}_2, t\right) d^3 r_2. \tag{19.33}$$

We have also the divergent of a part of current, $\mathbf{j}_1\left(\mathbf{r}_1\right)$, which is given by

$$\nabla_1\cdot\mathbf{j}_1\left(\mathbf{r}_1\right) = -\left(\frac{ie\hbar}{2m}\right)$$

$$\times \quad \nabla_1\cdot\int\left\{\nabla_1\Psi\left(\mathbf{r}_1, \mathbf{r}_2, t\right)\Psi^*\left(\mathbf{r}_1, \mathbf{r}_2, t\right) - \Psi\left(\mathbf{r}_1, \mathbf{r}_2, t\right)\nabla_1\Psi^*\left(\mathbf{r}_1, \mathbf{r}_2, t\right)\right\} d^2 r_2. \tag{19.34}$$

In order to have conservation law for the charge, in addition to $\mathbf{j}_1\left(\mathbf{r}_1\right)$ we must have a current $\mathbf{j}_2\left(\mathbf{r}_1\right)$ where

$$\nabla_1\cdot\mathbf{j}_2\left(\mathbf{r}_1\right) = \left(\frac{ie}{2\hbar}\right)\int \mathcal{V}\left(\mathbf{r}_1, \mathbf{r}_2; \mathbf{r}_3, \mathbf{r}_4\right)$$

$$\times \quad \left[\Psi\left(\mathbf{r}_3, \mathbf{r}_4, t\right)\Psi^*\left(\mathbf{r}_1, \mathbf{r}_2, t\right) - \Psi^*\left(\mathbf{r}_3, \mathbf{r}_4, t\right)\Psi\left(\mathbf{r}_1, \mathbf{r}_2, t\right)\right] d^3 r_2 d^3 r_3 d^3 r_4. \tag{19.35}$$

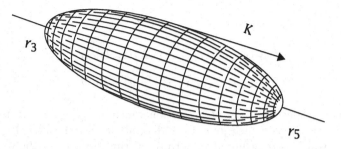

Figure 19.1: An elementary solution of Eq. (19.37). For any point on the surface, $\mathbf{K}$ is a unit vector in the direction of current multiplied by a two-dimensional delta function in the orthogonal plane.

From $\nabla_1 \cdot \mathbf{j}_2(\mathbf{r}_1)$ we find $\mathbf{j}_2(\mathbf{r}_1)$ to be [17]

$$\mathbf{j}_2\left(\mathbf{r}_1\right) = \left(\frac{ie}{2\hbar}\right) \int \mathbf{K}\left(\mathbf{r}_1 - \mathbf{r}_3; \mathbf{r}_1 - \mathbf{r}_5\right) \mathcal{V}\left(\mathbf{r}_2, \mathbf{r}_3; \mathbf{r}_4, \mathbf{r}_5\right)$$
$$\times \quad \Psi^*\left(\mathbf{r}_3, \mathbf{r}_2, t\right) \Psi\left(\mathbf{r}_4, \mathbf{r}_5, t\right) d^3r_2 d^3r_3 d^3r_4 d^3r_5, \qquad (19.36)$$

where

$$\nabla_1 \cdot \mathbf{K}\left(\mathbf{r}_1 - \mathbf{r}_3; \mathbf{r}_1 - \mathbf{r}_5\right) = \delta\left(\mathbf{r}_1 - \mathbf{r}_3\right) - \delta\left(\mathbf{r}_1 - \mathbf{r}_5\right). \qquad (19.37)$$

Note that the Hermiticity requires that

$$\mathbf{K}^*\left(\mathbf{r}_1 - \mathbf{r}_3; \mathbf{r}_1 - \mathbf{r}_5\right) = -\mathbf{K}\left(\mathbf{r}_1 - \mathbf{r}_5; \mathbf{r}_1 - \mathbf{r}_3\right). \qquad (19.38)$$

For a local potential from (19.30) we have

$$\mathcal{V}\left(\mathbf{r}_1, \mathbf{r}_2; \mathbf{r}_1', \mathbf{r}_2'\right) = \delta\left(\mathbf{r}_1 - \mathbf{r}_1'\right) \delta\left(\mathbf{r}_2 - \mathbf{r}_2'\right) V\left(|\mathbf{r}_1 - \mathbf{r}_2|\right), \qquad (19.39)$$

and therefore $\mathbf{j}_2(\mathbf{r}_1)$ is zero. Having obtained the total current $\mathbf{j}_1\left(\mathbf{r}_1\right) + \mathbf{j}_2\left(\mathbf{r}_1\right)$ we can express the conservation law as

$$\frac{\partial \rho}{\partial t} + \nabla_1 \cdot \mathbf{j}(\mathbf{r}_1) = 0, \qquad (19.40)$$

where

$$\mathbf{j}(\mathbf{r}_1) = \mathbf{j}_1\left(\mathbf{r}_1\right) + \mathbf{j}_2\left(\mathbf{r}_1\right). \qquad (19.41)$$

The interaction between the charged particle and the electromagnetic field can be written as H_{int} where

$$H_{int} = \int \left[\rho(\mathbf{r}_1)\phi(\mathbf{r}_1, t) - \frac{1}{c}\mathbf{j}(\mathbf{r}_1) \cdot \mathbf{A}_1(\mathbf{r}_1, t)\right] d^3r_1. \qquad (19.42)$$

In this equation $\phi(\mathbf{r}_1, t)$ and $\mathbf{A}_1(\mathbf{r}_1, t)$ are the electric and magnetic potentials respectively.

If $\phi(\mathbf{r}_1, t)$ is zero, then H_{int} becomes

$$H_{int} = -\frac{1}{c} \int \left[\Psi^*\nabla_1\Psi - \Psi\nabla_1\Psi^*\right] \cdot \mathbf{A}_1 d^3r_1 - H_I, \qquad (19.43)$$

where H_I is defined by

$$
\begin{aligned}
H_I \;=\; &\frac{1}{c}\int \mathbf{K}\,(\mathbf{r}_1 - \mathbf{r}_3; \mathbf{r}_1 - \mathbf{r}_5) \cdot \mathbf{A}_1(\mathbf{r}_1, t)\mathcal{V}\,(\mathbf{r}_2, \mathbf{r}_3; \mathbf{r}_4, \mathbf{r}_5) \\
&\times\; \Psi^*\,(\mathbf{r}_2, \mathbf{r}_3, t)\,\Psi\,(\mathbf{r}_4, \mathbf{r}_5, t)\, d^3 r_1 d^3 r_2 d^3 r_3 d^3 r_4 d^3 r_5.
\end{aligned}
\tag{19.44}
$$

Let us consider a surface of revolution which is obtained by rotating an arbitrary smooth curve which joins $\mathbf{r}_3$ and $\mathbf{r}_5$ about $\mathbf{r}_3 - \mathbf{r}_5$ axis (Fig. 19.1). Any current which is distributed uniformly on this surface will be a solution of (19.37). In particular we can join $\mathbf{r}_3$ and $\mathbf{r}_5$ by a straight line. This is the case of "minimal coupling", which corresponds to the shortest path connecting these two points. Thus if the effect of an arbitrary nonlocal potential (19.30) is expressed as a superposition of displacement operators, and if each $-i\hbar\nabla_r$ is replaced by $-i\hbar\nabla_r - \frac{ie}{c}\mathbf{A}(r)$, then the resulting electromagnetic current is the one where $\mathbf{K}$ follows a straight line from r to r' [15].

A simple solution of (19.37) corresponding to minimal coupling can be found if we choose $\mathbf{K}$ to be the unit vector in the direction of $\mathbf{r}_3 - \mathbf{r}_5$ multiplied by a two-dimensional δ-function in the plane orthogonal to $\mathbf{r}_3 - \mathbf{r}_5$, i.e.

$$
\mathbf{K} = \frac{\mathbf{r}_3 - \mathbf{r}_5}{|\mathbf{r}_3 - \mathbf{r}_5|}\delta^2(\mathbf{X}), \quad \mathbf{X}\cdot(\mathbf{r}_3 - \mathbf{r}_5) = 0.
\tag{19.45}
$$

Next let us study the "maximal coupling" solution. We can find the solution of (19.37) by analogy with the equation for the electric field due to a dipole with charges located at $\mathbf{r}_3$ and $\mathbf{r}_5$. Thus we write $\mathbf{K}$ as the sum of two vectors

$$
\mathbf{K} = \nabla_r \Lambda + \nabla_r \wedge \boldsymbol{\Gamma},
\tag{19.46}
$$

and substitute it in (19.37) to obtain

$$
\nabla_1 \cdot \mathbf{K}\,(\mathbf{r}_1 - \mathbf{r}_3; \mathbf{r}_1 - \mathbf{r}_5) = \nabla_r^2 \Lambda = \delta\,(\mathbf{r}_1 - \mathbf{r}_3) - \delta\,(\mathbf{r}_1 - \mathbf{r}_5).
\tag{19.47}
$$

Equation (19.47) can be solved for Λ subject to the boundary condition that Λ should go to zero as $r \to \infty$ with the result that

$$
\Lambda = -\frac{1}{4\pi}\left(\frac{1}{|\mathbf{r} - \mathbf{r}_3|} - \frac{1}{|\mathbf{r} - \mathbf{r}_5|}\right).
\tag{19.48}
$$

From Eqs. (19.46) and (19.48) we find $\mathbf{K}$:

$$
\mathbf{K}(\mathbf{r} - \mathbf{r}_3; \mathbf{r} - \mathbf{r}_5) = \frac{1}{4\pi}\left(\frac{\mathbf{r}_1 - \mathbf{r}_3}{|\mathbf{r}_1 - \mathbf{r}_3|^3} - \frac{\mathbf{r} - \mathbf{r}_5}{|\mathbf{r} - \mathbf{r}_5|^3}\right) + \nabla_r \wedge \boldsymbol{\Gamma}.
\tag{19.49}
$$

The last term in (19.49) is arbitrary but it must satisfy the same boundary condition as Λ does.

19.3 Damped Harmonic Oscillator and Optical Potential

In Section 4.4 we found the Lagrangian (4.64) for the motion of two harmonic oscillators. Let us now consider the quantum mechanical version of that problem where the optical potential is given by $\frac{m}{2}\omega^2 (1 - i\alpha)^2 z^2$. The classical Lagrangian for this system can be written either as (4.64) or as a real Lagrangian

$$L = \frac{m}{4}\left[\left(\frac{dz}{dt}\right)^2 + \left(\frac{dz^*}{dt}\right)^2\right] - \frac{m\omega^2}{4}\left[(1 - i\alpha)^2 z^2 + (1 + i\alpha)^2 z^{*2}\right]. \quad (19.50)$$

The Hamiltonian for this system is

$$H = \frac{1}{4m}\left(p_z^2 + p_{z^*}^2\right) + \frac{m\omega^2}{4}\left[(1 - i\alpha)^2 z^2 + (1 + i\alpha)^2 z^{*2}\right]. \quad (19.51)$$

Expressing H in terms of the original variables x and y we have

$$H = \frac{1}{2m}\left(p_x^2 - p_y^2\right) + \frac{m\omega^2}{2}\left[(1 - \alpha^2)\left(x^2 - y^2\right) + 4\alpha xy\right]. \quad (19.52)$$

Now the Hamiltonian (19.52) can be quantized with the resulting wave equation

$$\left(\frac{\partial^2}{\partial x^2} - \frac{\partial^2}{\partial y^2}\right)\psi + \left[\mathcal{E} - \beta^2\left\{(1 - \alpha^2)\left(x^2 - y^2\right) + 4\alpha xy\right\}\right]\psi = 0, \quad (19.53)$$

where

$$\mathcal{E} = \frac{2mE}{\hbar^2}, \quad \text{and} \quad \beta = \frac{m\omega}{\hbar}, \quad (19.54)$$

and E is the energy eigenvalue.

Equation (19.53) becomes separable if $\alpha = 0$, and then $\psi(x, y)$ will be a product of the wave functions for two harmonic oscillators. For the damped oscillator we expand $\psi(x, y)$ in terms of the normalized harmonic oscillator wave function $\psi_n(y)$,

$$\psi(x, y) = \sum_n X_n(x)\psi_n(y). \quad (19.55)$$

By substituting (19.55) in (19.53) and using the relations

$$\int_\infty^\infty \psi_j^*(y)y\psi_n(y)dy = \sqrt{\left(\frac{j + 1}{2\beta}\right)}\delta_{n,j+1} + \sqrt{\left(\frac{j}{2\beta}\right)}\delta_{n,j-1}, \quad (19.56)$$

and

$$\int_\infty^\infty \psi_j^*(y)y^2\psi_n(y)dy = \left(\frac{j + 1}{2\beta}\right)\delta_{n,j}, \quad (19.57)$$

we find that $X_n(x)$ satisfies the differential-difference equation

$$\frac{d^2 X_n(x)}{dx^2} + \left[\mathcal{E} + \beta\left(n + \frac{1}{2}\right)(2 - \alpha^2) - \beta^2\left(1 - \alpha^2\right)x^2\right] X_n(x)$$
$$- (2\beta)^{\frac{3}{2}}\alpha x \left(\sqrt{n+1}X_{n+1}(x) + \sqrt{n}X_{n-1}\right) = 0. \qquad (19.58)$$

This equation with the boundary conditions

$$X_n(x) \to 0, \quad \text{as} \quad x \to \pm\infty, \qquad (19.59)$$

is an eigenvalue equation with the solution $X_{nj}(x)$ for the characteristic values $\mathcal{E}_{nj}$.

Earlier in §18.4 we showed that the motion of a single particle coupled to a many-body system can be reduced to the motion of that particle in an optical potential. The present case shows that the motion in an optical potential field can be viewed as the interaction between a particle and a many-channel problem.

In the optical potential model that we have considered so far, a state or a particle decays into other states or particles, therefore the norm of the central particle (or state) is not preserved. We can use a model similar to Gisin's which preserves the norm of the wave function for the central particle (or state).

In the present problem the potential is complex, but we use it in a non-linear evolution equation [18] [19]

$$i\frac{\partial\psi}{\partial t} = H\psi + i\left[\mathcal{V}(x) - \langle\psi|\mathcal{V}(x)|\psi\rangle\right]\psi, \qquad (19.60)$$

where we have set $\hbar = 1$. In Eq. (19.60), H is a Hermitian Hamiltonian. From (19.60) and its complex conjugate we find that

$$\frac{\partial}{\partial t}\langle\psi|\psi\rangle = 0, \qquad (19.61)$$

and therefore ψ can be normalized.

We now change the wave function ψ to ϕ where

$$\psi(x, t) = \phi(x, t)\exp\left[-\int_0^t \langle\psi|\mathcal{V}(x)|\psi\rangle dt\right]. \qquad (19.62)$$

By substituting (19.62) in (19.60) we find the partial differential equation for ϕ

$$i\frac{\partial\phi}{\partial t} = H\phi + i\mathcal{V}(x)\phi. \qquad (19.63)$$

Next we define the set of complex eigenfunctions $\phi_n(x)$ by the relation

$$[H + i\mathcal{V}]\phi_n(x) = \left(E_n - \frac{i}{2}\Gamma_n\right)\phi_n(x). \qquad (19.64)$$

These ϕ_n s satisfy the orthogonality condition

$$\int_{-\infty}^{\infty} \phi_n(x)\phi_j(x)dx = 0, \quad n \neq j, \tag{19.65}$$

but since ϕ_j s are complex functions we do not have an orthonormal set. We write $\phi(x,t)$ in terms of $\phi_n(x)$;

$$\phi(x,t) = \sum_n C_n \exp\left[-i\left(E_n - \frac{i}{2}\Gamma_n\right)t\right]\phi_n(x), \tag{19.66}$$

where C_n s are the coefficients of the expansion and are related to $\psi(x,0)$ by

$$C_n = \frac{\int_{-\infty}^{\infty} \psi(x,0)\phi_n(x)dx}{\int_{-\infty}^{\infty} (\phi_n(x))^2 dx}. \tag{19.67}$$

Using the time-dependent wave packet we can determine the position of the center of the wave packet as a function of time, i.e.

$$\langle x(t) \rangle = \frac{\int_{-\infty}^{\infty} x|\psi(x,t)|^2 dx}{\int_{-\infty}^{\infty} |\psi(x,t)|^2 dx}. \tag{19.68}$$

Substituting for $\psi(x,t)$ we have

$$\langle x(t) \rangle = \frac{\sum_{n,j}\langle \phi_j|x|\phi_n \rangle C_k C_j^* \exp(-i\omega_{nj}t)\exp[-\frac{1}{2}(\Gamma_n + \Gamma_j)t]}{\sum_{n,j}\langle \phi_j|\phi_n \rangle C_k C_j^* \exp(-i\omega_{nj}t)\exp[-\frac{1}{2}(\Gamma_n + \Gamma_j)t]}, \tag{19.69}$$

where

$$\omega_{nj} = E_n - E_j. \tag{19.70}$$

If Γ_i is the smallest of the set of Γ_n's then as $t \to \infty$, $\langle x(t) \rangle$ tends to the limit

$$\langle x(t) \rangle \to \langle \phi_i|x|\phi_i \rangle, \tag{19.71}$$

and thus the final state of the system is given by the wave function $\phi_j(x)$. As an example of the model that we discussed in this section let us consider an oscillator with the harmonic optical potential $v(x) + iV(x) = \frac{m}{2}\omega^2(1 - i\alpha)^2 x^2$ where $\alpha = \lambda/2\omega$ (see Eq. (5.26). The Schrödinger equation for this potential can be written as

$$\frac{d^2\psi}{dx^2} + [\varepsilon - \beta^2 x^2]\psi = 0, \tag{19.72}$$

with

$$\beta = m(1 - i\alpha)\omega, \tag{19.73}$$

and the energy eigenvalues

$$E = \left(\frac{\omega}{2\beta}\right)\varepsilon. \tag{19.74}$$

The solution of (19.72) is given by the parabolic cylinder function

$$\psi_\nu(x) = \frac{1}{\sqrt{n!}} \left(\frac{\beta}{\pi}\right)^{\frac{1}{4}} D_\nu\left(x\sqrt{2\beta}\right).$$ (19.75)

This function has the asymptotic property [20]

$$D_\nu\left(x\sqrt{2\beta}\right) \rightarrow \begin{cases} \left(x\sqrt{2\beta}\right)^\nu e^{-\frac{1}{2}\beta x^2} & \text{as } x \to \infty \\ \frac{\sqrt{2\pi}}{\Gamma(-\nu)}\left[\frac{e^{\frac{1}{2}\beta x^2}}{\left(-x\sqrt{2\beta}\right)^{\nu+1}}\right] & \text{as } x \to -\infty \end{cases},$$ (19.76)

and ν is related to ε by

$$\nu = \frac{\varepsilon}{2m(1-i\alpha)\omega} - \frac{1}{2},$$ (19.77)

and $\Gamma(z)$ denotes the Gamma function. Now for $\psi(x)$ to be square integrable we must have

$$\Gamma(-\nu) = \infty, \quad \text{or} \quad \nu = n,$$ (19.78)

where n is an integer. Thus for negative values of x we have

$$D_n\left(-x\sqrt{2\beta}\right) = (-1)^n D_n\left(x\sqrt{2\beta}\right).$$ (19.79)

From Eqs. (19.74), (19.77) and (19.78) we find the energy eigenvalues to be

$$E_n - \frac{i}{2}\Gamma_n = \left(n+\frac{1}{2}\right)(1-i\alpha)\,\omega = \left(n+\frac{1}{2}\right)\left(\omega - \frac{i\lambda}{2}\right).$$ (19.80)

This result is similar to the energy eigenvalues of H_n^-, Eq. (12.114).

For the initial wave packet of the form of a displaced Gaussian,

$$\psi(x,0) = \left(\frac{\beta^2}{\pi}\right)^{\frac{1}{4}} \exp\left[-\frac{\beta}{2}(x-x_0)^2\right],$$ (19.81)

we find that the center of the wave packet oscillates with decreasing amplitude as is shown in Fig. (19.2). The time-dependence of $\langle x(t)\rangle$ is found from Eq. (19.69).

19.4   Quantum Mechanical Analogue of the Raleigh Oscillator

We have seen that the motion of a tuning fork in a viscous fluid can be described by a velocity-dependent frictional force and by the fluid accreting to the oscillator and thus changing its effective mass (Chapter 2). The quantum analogue

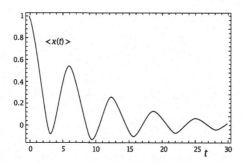

Figure 19.2: Damped motion of the center of a displaced Gaussian wave packet calculated from Eq. (19.69).

of this problem can be formulated in the following way [21]:

We write the total Hamiltonian as the sum of two terms $\hat{H} = \hat{H}_0 + \hat{H}'$, where

$$\hat{H}_0 = \frac{\hat{p}^2}{2m} + \frac{1}{2}\omega_0^2 x^2 \tag{19.82}$$

and

$$\hat{H}' = \frac{\hat{p}^2}{2}\left(\frac{1}{m'} - \frac{1}{m}\right) - \frac{i\lambda\hbar}{4}. \tag{19.83}$$

In these relations λ is the damping constant and $\hat{p} = -i\hbar\frac{\partial}{\partial x}$ is the linear momentum operator and m' is defined by

$$\frac{m}{m'} = 1 - \frac{\lambda^2}{4\omega_0^2}. \tag{19.84}$$

The total Hamiltonian is that of an optical potential Hamiltonian with a constant $\mathcal{V}$.

We note that the classical motion generated by the Hamiltonian H does not depend on λ. But the imaginary part of $\hat{H}$ changes the eigenvalues of (Re H) by the amount $\left(\frac{-i\hbar\lambda}{4}\right)$. If we denote the initial eigenstates of Re $\hat{H}$ by $\phi_n(x)$, then the time evolution of the wave function is expressed by

$$\psi_n(x,t) = \exp\left(-\frac{iHt}{\hbar}\right)\phi_n(x) = \exp\left(-\frac{\lambda t}{4}\right)\exp\left(-\frac{i\mathcal{E}_n t}{\hbar}\right)\phi_n(x). \tag{19.85}$$

Here $\mathcal{E}_n$ is the n-th eigenvalue of Re $\hat{H}$, and is given by

$$\mathcal{E}_n = \hbar\omega\left(n + \frac{1}{2}\right), \quad \omega = \left(\omega_0^2 - \frac{\lambda^2}{4}\right)^{\frac{1}{2}}. \tag{19.86}$$

The expectation value of $x(t)$ for any arbitrary time-dependent wave function $\Psi(x,t)$ can be found from (19.85) and is given by

$$\langle\Psi|x|\Psi\rangle = e^{\frac{-\lambda t}{2}}\int_{-\infty}^{\infty}\phi^*(x,t)x\phi(x,t)dx, \tag{19.87}$$

where $\phi(x,t)$ is the solution of

$$-i\hbar\frac{\partial\phi(x,t)}{\partial t} = \left[\text{Re}\hat{H}\right]\phi(x,t). \tag{19.88}$$

From this expression we can verify that the expectation value satisfies the equation of motion

$$m'\langle\ddot{x}(t)\rangle + \lambda m'\langle\dot{x}(t)\rangle + m'\omega_0^2\langle x(t)\rangle = 0. \tag{19.89}$$

We can also find Eq. (19.89) from the Ehrenfest theorem for complex potentials, Eqs. (19.16) and (19.17).

As we noted in the general formulation of the optical potential problem, this model can be viewed as the coupling of the oscillator to an infinite number of channels or to a sink. A sink here represents any process by which the quantum system can leak out from the original state. The probability of the system remaining in its initial state being proportional to $e^{-\lambda t}$.

Bibliography

[1] H. Feshbach, Ann. Phys. (NY) 5, 357 (1958).

[2] L.L. Foldy and J.D. Walecka, Ann. Phys. (NY) 54, 447 (1969).

[3] P.E. Hodgson, *Nuclear Reactions and Nuclear Structure*, (Oxford University Press, London, 1971). Chapter 6.

[4] G.R. Satchler, *Introduction to Nuclear Reactions*, (MacMillan, London, 1980) Chapter 4.

[5] N.F. Mott and H.S.W. Massey, *Theory of Atomic Collisions*, Third Edition, (Oxford University Press, London, 1965) p. 181.

[6] M. Razavy, Phys. Rev. 125, 269 (1962).

[7] M. Razavy, Phys. Rev. 171, 1201 (1968).

[8] J.S. Eck and W.J. Thompson, Am. J. Phys. 45, 161 (1977).

[9] See for instance A. Messiah, *Quantum Mechanics*, vol. I (John Wiley & Sons, New York, 1958) Chapter VI.

[10] Y. Takahashi and T. Toyoda, Physica, 138A, 501 (1986).

[11] M. Razavy, Hadronic J. 11, 75 (1988).

[12] E. Madelung, Z. Physik, 40, 322 (1926).

[13] D. Schuch and K.M. Chung, Int. J. Quantum Chem. xxix, 1561 (1986).

[14] Y. Yamaguchi, Phys. Rev. 95, 1628 (1954).

[15] R.G. Sachs, Phys. Rev. 74, 433 (1948).

[16] K.K. Osborne and L.L. Foldy, Phys. Rev. 79, 795 (1948).

[17] L. Heller in *Symposium on the Two-Body Forces in Nuclei*, edited by S.M. Austin and C.M. Cawley, (Plenum Press, New York, 1972).

[18] M. Razavy and A. Pimpale, Phys. Rep. 168, 305 (1988).

[19] M. Razavy, Can. J. Phys. 73, 131 (1994).

[20] P.M. Morse and H. Feshbach, *Methods of Theoretical Physics*, Part II (McGraw-Hill, New York, 1953) p. 1641.
[21] R.L. Anderson, Am. J. Phys. 61, 343 (1993).

Index